T0185377

Lecture Notes in Computer Science 12421

Founding Editors

Gerhard Goos
Karlsruhe Institute of Technology, Karlsruhe, Germany
Juris Hartmanis
Cornell University, Ithaca, NY, USA

Editorial Board Members

Elisa Bertino
Purdue University, West Lafayette, IN, USA
Wen Gao
Peking University, Beijing, China
Bernhard Steffen
TU Dortmund University, Dortmund, Germany
Gerhard Woeginger
RWTH Aachen, Aachen, Germany
Moti Yung
Columbia University, New York, NY, USA

More information about this series at http://www.springer.com/series/7407

Marco Dorigo · Thomas Stützle ·
Maria J. Blesa · Christian Blum ·
Heiko Hamann · Mary Katherine Heinrich ·
Volker Strobel (Eds.)

Swarm Intelligence

12th International Conference, ANTS 2020
Barcelona, Spain, October 26–28, 2020
Proceedings

 Springer

Editors
Marco Dorigo ⓘ
Université Libre de Bruxelles
Brussels, Belgium

Thomas Stützle ⓘ
Université Libre de Bruxelles
Brussels, Belgium

Maria J. Blesa ⓘ
Universitat Politècnica de Catalunya
Barcelona, Spain

Christian Blum ⓘ
Artificial Intelligence Research Institute
Bellaterra, Spain

Heiko Hamann ⓘ
University of Lübeck
Lübeck, Germany

Mary Katherine Heinrich ⓘ
Université Libre de Bruxelles
Brussels, Belgium

Volker Strobel ⓘ
Université Libre de Bruxelles
Brussels, Belgium

ISSN 0302-9743   ISSN 1611-3349 (electronic)
Lecture Notes in Computer Science
ISBN 978-3-030-60375-5   ISBN 978-3-030-60376-2 (eBook)
https://doi.org/10.1007/978-3-030-60376-2

LNCS Sublibrary: SL1 – Theoretical Computer Science and General Issues

© Springer Nature Switzerland AG 2020
This work is subject to copyright. All rights are reserved by the Publisher, whether the whole or part of the material is concerned, specifically the rights of translation, reprinting, reuse of illustrations, recitation, broadcasting, reproduction on microfilms or in any other physical way, and transmission or information storage and retrieval, electronic adaptation, computer software, or by similar or dissimilar methodology now known or hereafter developed.
The use of general descriptive names, registered names, trademarks, service marks, etc. in this publication does not imply, even in the absence of a specific statement, that such names are exempt from the relevant protective laws and regulations and therefore free for general use.
The publisher, the authors and the editors are safe to assume that the advice and information in this book are believed to be true and accurate at the date of publication. Neither the publisher nor the authors or the editors give a warranty, expressed or implied, with respect to the material contained herein or for any errors or omissions that may have been made. The publisher remains neutral with regard to jurisdictional claims in published maps and institutional affiliations.

This Springer imprint is published by the registered company Springer Nature Switzerland AG
The registered company address is: Gewerbestrasse 11, 6330 Cham, Switzerland

Preface

These proceedings contain the papers presented at the 12th International Conference on Swarm Intelligence (ANTS 2020), which took place during October 26–28, 2020. The conference would have been held at the Spanish National Research Council (CSIC), Barcelona, Spain, but instead was held as an online conference due to the COVID-19 pandemic. The ANTS series started in 1998 with the First International Workshop on Ant Colony Optimization (ANTS 1998). Since then ANTS, which is held bi-annually, has gradually become an international forum for researchers in the wider field of swarm intelligence. In 2004, this development was acknowledged by the inclusion of the term "Swarm Intelligence" (next to "Ant Colony Optimization") in the conference title. Starting in 2010, the ANTS conference has been officially devoted to the field of swarm intelligence as a whole, without any bias towards specific research directions. This is reflected in the current title of the conference: "International Conference on Swarm Intelligence."

This volume contains 28 papers selected from 50 initial submissions. Of these, 20 were accepted as full-length papers, and 8 were accepted as short papers. This corresponds to an overall acceptance rate of 56%. Also included in this volume are five extended abstracts.

All contributions were presented online. Extended versions of the best papers presented at the conference will be published in a special issue of the journal *Swarm Intelligence*.

We take this opportunity to thank the large number of people that were involved in making this conference a success. We express our gratitude to the authors who contributed their work, to the members of the international Program Committee, and to the additional referees for their qualified and detailed reviews.

We hope the reader will find this volume useful both as a reference disseminating current research in swarm intelligence and as a starting point for future work.

August 2020

Marco Dorigo
Thomas Stützle
Maria J. Blesa
Christian Blum
Heiko Hamann
Mary Katherine Heinrich
Volker Strobel

Organization

Organizing Committee

General Chairs

Marco Dorigo — Université Libre de Bruxelles, Belgium
Thomas Stützle — Université Libre de Bruxelles, Belgium

Local Organization and Publicity Chairs

Maria J. Blesa — Universitat Politècnica de Catalunya, Spain
Christian Blum — Artificial Intelligence Research Institute (IIIA-CSIC), Spain

Technical Program Chairs

Christian Blum — Artificial Intelligence Research Institute (IIIA-CSIC), Spain
Heiko Hamann — University of Lübeck, Germany

Publication Chair

Mary Katherine Heinrich — Université Libre de Bruxelles, Belgium

Paper Submission Chair

Volker Strobel — Université Libre de Bruxelles, Belgium

Program Committee

Ashraf Abdelbar — Brandon University, Canada
Michael Allwright — Université Libre de Bruxelles, Belgium
Martyn Amos — Northumbria University, UK
Jacob Beal — BBN Technologies, USA
Giovanni Beltrame — Polytechnique Montréal, Canada
Spring Berman — Arizona State University, USA
Tim Blackwell — Goldsmiths, University of London, UK
Wei-Neng Chen — South China University of Technology, China
Maurice Clerc — Independent Consultant on Optimization, France
Leandro dos Santos Coelho — Pontifícia Universidade Católica do Paraná, Brazil
Carlos Coello Coello — CINVESTAV-IPN, Mexico
Oscar Cordon — University of Granada, Spain
Guido De Croon — Delft University of Technology, The Netherlands
Gianni Di Caro — Carnegie Mellon University in Qatar, Qatar
Swagatam Das — Indian Statistical Institute, India

Luca Di Gaspero	University of Udine, Italy
Eliseo Ferrante	Vrije Universiteit Amsterdam, The Netherlands
Ryusuke Fujisawa	Kyushu Institute of Technology, Japan
Luca Maria Gambardella	Istituto Dalle Molle di Studi sull'Intelligenza Artificiale, Switzerland
José García-Nieto	University of Málaga, Switzerland
Simon Garnier	New Jersey Institute of Technology, USA
Morten Goodwin	University of Agder, Norway
Roderich Gross	The University of Sheffield, UK
Julia Handl	The University of Manchester, UK
Yara Khaluf	Ghent University, Belgium
Xiaodong Li	RMIT University, Australia
Simone Ludwig	North Dakota State University, USA
Manuel López-Ibáñez	The University of Manchester, UK
Vittorio Maniezzo	University of Bologna, Italy
Massimo Mastrangeli	Delft University of Technology, The Netherlands
Bernd Meyer	Monash University, Australia
Martin Middendorf	University of Leipzig, Germany
Marco Montes de Oca	Northeastern University, USA
Melanie Moses	University of New Mexico, USA
Radhika Nagpal	Harvard University, USA
Kazuhiro Ohkura	Hiroshima University, Japan
Konstantinos Parsopoulos	University of Ioannina, Greece
Orit Peleg	University of Colorado Boulder, USA
Paola Pellegrini	IFSTTAR, France
Carlo Pinciroli	Worcester Polytechnic Institute, USA
Günther Raidl	Vienna University of Technology, Austria
Andreagiovanni Reina	The University of Sheffield, UK
Pawel Romanczuk	Humboldt University of Berlin, Germany
Mike Rubenstein	Northwestern University, USA
Roberto Santana	University of the Basque Country, Spain
Thomas Schmickl	University of Graz, Austria
Kevin Seppi	Brigham Young University, USA
Dirk Sudholt	The University of Sheffield, UK
Munehiro Takimoto	Tokyo University of Science, Japan
Danesh Tarapore	University of Southampton, UK
Guy Theraulaz	Université Paul Sabatier, France
Dhananjay Thiruvady	Deakin University, Australia
Vito Trianni	Italian National Research Council, Italy
Elio Tuci	University of Namur, Belgium
Ali Emre Turgut	Middle East Technical University, Turkey
Gabriele Valentini	Arizona State University, USA
Justin Werfel	Harvard University, USA
Masahito Yamamoto	Hokkaido University, Japan

Cheng-Hong Yang   National Kaohsiung University of Science
                                  and Technology, Taiwan
Zhi-Hui Zhan   South China University of Technology, China

Additional Reviewers

Sameera Abar   Loughborough University, UK
David Askay   California Polytechnic State University, USA
Filippo Bistaffa   Artificial Intelligence Research Institute (IIIA-CSIC),
                                  Spain
Denis Boyer   Universidad Nacional Autónoma de México, Mexico
Christian Leonardo   Université Libre de Bruxelles, Belgium
  Camacho Villalón
Anders L. Christensen   University of Southern Denmark, Denmark
Wilfried Elmenreich   Alpen-Adria-Universität Klagenfurt, Austria
Di Liang   South China University of Technology, China
Run-Dong Liu   South China University of Technology, China
Louis Rosenberg   Unanimous AI, USA
Melanie Schranz   Lakeside Labs GmbH, Austria
Shu-Zi Zhou   South China University of Technology, China

Contents

Extended Abstracts

Full Papers

A Blockchain-Controlled Physical Robot Swarm Communicating via an Ad-Hoc Network

Alexandre Pacheco⬤, Volker Strobel(✉)⬤, and Marco Dorigo⬤

IRIDIA, Université Libre de Bruxelles, Brussels, Belgium
{alexandre.melo.pacheco,vstrobel,mdorigo}@ulb.ac.be

Abstract. We present a robot swarm composed of Pi-puck robots that maintain a blockchain network. The blockchain serves as security layer to neutralize Byzantine robots (faulty, malfunctioning, or malicious robots). In the context of this work, we implemented a framework for high-throughput communication using a decentralized mobile ad-hoc network. This work serves as a building block for secure real-world deployments of robot swarms. Our results show that the use of a blockchain is feasible and warranted in embodied robot swarm deployments.

1 Introduction

In real-world deployments, robot swarms will face a multitude of security challenges that are rarely taken into consideration in the swarm robotics field [7]. In particular, the presence of *Byzantine* robots, that is, malfunctioning, faulty, or malicious robots, might lead to a discrepancy between the *intended* and the *actual* behavior of a swarm. In a recent article, Strobel et al. [16] deliver a comprehensive proof-of-concept for a blockchain-based approach that greatly limits, in a fully decentralized manner, the impact of Byzantine robots on the robot swarm behavior.

The field of Blockchain Technology was initiated through the digital currency Bitcoin [12], in which the blockchain serves as a decentralized *ledger* for storing financial transactions. Later blockchain frameworks, such as Ethereum [2] (used in this work), extended the capability of blockchains to be decentralized *computing platforms*. Put simply, in the Ethereum blockchain the network participants can run programming code on the blockchain and agree on the outcome of the programs, without the need for supervision or mutual trust. It is precisely these decentralized programs that have proven to be useful in robot swarms, where a blockchain can serve as a secure decentralized coordinator and database [14–17]. To the best of our knowledge, all existing multi-robot systems that use blockchain and smart contracts technology have been demonstrated on simulated robots: the present paper is the first successful implementation of this technology in a physical robot swarm.

The replication of simulation results with physically embodied robots is crucial to convince the robotics community of the feasibility of blockchain and smart

© Springer Nature Switzerland AG 2020
M. Dorigo et al. (Eds.): ANTS 2020, LNCS 12421, pp. 3–15, 2020.
https://doi.org/10.1007/978-3-030-60376-2_1

contracts technology as a tool for solving security issues in robot swarms. However, moving from simulated to real robots is not straightforward and involves, among other things, the choice of adequate robot platforms, communication protocols, and blockchain frameworks. Unfortunately, in previous works addressing the topic, many questions whose answer would be of paramount importance for a successful real robot implementation were not addressed. Examples are:

- which lightweight robot platform can provide the requirements for running a blockchain framework?
- which blockchain consensus protocols are appropriate for robot swarms?
- which communication infrastructure should be used?

In this work we give a first answer to the above questions by presenting an experimental setup consisting of Pi-puck robots [10] which maintain a proof-of-authority blockchain network and communicate through a mobile ad-hoc network. The chosen robot platform—the Pi-puck—is a reasonable choice due to the low cost and easy availability of the Pi-pucks, and to their support for Linux that allows the easy installation of the blockchain software—Ethereum in our case. The choice of the proof-of-authority protocol is motivated by its low computational cost that allows for running it efficiently on the Pi-puck processor. Finally, we chose to implement a mobile ad-hoc network as communication infrastructure. This choice is consistent both with the standard swarm robotics requirements of *decentralized* and *local/peer-to-peer* communication (see e.g., [1,4,6]), and with the throughput required for blockchain synchronization.

The field of blockchain-based swarm robotics was set out in 2016 by [3]; the paper describes several use cases for blockchain-controlled robot swarms. Since 2018 there has been a number of simulation results for blockchain applications in robots such as: the achievement of consensus in robot swarms in the presence of Byzantine robots [15]; the improvement of communications and performance in industrial robots [5]; the formation of coalitions in cyber-physical systems [9]; the management of collaboration in heterogeneous multirobot systems [14]; the secure collection of data from robots [20]; and path planning in multi-robot systems [11]. Hence, research addressing the application of blockchain technology to robotics is an active research area and, as such, there is a high demand for platforms to run blockchain experiments on physical robots.

The remainder of this paper is structured as follows. Section 2 describes the robot hardware and control routines, the used blockchain software, and the setup of the experiments. Section 3 presents the experimental results. Finally, Sect. 4 concludes the paper and indicates directions for future research.

2 Methods

2.1 Experimental Scenario

In this paper, we consider a scenario where a robot swarm is given the task to determine the fraction of white tiles in a checkerboard environment (Fig. 1).

The difficulty of this task can be increased by increasing the number of Byzantine robots that distribute false information, or by decreasing the size of the swarm, which leads to lower coverage of the map and lower network connectivity. Our goal is to study how a blockchain—and its relevant components, such as smart contracts, cryptotokens and protocols for consensus—can be used to counteract the negative influence of Byzantine robots when a swarm is trying to achieve swarm wide consensus. As this is the first implementation of a physical robot swarm that uses blockchain and smart contracts technology, we provide guidelines for the establishment of such a system, and insights into some details of its operation.

Fig. 1. The Pi-puck robots move in an $1 \times 1\,\mathrm{m}^2$ arena covered by 68 black and 32 white tiles. The robots' goal is to determine the fraction of white tiles by using their ground sensors. A part of the swarm act as Byzantine robots that disseminate wrong estimates. Using their LEDs, the robots communicate events—such as the receipt of a new block or the receipt of the consensus signal—to the experimenter for visual analysis.

The Robot Platform. The experiments are conducted using a swarm of up to $N = 10$ Pi-puck robots [10]. Pi-pucks are e-puck robots extended by a Raspberry Pi Zero W single board computer. Compared to the e-puck robot, the extension board improves the robots' communication capabilities and computational power. The Raspberry Pi Zero W has a 1 GHz processor with 512 MB of RAM. Hence, the use of the Raspberry Pi Zero W extension board allows for the implementation of more complex algorithms compared to previous e-puck robot versions.

The Arena. The arena has a dimension of $1 \times 1\,\mathrm{m}^2$. Its plywood floor is covered with 68 black and 32 white square tiles; therefore, the fraction of white tiles is 0.32. Each tile is $10 \times 10\,\mathrm{cm}^2$. The arena is bounded on each side by a wooden barrier which can be detected by the robots' obstacle avoidance sensors.

2.2   Control Routines

In the following, a high-level overview of the software that is executed on each robot is given. A more in-depth description is given in a separate technical report [13], in which we provide a detailed description of all the steps needed to set up and replicate our experiments using the Pi-puck robot.

The control for each robot is composed of five high-level routines that are executed in parallel at different frequencies:

- *Random-walk with obstacle avoidance* (frequency: 10 Hz): at each step the robot can perform: a) straight movement, b) rotation on site, or c) obstacle avoidance. If the robot's infrared (IR) sensors detect an obstacle, phase c) is selected in order to prevent collisions; otherwise, the robot alternates between the random-walk phases a) and b); the duration of each phase is sampled from an exponential distribution (we use the same parameters as [19]).
- *Estimation* (frequency: 1 Hz): the robots calculate a local estimate of the fraction of white tiles in the environment by dividing the number of white ground sensor readings by the total number of readings; to reduce noise, the sensors sample the floor at a rate 20 Hz and the average of these samples is used in this routine.
- *Peering* (frequency: 3 Hz): each robot uses its range-and-bearing IR actuators/sensors to simultaneously transmit its ID, and listen for other IDs within a fixed range (approx. 10 cm). After an ID is received, the robot executes a TCP request to obtain the Enode—a unique identifier of a blockchain node, used for the peering calls. After 2 s without receiving close range IR messages, a peer is removed from the blockchain and all information regarding that peer is deleted.
- *Local estimate dissemination* (frequency: $\frac{1}{45}$ Hz): Every 45 s, a robot sends its local estimate to the smart contract where local estimates are stored, aggregated, and refined to generate a *shared* estimate.
- *Block sealing* (frequency: varying—see below): On each robot, an instance of the Ethereum software *geth* is executed during the entire course of the experiment. In order to create new blocks on the blockchain, each robot acts as a sealer in this background process.

With the exception of the Ethereum client, which is implemented in the Go language [2], we implemented each control routine in Python[1]; in order to run the routines in parallel and guarantee the specified frequencies, the multi-threading package is used.

2.3   Blockchain Technology

For fundamentals of blockchain technology, we refer the reader to [16, Section 2] and to the original papers on Bitcoin [12] and Ethereum [2]. Here, we limit ourselves to the description of those aspects of blockchain technology that are most relevant for our setup—namely, the used *proof-of-authority* consensus protocol, as well as the concepts of *smart contract* and *cryptotoken*.

---

[1] Project repository: https://github.com/teksander/geth-pi-pucks.

Consensus Protocol. The decentralized nature of blockchains can result in conflicting situations (e.g., a different order of transactions in different versions of the blockchain). To resolve these conflicts and agree on a common order of transactions, a consensus protocol is needed. In this work, we use proof-of-authority (see [18] for the full specifications), an alternative to the original and most commonly used blockchain consensus protocol known as proof-of-work [12]. In contrast to the computationally expensive proof-of-work, proof-of-authority requires a majority of preselected nodes (i.e., in this work, a majority of robots) to agree on the state of the blockchain database. In the proof-of-authority protocol there are two kinds of blockchain nodes: normal blockchain nodes and *sealer* nodes, which are analogous to *miners* in proof-of-work and that are able to create new blocks by signing them. For each block, there is a preferred sealer, chosen in a round-robin fashion. If the preferred sealer signs the block, it is called an *in-turn* signature, if another sealer signs it, it is called an *out-of-turn* signature. The sealers can sign new blocks anytime they want, but in order for a new block to be valid:

- the timestamp of the new block must be at least $t = 15\,\mathrm{s}$ after the previous block (also known as the block time);
- a sealer can only sign one block in $\lfloor \frac{N}{2} \rfloor + 1$ blocks (to guarantee majority voting);
- a sealer must create a correct signature using its private key and sign the hash of the current block.

As soon as a sealer has signed a block, it disseminates the block in the network. The other nodes verify the signature and the validity of the block. The nodes agree on the *strongest* chain, that is the chain with the highest difficulty. The difficulty for in-turn signature is 2, and the one for out-of-turn signature is 1.

The major advantage of proof-of-authority is that it does not depend on solving computationally complex mathematical puzzles (as the standard proof-of-work-based consensus protocol does). Even though in this work our swarms contain a fixed number of robots (from 5 to 10), proof-of-authority also works with swarms with varying number of robots by either keeping a core of trusted sealers or adding and removing sealers based on majority vote.

Blockchain-Based Smart Contract. A blockchain-based smart contract is a piece of programming code that is stored on the blockchain. The smart contract encapsulates functions and variables, and participants of the blockchain network are able to alter its state by sending transactions to its functions. Blockchains additionally store the amount of "cryptotokens"—that is, immutable shares of a digital currency—that each participant possesses (see below).

The smart contract used in our research has four functions with which the robots can interact:

1. `sendEstimate(localEstimate)`: this function enables the robots to store their local estimates on the blockchain. In order to store an estimate, robots

have to send 40 ether. *Ether* is a scarce cryptotoken, and the fact that sending transactions requires ether effectively limits the number of transactions a robot can send;

2. `askForUBI()`: by sending a transaction to this function, robots can make a request for the universal basic income (see below for a description);
3. `getEstimate()`: this function returns the aggregated estimate of the fraction of white tiles, as determined by the blockchain-based smart contract;
4. `hasConverged()`: this function checks if the smart contract has reached convergence on an estimate (i.e., it determines if the absolute difference between the previous and the current value of the shared estimate is smaller than $\tau = 0.01$), in which case it returns 'true';
5. `registerRobot()` this function is called by each robot at least once, and is required before a robot is allowed to use any other function of the smart contract. This function allows using the same smart contract independently of the number of robots N.

The flow of information in the smart contract works as follows. After a robot registers itself, it can begin sending transactions that contain local estimates. Sent local estimates are stored in a list of proposals in the smart contract. As soon as N proposals are received, the smart contract performs a simple outlier detection, where all proposals with an absolute difference to the current blockchain estimate larger than $\delta = 0.2$ are discarded (except for the very first N proposals that are all accepted). The accepted proposals are used to update the estimate in the blockchain, which is the arithmetic mean of all accepted proposals. All robots that sent an accepted proposal get back their 40 ether plus a bonus consisting of a share of the non-repaid ethers of the discarded proposals.

Cryptotokens. In order to store their local estimate in the blockchain, robots send `sendEstimate` transactions accompanied by a fixed amount of *cryptotokens*. Cryptotokens are an immutable and scarce asset which is stored on a blockchain ledger. A digital asset with these properties is a key component to limiting the number of transactions robots can send, and thus prevent Sybil attacks. Robots can obtain tokens in two ways:

- by being reimbursed when sending accepted proposals (that is, by sending useful information);
- by receiving the universal basic income (UBI).

The UBI is an economy mechanism we established within the smart contract to allow the fair distribution of tokens between the robots. It functions as follows: at block numbers which are a power of 2 (i.e., in the blocks $2, 4, 8, \cdots$), the smart contract grants 20 ether to each robot in the swarm. This exponential scheme makes sure that in the beginning of an experiment every robot receives enough ether to be able to send its local estimate; however, over time sending useful information becomes the main means to receive additional ethers and to be able to continue participating in the experiment. Using this scheme, we can take advantage of the immutability and scarcity of blockchain cryptotokens to filter Byzantine robots out and limit their influence on the smart contract estimate.

2.4 Ad-Hoc Network

In order to exploit the Wi-Fi communication abilities of the Raspberry Pi Zero W without compromising the decentralization of the robot swarm, we establish a Mobile Ad-hoc Mesh Network using the B.A.T.M.A.N. routing protocol [8]. The advantage of such a network is that it does not rely on any central hubs (such as routers or master servers) nor does it assume global connectivity. Instead, each node participates in routing by forwarding data of other nodes.

In our experiments, the communication range of the ad-hoc network is additionally constrained by the range-and-bearing (RAB) board of the robots. We do this to enforce the swarm robotics core assumption of local communication, or, otherwise, the small arena size would lead to global communication. By tuning the power allocated to the RAB board it is possible to physically limit the communication range to approximately 5 cm. Robots broadcast the last 8 bits of their IP address in this fashion, and once an exchange has taken place, the connection to this IP address is established via the Ad-hoc network. Then, the robots exchange their enodes (an enode is a unique identifier for each Ethereum node) using TCP, in order to connect their Ethereum nodes and begin synchronization of the blockchain. From the moment a robot stops receiving signals from the RAB of a blockchain peer, a 2-s grace period is started after which the peer is removed.

2.5 Experiment Setup and Evaluation

Initialization and Termination. At the start of each experimental run, the robots are randomly distributed in the arena by the experimenter. Then, all robots connect to the Ethereum process of a bootstrap node (a desktop PC) and wait for a signal to start executing the parallel routines of Sect. 2.2. The experimenter sends the start signal from the bootstrap node, which consists of broadcasting a transaction containing the smart contract. An experimental run is stopped after all robots have received 'true' when querying `hasConverged()`, at which time they turn on their green LEDs.

Independent Variables. Each experiment may differ in (i) the total number of robots in the swarm; or (ii) the number of Byzantine robots.

(i) Swarm Size: Changing the swarm size allows for analyzing our platform in terms of two key features of robot swarms: *scalability*, i.e., the ability of the system to maintain or improve performance as the swarm size increases; and *partition-tolerance*, i.e., the ability of the system to reach consensus when there is reduced network connectivity (in this case, induced by a more sparse distribution of robots in the arena). As mentioned before, the dynamic addition and removal of block sealers is a feature in proof-of-authority; however, it has not been exploited in this work, and instead, all sealers are included on the genesis block in each experiment.

(ii) Byzantine Robots: To study the performance of our approach for increasing numbers of Byzantine robots, we model a Byzantine robot as a robot that disables its ground sensor, keeps a local estimate $\hat{\rho} = 0.0$, and sends this faulty estimate to the smart contract. This failure mode is well-motivated by our tests with physical robots and can occur in several situations: (1) a robot gets stuck on a tile during the course of the experiments, for example, due to a broken motor; (2) a robot's ground sensor does not have the correct distance from the floor, for example, due to a loose screw; (3) the communication to the ground sensor is broken, for example, due to a crash of the I^2C communication protocol; or (4) the robot is controlled by a malicious entity that tries to work against the goal of the swarm. Byzantine robots are selected randomly by the experimenter at the start of the experiment.

Table 1. Overview of the experiments and their parameters

No.	Experiment name	Swarm size	# Byzantine robots
1	Increasing Byzantines	10	$0, 1, 2, 3, 4$
2	Increasing swarm size (no Byzantine robots)	$5, 6, 7, 8, 9, 10$	0
3	Increasing swarm size (20% Byzantine robots)	$5, 10$	20%

Metrics. The performance of our approach is evaluated by comparing the error between the actual fraction ρ of black tiles to the blockchain estimate of a randomly selected robot at the end of each run; and the time required for all robots to receive the consensus signal. In addition, we record the size of the blockchain in MB and we use it to draw conclusions regarding the scalability of the approach.

3   Results

In order to evaluate the presented approach with physical robots, we conduct three experiments (Table 1). For each setting of each experiment we conduct 10 repetitions.

3.1   Experiment 1: Increasing Byzantines

The first experiment studies the impact of Byzantine robots in a swarm of fixed size $N = 10$. The number of Byzantines is increased from 0 to 5. Our hypothesis is that the approach has the lowest absolute error when no Byzantines are part of the swarm. We expect the Byzantines to have little effect up to a crucial point where they have a strong adverse effect on the estimate.

Results and Short Discussion. Figure 2 shows the results obtained. The Byzantine robots have a small impact when their number is between 0 and 3 (median of absolute error <5%) because their estimates are rejected by the smart contract and therefore they eventually run out of cryptotokens. As soon as four Byzantines are part of the swarm, the median error becomes significantly larger as Byzantines begin to collect rewards and therefore to have a stronger influence on the estimate. We also observe a high variability that is due to the fact that Byzantine robots may or may not become the dominant party—the estimate may therefore sway in either direction: reality, or zero. The estimate variability decreases with five Byzantines because in this case the Byzantines become dominant as they always send the same 0% estimate, and are therefore able to consistently collect rewards and steer the estimate towards the wrong value.

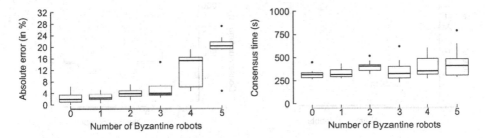

Fig. 2. Experiment 1 – Increasing Byzantines (10 robots in total, 0 to 5 Byzantines). *Left*: The median of the absolute error stays below 5% for 0 to 3 Byzantine robots. With 4 and 5 Byzantine robots the estimate error becomes much larger as the Byzantine robots are able to steer the estimate towards the wrong value. *Right*: There are no statistically significant differences in the consensus time when the number of Byzantine robots increases, even though the variability tends to increase.

3.2 Experiment 2: Increasing Swarm Size (No Byzantine Robots)

In Experiment 2 we investigate to what extent the size of the swarm has an influence on the consensus time as well as on the blockchain size. To this end, we increase the swarm size from 5 to 10 robots.

Results and Short Discussion. The median of the absolute error is below 5% for all swarm sizes and independent of the swarm size (results not shown). As expected, the consensus time decreases with an increasing swarm size (Fig. 3, *left*). The consensus time is influenced by several variables, such as the number of transactions and the average connectivity of the network, which is higher when there are more robots distributed in the arena. The blockchain size grows linearly in time. To obtain the growth rate for the different swarm sizes, a linear regression was performed using time as a predictor of blockchain size (Fig. 3, *right*). The larger the swarm size, the more transactions are created, thus the faster the blockchain size grows.

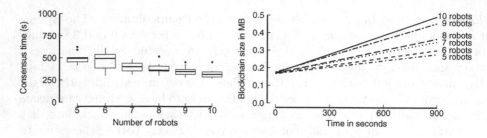

Fig. 3. Experiment 2 – Increasing swarm size (5 to 10 robots, no Byzantine Robots). The consensus time (*left*) decreases approximately linearly with the number of robots. The blockchain size (*right*) increases linearly over time (values obtained by linear regression for each swarm size).

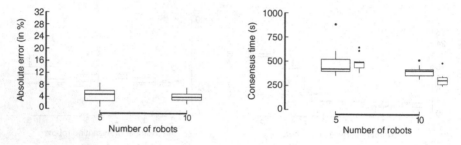

Fig. 4. Experiment 3 – Increasing swarm size (5 or 10 robots, 20% Byzantine robots). *Left*: The absolute error is not influenced by the number of robots. *Right*: With 5 robots the consensus time with (large boxplots) or without (small boxplots) Byzantine robots is not significantly different; with 10 robots the presence of Byzantines significantly increase the consensus time.

3.3 Experiment 3: Increasing Swarm Size (20% Byzantines)

In the third and final experiment, we study how different swarm sizes deal with a fixed fraction of 20% Byzantine robots. We perform Experiment 3 exclusively with 5 and 10 robots, because only these two swarm sizes permit to have exactly 20% Byzantine robots.

Results and Short Discussion. Figure 4 (left) shows that, as without Byzantines, the absolute error is independent of the swarm size. Figure 4 (right) shows that, when comparing the results to Experiment 2, with 5 robots the consensus time with or without Byzantines is not significantly different, while with 10 robots the presence of Byzantines significantly increases the consensus time.

4   Conclusions

In our research we are interested in developing robot swarms that present a high level of security against the possible presence of Byzantine robots.

The blockchain protocol uses a set of technologies to generate secure and tamper-proof knowledge shared by a network of mutually untrusting agents (robots in our case): public-key cryptography, digital signatures, consensus protocols, decentralized databases, and smart contracts. By using these technologies within the blockchain protocol it becomes possible to protect a robot swarm from Sybil attacks [16] and to reduce the influence that Byzantine robots can have on the overall swarm behavior. Additionally, it makes it possible to let the swarm reach consensus about the overall status of the system as stored in a decentralized ledger that can also be used as a tamper-proof register of events, accessible during, or after, an experiment.

While we had already demonstrated the feasibility of using the blockchain protocol in a robot swarm in previous work [15–17], this was done only in simulation. One important question when moving to real robots is whether the computations and communications required by the blockchain protocol are still feasible in a system composed of agents (the robots) that have limited computational power and only local communication (i.e., they can only communicate with neighbour robots)—as opposed to standard implementations of the blockchain protocol where the individual nodes are powerful computers that are fully connected to each other.

In this paper we have demonstrated the first example of a physical robot swarm that uses the blockchain protocol and smart contracts. In particular, we have showed that it is possible to do so in a swarm of not-so-powerful robots using a low-cost Raspberry Pi Zero W as onboard computer. To get these results, we have used the proof-of-authority protocol and our results show that the Ethereum client *geth* running on our robots uses, regardless of experimental parameters, about 13.7% of the available Raspberry Pi's CPU power.

Our results show that the blockchain approach is feasible also in terms of data storage: the size of the blockchain data folder in our robots grows linearly over time, and remains under 0.5 MB at the end of a 15-min run (Fig. 3). Therefore, the same experiment could be executed for about one year using a 16 GB SD card as data storage.

In conclusion, the reader should note that our goal for this article was to show that our previous results obtained in simulation would carry over to a swarm of real robots and consequently that the use of a blockchain is warranted for real-world deployment of secure robot swarms. We did not, however, intend to provide the most efficient possible implementation, nor fine tune parameters in order to achieve the best possible results: these aspects are left for future work.

Acknowledgements. Alexandre Pacheco acknowledges support via a fellowship from the Faculty of Applied Sciences of the Université Libre de Bruxelles. Volker Strobel and Marco Dorigo acknowledge support from the Belgian F.R.S.-FNRS, of which they are a Research Fellow and a Research Director respectively.

References

1. Brambilla, M., Ferrante, E., Birattari, M., Dorigo, M.: Swarm robotics: a review from the swarm engineering perspective. Swarm Intell. **7**(1), 1–41 (2013). https://doi.org/10.1007/s11721-012-0075-2
2. Buterin, V.: A next-generation smart contract and decentralized application platform. Ethereum project white paper. Technical report, Ethereum Foundation (2014). https://github.com/ethereum/wiki/wiki/White-Paper. Accessed 18 July 2019
3. Castelló Ferrer, E.: The blockchain: a new framework for robotic swarm systems. e-print (2016). arXiv:1608.00695v3
4. Dorigo, M., Birattari, M., Brambilla, M.: Swarm robotics. Scholarpedia **9**(1), 1463 (2014)
5. Fernandes, M., Alexandre, L.A.: Robotchain: using Tezos technology for robot event management. Ledger **4**(Suppl. 1) (2019). https://doi.org/10.5195/ledger.2019.175
6. Garattoni, L., Birattari, M.: Swarm robotics. In: Webster, J.G. (ed.) Wiley Encyclopedia of Electrical and Electronics Engineering. Wiley, Hoboken (2016)
7. Higgins, F., Tomlinson, A., Martin, K.M.: Survey on security challenges for swarm robotics. In: Proceedings of the Fifth International Conference on Autonomic and Autonomous Systems, pp. 307–312. IEEE Press (2009). https://doi.org/10.1109/ICAS.2009.62
8. Johnson, D., Ntlatlapa, N., Aichele, C.: A simple pragmatic approach to mesh routing using BATMAN. In: Proceedings of the 2nd IFIP International Symposium on Wireless Communications and Information Technology in Developing Countries (WCTID 2008) (2008)
9. Kashevnik, A., Teslya, N.: Blockchain-oriented coalition formation by CPS resources: ontological approach and case study. Electronics **7**, 66 (2018). https://doi.org/10.3390/electronics7050066
10. Millard, A.G., et al.: The Pi-puck extension board: a Raspberry Pi interface for the e-puck robot platform. In: 2017 IEEE/RSJ International Conference on Intelligent Robots and Systems (IROS), pp. 741–748. IEEE Press (2017)
11. Mokhtar, A., Murphy, N., Bruton, J.: Blockchain-based multi-robot path planning. In: 2019 IEEE 5th World Forum on Internet of Things, pp. 584–589 (2019). https://doi.org/10.1109/WF-IoT.2019.8767340
12. Nakamoto, S.: Bitcoin: A peer-to-peer electronic cash system. Technical report (2008). https://bitcoin.org/bitcoin.pdf. Accessed 11 Aug 2018
13. Pacheco, A., Strobel, V., Dorigo, M.: A framework for swarm robotics experimentation with Pi-puck robots and an Ethereum-based blockchain. Technical report TR/IRIDIA/2020-001, IRIDIA, Université Libre de Bruxelles, Brussels, Belgium (2020)
14. Queralta, J.P., Westerlund, T.: Blockchain-powered collaboration in heterogeneous swarms of robots. e-print (2019). arXiv:1912.01711v2
15. Strobel, V., Castelló Ferrer, E., Dorigo, M.: Managing Byzantine robots via blockchain technology in a swarm robotics collective decision making scenario. In: Dastani, M., Sukthankar, G., André, E., Koenig, S. (eds.) Proceedings of the 17th International Conference on Autonomous Agents and MultiAgent Systems (AAMAS 2018), Richland, SC, USA , pp. 541–549. International Foundation for Autonomous Agents and Multiagent Systems (2018)

16. Strobel, V., Castelló Ferrer, E., Dorigo, M.: Blockchain technology secures robot swarms: a comparison of consensus protocols and their resilience to Byzantine robots. Front. Robot. AI **7**, 54 (2020). https://doi.org/10.3389/frobt.2020.00054
17. Strobel, V., Dorigo, M.: Blockchain technology for robot swarms: a shared knowledge and reputation management system for collective estimation. In: Dorigo, M., Birattari, M., Blum, C., Christensen, A.L., Reina, A., Trianni, V. (eds.) Swarm Intelligence – Proceedings of ANTS 2018 – Eleventh International Conference. LNCS, vol. 11172, pp. 425–426. Springer, Cham (2018). https://doi.org/10.1007/978-3-030-00533-7
18. Szilágyi, P.: EIP 225: Clique proof-of-authority consensus protocol (2017). https://github.com/ethereum/EIPs/issues/225. Accessed 10 May 2020
19. Valentini, G., Brambilla, D., Hamann, H., Dorigo, M.: Collective perception of environmental features in a robot swarm. In: Dorigo, M., Birattari, M., Li, X., López-Ibáñez, M., Ohkura, K., Pinciroli, C., Stützle, T. (eds.) ANTS 2016. LNCS, vol. 9882, pp. 65–76. Springer, Cham (2016). https://doi.org/10.1007/978-3-319-44427-7_6
20. White, R., Caiazza, G., Cortesi, A., Cho, Y., Christensen, H.: Black block recorder: immutable black box logging for robots via blockchain. IEEE J. Robot. Autom. **4**, 3812–3819 (2019). https://doi.org/10.1109/LRA.2019.2928780

A New Approach for Making Use of Negative Learning in Ant Colony Optimization

Teddy Nurcahyadi$^{(\boxtimes)}$ ⓘ and Christian Blum ⓘ

Artificial Intelligence Research Institute (IIIA-CSIC),
Campus of the UAB, Bellaterra, Spain
{teddy.nurcahyadi,christian.blum}@iiia.csic.es

Abstract. The overwhelming majority of ant colony optimization approaches from the literature is exclusively based on learning from positive examples. Natural examples from biology, however, indicate the potential usefulness of negative learning. Several research works have explored this topic over the last two decades in the context of ant colony optimization, with limited success. In this work we present an alternative proposal for the incorporation of negative learning in ant colony optimization. The results obtained for the capacitated minimum dominating set problem indicate that this approach can be quite useful. More specifically, our extended ant colony algorithm clearly outperforms the standard approach. Moreover, we were able to improve the current state-of-the-art results in 10 out of 36 cases.

1 Introduction

Combinatorial optimization (CO) problems are of utmost importance in many real-life scenarios. Large-scale instances of hard CO problems are often solved by heuristic methods. The family of metaheuristics [2] includes techniques based on local search (such as tabu search) and it includes a whole range of bio-inspired techniques such as ant colony optimization and evolutionary algorithms. In this paper we deal with the metaheuristic ant colony optimization (ACO) [5,6], whose development was inspired by the shortest path finding behavior of natural ant colonies. ACO, which is a metaheuristic based on learning, works as follows. At each iteration, a number of artificial ants generate solutions to the tackled optimization problem in a probabilistic way. This is done based on two types of information: greedy information and pheromone information. Then, the best ones of these solutions are used to update the pheromone values, with the aim of moving the probability distribution used for generating solutions to areas of the search space in which high-quality solutions can be found.

As in most metaheuristics based on learning, the type of learning generally used in ACO is *positive learning*, that is, the algorithm tries to learn which components are necessary for assembling high-quality solutions. Nevertheless, learning from negative examples (negative learning) seems to play an important

© Springer Nature Switzerland AG 2020
M. Dorigo et al. (Eds.): ANTS 2020, LNCS 12421, pp. 16–28, 2020.
https://doi.org/10.1007/978-3-030-60376-2_2

role in biological self-organizing systems. Pharaoh ants (*Monomorium pharaonis*), for example, make use of negative trail pheromone in order to deploy 'no entry' signals to mark unrewarding foraging paths [15]. Another example is the use of anti-pheromone hydrocarbons produced by male tsetse flies. These anti-pheromones play an important role in tsetse communications [17]. As already noted in [18], it might therefore be possible to boost the performance of ACO with an additional mechanism that learns (or marks) *undesirable components* by means of a negative feedback mechanisms.

The research community has made several attempts to take benefit from negative learning. Maniezzo [11] and Cordón et al. [4] were the first ones to introduce an active decrease of pheromone values based on low-quality solutions. Montgomery and Randall [12] proposed three different anti-pheromone strategies, partially inspired by previous works [4,8]. In their first approach some amount of pheromone is removed from the solution components of the worst solution in each iteration. Their second approach makes explicit use of negative pheromone in addition to the standard pheromone. Finally, their third approach allocates a small number of ants at each iteration to explore the use of solution components with lower pheromone values, without adding dedicated anti-pheromones. Unfortunately, the experimental evaluation did not show a clear advantage of any of the three strategies over standard ACO. Simons and Smith [19] explored different extensions of [12]. They state, however, that nearly all their approaches were proved counter-intuitive by the results. The only approach that showed some usefulness was to make use of a high amount of anti-pheromone in the very early stages of the search process. In [16], Rojas-Morales et al. propose an extension of an ACO algorithm for the multidimensional knapsack problem based on opposite learning. In a first phase, the algorithm builds anti-pheromone values whose intention it is to repel the ACO algorithm during the second phase from solution components that seem locally attractive (due to a rather high heuristic value) but that lead to low-quality solutions. Unfortunately, the results do not show a consistent improvement over standard ACO. Finally, note that earlier strategies based on opposition-based learning were tested on four small TSP instances in [10].

In this paper we introduce a conceptually new way of making use of negative learning in ACO, in the context of the so-called capacitated minimum dominating set (CapMDS) problem [14]. Our results show that the performance of the standard ACO algorithm (without negative learning) is significantly improved for most of the considered problem instance types. Moreover, the current state-of-the-art algorithm is improved in the context of 10 out of 36 cases.

2   The CapMDS Problem

Before introducing the CapMDS problem and the developed algorithms, let us briefly recall some necessary definitions and notions from graph theory. Henceforth, $G = (V, E)$ denotes an undirected graph with a set $V = \{v_1, v_2, \cdots, v_n\}$ of n vertices, and a set E of edges. We assume that the given graph neither contains self-loops nor multi-edges. Two vertices $u, v \in V$ are called neighbors—that is, they are adjacent—if and only if $(u, v) = (v, u) \in E$. Furthermore, $N(v) := \{u \in V \mid (v, u) \in E\}$ is called the *(open) neighborhood* of v and denotes the set of neighbors of $v \in V$. In contrast, the *closed neighborhood* $N[v]$ of a vertex $v \in V$ is $N[v] := N(v) \cup \{v\}$. The *degree* $\deg(v)$ of v is defined as the cardinality of the set of neighbors of v, that is, $\deg(v) = |N(v)|$. Any subset $S \subseteq V$ is called a *dominating set* of G if each vertex $v \in V \setminus S$ is adjacent to at least one vertex from S. A vertex from S is called a *dominator*. Given an undirected graph $G = (V, E)$, the classical minimum dominating set (MDS) problem asks to find a smallest-size dominating set $S \subseteq V$.

A problem instance of the CAPMDS problem is given by a tuple (G, Cap) that consists of an undirected (simple) graph $G = (V, E)$ and a capacity function $Cap : V \to \mathbb{N}$. This capacity function assigns a positive integer $Cap(v) > 0$ to each vertex $v \in V$, indicating the maximum number of adjacent vertices this vertex is allowed to dominate in a valid solution.

A solution S to an instance (G, Cap) is a tuple $(D^S, \{C^S(v) \mid v \in D^S\})$, where $D^S \subseteq V$ is the set of selected dominators, and $\{C^S(v) \mid v \in D^S\}$ is a set that contains for each dominator $v \in D^S$ the (sub-)set $C^S(v) \subseteq N(v)$ of those of its neighbors that are (chosen to be) dominated by v. The following conditions have to be fulfilled in order for S to be a valid solution:

1. $D^S \cup \left(\bigcup_{v \in D^S} C^S(v) \right) = V$, that is, all vertices from V are either chosen to be a dominator, or are dominated by at least one dominator.
2. $|C^S(v)| \leq Cap(v)$ for all $v \in D^S$, that is, all chosen dominators dominate at most $Cap(v)$ of their neighbors.

Finally, the objective function value (to be minimized) is defined as $f(S) := |D^S|$.

ILP Model for the CAPMDS Problem. The following integer linear program (ILP) is reproduced from [13]. The model is presented because it plays an important role for the negative learning mechanism that is presented later. It works on the following sets of binary variables. First, a binary variable x_v is associated to each vertex $v \in V$ indicating whether or not v is selected as a dominator. Second, the model contains for each edge $(v, v') \in E$ two binary variables $y_{v,v'}$ and $y_{v',v}$. Variable $y_{v,v'}$ takes value one if vertex v is chosen to dominate vertex v'; similarly for $y_{v',v}$. The CapMDS problem can then be stated as follows:

$$\text{minimize} \sum_{v \in V} x_v \tag{1}$$

$$\text{s.t.} \sum_{v' \in N(v)} y_{v',v} \geq 1 - x_v \quad \forall v \in V \tag{2}$$

$$\sum_{v' \in N(v)} y_{v,v'} \leq Cap(v) \quad \forall v \in V \tag{3}$$

$$y_{v,v'} \leq x_v \quad \forall v \in V, v' \in N(v) \tag{4}$$

$$x_v, y_{v,v'} \in \{0, 1\} \tag{5}$$

Hereby, constraint (2) ensures that all non-chosen vertices must be dominated by at least one dominator, whereas constraint (3) limits the total number of vertices dominated by a particular vertex v to $Cap(v)$. Consequently, a dominator v can dominate at most $Cap(v)$ vertices from its (open) neighborhood.

3   Proposed Approach

First of all, we present a standard ACO approach (without negative learning) for the CapMDS problem. We chose a $\mathcal{MAX}$-$\mathcal{MIN}$ Ant System (MMAS) implemented in the Hypercube Framework [1] for this purpose. In the context of this algorithm, the construction of a solution S is done in a step-by-step manner. At each construction step, first, exactly one new dominator $v \in V \setminus D^S$ is chosen. In the second part of the construction step, it is decided which ones of the so-far non-dominated neighbors of v will be dominated by v. Therefore, the pheromone model $\mathcal{T}$ used by our algorithm consists of the following values:

1. A value τ_v for each $v \in V$. These values are used to choose dominators.
2. Values $\tau_{v,v'}$ and $\tau_{v',v}$ for each edge $(v, v') \in E$. These values are used in the second part of each construction step for deciding which ones of its neighbors a newly chosen dominator will dominate.

In general terms, a MMAS algorithm (when implemented in the Hypercube Framework) works as follows (see also Algorithm 1). At each iteration, first, n_a solutions are probabilistically generated both based on pheromone and on greedy information. Second, the pheromone values are modified using (at most) three solutions: (1) the iteration-best solution S^{ib}, (2) the restart-best solution S^{rb}, and (3) the best-so-far solution S^{bs}. The pheromone update is done with the aim to focus the search process of the MMAS algorithm on areas of the search space with high-quality solutions. Note that the algorithm also performs restarts when necessary—that is, a re-initializations of the pheromone values is performed once convergence is detected. Restarts are controlled by a convergence measure called the *convergence factor* (*cf*) and by a Boolean control variable called bs_update. The implementation of all these components for the CapMDS is detailed in the following.

InitializePheromoneValues(): In this function all pheromone values τ_v for $v \in V$ are initialized to 0.5. Moreover, all pheromone values $\tau_{v,v'}$ and $\tau_{v',v}$ for all $(v, v') \in E$ are equally initialized to 0.5.

Construct_Solution(): The construction of a solution starts with an empty solution $S = (D^S = \emptyset, \emptyset)$. Moreover, the set of non-dominated neighbors of each vertex $v \in V$, denoted by ND_v, is initialized to $N(v)$. At each construction step, first, one vertex v^* is chosen from a set O (options) that includes all those vertices v that still have non-dominated neighbors and that do not already form part of D^S:

$$O := \{v \in V \mid \text{ND}_v \neq \emptyset, v \notin D^S\} \tag{6}$$

Algorithm 1. MMAS for the CapMDS problem

1: **input:** a problem instance (G, Cap)
2: $S^{bs} :=$ NULL, $S^{rb} :=$ NULL, $cf := 0$, bs_update := FALSE
3: InitializePheromoneValues()
4: **while** termination conditions not met **do**
5:    $S^{\text{iter}} := \emptyset$
6:    **for** $k = 1, \ldots, n_a$ **do**
7:       $S^k :=$ Construct_Solution()
8:       $S^{\text{iter}} := S^{\text{iter}} \cup \{S^k\}$
9:    **end for**
10:   $S^{ib} := \operatorname{argmin}\{f(S) \mid S \in S^{\text{iter}}\}$
11:   **if** $f(S^{ib}) < f(S^{rb})$ **then** $S^{rb} := S^{ib}$
12:   **if** $f(S^{ib}) < f(S^{bs})$ **then** $S^{bs} := S^{ib}$
13:   ApplyPheromoneUpdate(cf, bs_update, S^{ib}, S^{rb}, S^{bs})
14:   $cf :=$ ComputeConvergenceFactor()
15:   **if** $cf > 0.9999$ **then**
16:      **if** bs_update = TRUE **then**
17:         $S^{rb} :=$ NULL, and bs_update := FALSE
18:         InitializePheromoneValues()
19:      **else**
20:         bs_update := TRUE
21:      **end if**
22:   **end if**
23: **end while**
24: **output:** S^{bs}, the best solution found by the algorithm

---

Note that the solution construction process stops once $O = \emptyset$. The greedy function value $\eta(v)$ of a vertex $v \in O$ is defined as $\eta(v) := \min\{Cap(v), |\text{ND}_v|\} + 1$. Based on this greedy function, the probability for a vertex $v \in O$ to be selected is determined as follows:

$$\mathbf{p}^{\text{step1}}(v) := \frac{\eta(v) \cdot \tau_v}{\sum_{v' \in O} \eta(v') \cdot \tau_{v'}} \tag{7}$$

Given the probabilities from Eq. (7), a vertex $v^* \in O$ is chosen in the following way. First a value $0 \leq r \leq 1$ is drawn uniformly at random. In case $r \leq d_{\text{rate}}$, the vertex with the highest probability is chosen deterministically. Otherwise, a vertex is chosen randomly according to the probabilities (roulette-wheel-selection). Hereby, the *determinism rate* $d_{\text{rate}} \leq 1$ is a parameter of the algorithm. Note that after choosing v^*, the sets of non-dominated neighbors of the neighbors of v^* are updated by removing v^*.

In the second part of each construction step, a set of $\min\{Cap(v^*), |\text{ND}_{v^*}|\}$ non-dominated neighbors of v^* is chosen and placed into $C^S(v^*)$ as follows. In case $|\text{ND}_{v^*}| \leq Cap(v^*)$, we set $C^S(v^*) := \text{ND}_{v^*}$. Otherwise, vertices are sequentially selected from ND_{v^*} in the following way. First, the probability for

Table 1. Setting of κ_{ib}, κ_{rb}, and κ_{bs} depending on the convergence factor cf and the Boolean control variable bs_update

	bs_update = FALSE				bs_update = TRUE
	$cf < 0.4$	$cf \in [0.4, 0.6)$	$cf \in [0.6, 0.8)$	$cf \geq 0.8$	
κ_{ib}	1	2/3	1/3	0	0
κ_{rb}	0	1/3	2/3	1	0
κ_{bs}	0	0	0	0	1

each vertex $v \in \mathrm{ND}_{v^*}$ to be selected is determined as follows:

$$\mathbf{p}^{\text{step2}}(v) := \frac{(|\mathrm{ND}_v| + 1) \cdot \tau_{v^*, v}}{\sum_{v' \in \mathrm{ND}_{v^*}} (|\mathrm{ND}_{v'}| + 1) \cdot \tau_{v^*, v'}} \tag{8}$$

Then, given the probabilities from Eq. (8), a vertex $\hat{v} \in \mathrm{ND}_{v^*}$ is chosen in the same way as outlined above in the context of the first part of the construction step. Vertex $\hat{v}$ is then added to an initially empty set $C^S(v^*)$, the respective ND-sets are updated, the probabilities from Eq. (8) are recalculated, and the next vertex from ND_{v^*} is chosen. This process stops once $\min\{Cap(v^*), |\mathrm{ND}_{v^*}|\}$ are selected. Finally, $C^S(v^*)$ is added to solution S, and the solution construction process proceeds with the next construction step.

ApplyPheromoneUpdate(cf, bs_update, S^{ib}, S^{rb}, S^{bs}): This is a standard procedure in any MMAS algorithm implemented in the Hypercube Framework. In particular, solutions S^{ib}, S^{rb}, and S^{bs} are used for the pheromone update. The influence of each of these solutions on the pheromone update is determined on the basis of the convergence factor (cf) and the value of bs_update (see Table 1). Each pheromone value τ_v is updated as follows: $\tau_v := \tau_v + \rho \cdot (\xi_v - \tau_v)$, where $\xi_v := \kappa_{ib} \cdot \Delta(S^{ib}, v) + \kappa_{rb} \cdot \Delta(S^{rb}, v) + \kappa_{bs} \cdot \Delta(S^{bs}, v)$. Hereby, κ_{ib} is the weight of solution S^{ib}, κ_{rb} the one of solution S^{rb}, and κ_{bs} the one of solution S^{bs}. Moreover, $\Delta(S, v)$ evaluates to 1 if and only if $v \in D^S$ (that is, v is chosen as a dominator). Otherwise, the function evaluates to 0. Note also that the three weights must be chosen such that $\kappa_{ib} + \kappa_{rb} + \kappa_{bs} = 1$. Finally, note that in the case of pheromone values $\tau_{v,v'}$, the pheromone update is the same, just that functions $\Delta(S, v)$ are replaced by functions $\Delta(S, v, v')$. Hereby, function $\Delta(S, v, v')$ evaluates to 1 if and only if $v \in D^S$ and $v' \in C^S(v)$ (that is, dominator v is chosen to dominate its neighbor v' in solution S). After the pheromone update, pheromone values that exceed $\tau_{\max} = 0.99$ are set back to $\tau_{\max}$, and pheromone values that have fallen below $\tau_{\min} = 0.01$ are set back to $\tau_{\min}$. This prevents the algorithm from reaching the state of complete convergence.

ComputeConvergenceFactor($\mathcal{T}$): The value of the convergence factor cf is computed, in a standard way, on the basis of the pheromone values:

$$cf := 2 \left(\left(\frac{\sum_{\tau \in \mathcal{T}} \max\{\tau_{\max} - \tau, \tau - \tau_{\min}\}}{|\mathcal{T}| \cdot (\tau_{\max} - \tau_{\min})} \right) - 0.5 \right) \tag{9}$$

Hereby, $\mathcal{T}$ stands for the set of all τ_v-values and all $\tau_{v,v'}$-values. With this formula, the value of cf results in zero, when all pheromone values are set to 0.5. In contrast, when all pheromone values have either value $\tau_{\min}$ or $\tau_{\max}$, the value cf evaluates to one. In all other cases, cf has a value between 0 and 1. This completes the description of all components of the proposed algorithm.

3.1 Adding Negative Learning

First of all, for all pheromone values τ_v ($v \in V$) we introduce a negative pheromone value τ_v^{neg}. Moreover, for all pheromone values $\tau_{v,v'}$ we also introduce the negative version $\tau_{v,v'}^{\text{neg}}$. In contrast to the standard pheromone values, these negative pheromone values are initialized to $\tau_{\min}$ at the start of the algorithm, and whenever the algorithm is restarted (which still depends exclusively on the standard pheromone values).

The negative pheromone values are used in the following way to change the probabilities in both phases of each step for the construction of a solution S. Remember that the first phase concerns the choice of the next dominator v^*, and the second phase concerns the choice of a set $C^S(v^*)$ of so-far uncovered neighbors of v^* that v^* will dominate. The updated formula for calculating the probabilities in the first phase is as follows (compare to Eq. 7):

$$\mathbf{p}^{\text{step1}}(v) := \frac{\eta(v) \cdot \tau_v \cdot (1 - \tau_v^{\text{neg}})}{\sum_{v' \in O} \eta(v') \cdot \tau_{v'} \cdot (1 - \tau_{v'}^{\text{neg}})} \tag{10}$$

In the second phase of each construction step $C^S(v^*)$ is sequentially filled with vertices taken from ND_{v^*} (the set of currently uncovered neighbors of v^*) in the following way. First, the probability for each vertex $v \in \text{ND}_{v^*}$ to be selected is determined as follows (compare to Eq. 8):

$$\mathbf{p}^{\text{step2}}(v) := \frac{(|\text{ND}_v| + 1) \cdot \tau_{v^*,v} \cdot (1 - \tau_{v^*,v}^{\text{neg}})}{\sum_{v' \in \text{ND}_{v^*}} (|\text{ND}_{v'}| + 1) \cdot \tau_{v^*,v'} \cdot (1 - \tau_{v^*,v'}^{\text{neg}})} \tag{11}$$

Another change in comparison to the standard way of generating solutions is that, during this second phase, only vertices whose probability $\mathbf{p}^{\text{step2}}(v)$ is greater or equal to 0.001 can be selected. This makes it possible to generate solutions in which a vertex selected as a dominator might not be chosen to dominate as many of its uncovered neighbors as possible in that moment.

As mentioned before, we strongly believe that the information that is used to determine the negative pheromone values should originate from an algorithmic component different to the ACO algorithm itself. In the context of the CapMDS problem we therefore propose the following. At each iteration of our MMAS algorithm, the set of solutions generated at the incumbent iteration ($\mathcal{S}^{\text{iter}}$) is used for generating a subinstance of the tackled problem instance. Such a subinstance I^{sub} is a tuple $(D^{\text{sub}}, \{C^{\text{sub}}(v) \mid v \in D^{\text{sub}}\})$ where

- $D^{\text{sub}} := \bigcup_{S \in \mathcal{S}^{\text{iter}}} D^S$

- $C^{\text{sub}}(v) := \bigcup_{\substack{S \in \mathcal{S}^{\text{iter}} \\ \text{s.t. } v \in D^S}} C^S(v)$

After generating the n_a solutions per iteration, the ILP solver CPLEX is used (with a time limit of t_{ILP} CPU seconds) to solve the corresponding subinstance (if possible) to optimality. Otherwise, the best solution found within the allotted computation time is returned. In any case, the returned solution is denoted by S^{ILP}. In order to solve the subinstance, the ILP model from Sect. 2 is used with the following additional restrictions. All variables x_v such that $v \notin D^{\text{sub}}$ are set to zero. Moreover, all variables $x_{v,v'}$ such that either $v \notin D$ or $v' \notin C^{\text{sub}}(v)$ are set to zero too.

After obtaining solution S^{ILP} both the standard pheromone value update and the update of the negative pheromone values is performed. The update of the negative pheromone values is done with the same formula as in the case of the standard pheromone update (see the description of function ApplyPheromone-Update(cf, bs_update, S^{ib}, S^{rb}, S^{bs})). Only the learning rate ρ is replaced by a *negative learning rate* ρ^{neg}, and the definition of the ξ_v (respectively $\xi_{v,v'}$) values changes. In particular, ξ_v is set to 1 for all $v \in D^{\text{sub}}$ with $v \notin D^{S^{\text{ILP}}}$. In all other cases ξ_v is set to 0. Moreover, $\xi_{v,v'}$ is set to 1, for all $v' \in C^{\text{sub}}(v)$ with $v' \notin C^{S^{\text{ILP}}}(v)$. In all other cases $\xi_{v,v'}$ is set to 0. In other words, those *solution components* that form part of the subinstance (and, therefore, form part of at least one of the solutions generated by MMAS) but that do not form part of the (possibly optimal) solution S^{ILP} to the subinstance, are penalized.

Note that—in contrast to the standard MMAS algorithm, which is simply denoted by ACO in the following section—the algorithm making use of negative learning is henceforth denoted by ACO_{neg}. Finally, not taking profit from solution S^{ILP} in an additional, more direct, way may result in wasting valuable information. Therefore, we also test an extended version of ACO_{neg}, henceforth denoted by $\text{ACO}_{\text{neg}}^+$, that updates solutions S^{rb} and S^{bs} at each iteration with solution S^{ILP} if appropriate.

4   Experimental Evaluation

All experiments concerning ACO, ACO_{neg} and $\text{ACO}_{\text{neg}}^+$ were performed on a cluster of machines with Intel® Xeon® CPU 5670 CPUs with 12 cores of 2.933 GHz and a minimum of 32 GB RAM. Moreover, for solving the subinstances in ACO_{neg} and $\text{ACO}_{\text{neg}}^+$ we used CPLEX 12.8 in one-threaded mode.

The proposed algorithms were evaluated on the largest ones of the general graphs benchmark set for the CapMDS problem from [14]. These graphs are

Table 2. Outcome of parameter tuning.

Algorithm	n_a	d_{rate}	ρ	ρ^{neg}	t_{ILP}
ACO	5	0.9	0.1	n.a	n.a.
ACO_{neg}	20	0.7	0.1	0.3	10.0
$\text{ACO}_{\text{neg}}^{+}$	20	0.6	0.1	0.2	5.0

characterized by a number of vertices (n), a number of edges (m), a vertex capacity type (uniform vs. variable), and a capacity. In the case of uniform capacities, graphs with three different capacities (2, 5 and α) exist. Hereby, α refers to the average degree of the corresponding graph. In the case of variable capacities, the vertex capacities are—for each vertex—randomly chosen from the following three intervals: $(2,5)$, $(\alpha/5, \alpha/2)$ and $[1, \alpha]$. For each combination of these graph characteristics, the benchmark set consists of 10 randomly generated graphs.

Algorithm Tuning. All three algorithm variants require parameter values to be set to well-working options. In particular, all three algorithm versions need parameter values for n_a (the number of solutions per iteration), d_{rate} (the determinism rate for solution construction), and ρ (the learning rate). Additionally, ACO_{neg} and $\text{ACO}_{\text{neg}}^{+}$ require values for parameters ρ^{neg} (the negative learning rate) and t_{ILP} (the time limit, in CPU seconds, for CPLEX at each iteration). For the purpose of parameter tuning we made use of irace [9], which is a scientific tool for parameter tuning. This tool was used for generating one single parameter setting for each algorithm. As tuning instances we chose the first (out of 10) instances for each combination of the four input graph characteristics. Moreover, a budget of 2000 applications was given to irace. The parameter value domains were fixed as follows: $n_a \in \{3, 5, 10, 20\}$, $d_{\text{rate}} \in \{0.1, 0.2, \ldots, 0.8, 0.9\}$, $\rho, \rho^{\text{neg}} \in \{0.1, 0.2, 0.3\}$, and $t_{\text{ILP}} \in \{2.0, 3.0, 5.0, 10.0\}$ (in seconds). The parameter value settings determined by irace are shown in Table 2.

Numerical Results. Each algorithm was applied exactly once (with a time limit of 1000 CPU seconds) to each problem instance. The results, averaged over 10 instances per table row, are shown in Table 3 (uniform capacity graphs) and in Table 4 (variable capacity graphs). While the two tables separate the instances with respect to the vertex capacity type (uniform vs. variable), the first three columns of each table provide information about the remaining three input graph characteristics (n, m, and vertex capacity). The fourth table column provides information about the best result known from the literature, while the fifth and sixth table columns present the results of CMSA, which is the current state-of-the-art algorithm from [13]. Both the results of CMSA and of the three ACO versions are shown by means of the average solution quality and the average computation time needed for producing these results.

Table 3. Results for general graphs with uniform capacity.

n	m	Cap.	Best known	CMSA		ACO		ACO_{neg}		ACO_{neg}^+	
				Avg.	Time	Avg.	Time	Avg.	Time	Avg.	Time
800	1000	2	267.0	**267.0**	3.6	285.3	136.2	**267.0**	8.4	**267.0**	2.5
800	2000	2	267.0	**267.0**	3.9	269.4	80.3	269.3	129.3	**267.0**	67.8
800	5000	2	267.0	**267.0**	3.2	**267.0**	59.0	271.1	192.1	**267.0**	119.4
1000	1000	2	334.0	**334.0**	7.9	364.0	157.1	**334.0**	7.2	**334.0**	0.6
1000	5000	2	334.0	**334.0**	6.5	334.2	88.0	384.6	50.3	**334.0**	126.9
1000	10000	2	334.0	**334.0**	5.8	**334.0**	32.4	379.5	176.7	337.2	136.7
800	1000	5	242.5	243.1	205.6	262.8	113.7	245.5	89.2	244.4	76.1
800	2000	5	162.8	162.8	574.7	177.0	116.1	163.2	61.0	**161.9***	79.1
800	5000	5	134.0	**134.0**	4.7	135.3	72.4	158.7	6.3	**134.0**	160.2
1000	1000	5	333.7	**333.7**	10.5	362.8	141.2	**333.7**	8.8	**333.7**	0.6
1000	5000	5	167.0	**167.0**	40.8	172.2	101.1	206.3	61.4	**167.0**	173.6
1000	10000	5	167.0	**167.0**	3.7	167.8	67.3	188.4	8.6	**167.0**	102.7
800	1000	α	267.0	**267.0**	4.6	284.0	153.8	**267.0**	10.1	**267.0**	2.8
800	2000	α	162.8	162.8	537.3	178.8	93.0	163.4	73.8	**162.0***	69.7
800	5000	α	91.1	93.0	717.9	92.9	62.8	90.9	74.0	**89.2***	104.3
1000	1000	α	334.0	**334.0**	13.7	365.1	175.2	**334.0**	6.9	**334.0**	0.6
1000	5000	α	132.5	135.0	782.9	137.3	82.0	131.6	65.4	**127.3***	116.3
1000	10000	α	81.3	86.8	518.7	82.6	67.9	87.9	98.4	**80.7***	133.1

In order to facilitate an interpretation of these results we provide the corresponding *critical difference* (CD) plots [3]. First, the Friedman test was used to compare the three approaches simultaneously. As a consequence of the rejection of the hypothesis that the techniques perform equally, the corresponding pairwise comparisons were performed using the Nemenyi post-hoc test [7]. The obtained results are graphically shown by means of the above-mentioned CD plots in Fig. 1. In these plots, each considered algorithm variant is placed on the horizontal axis according to its average ranking for the considered subset of problem instances. The performances of those algorithm variants that are below the critical difference threshold (computed with a significance level of 0.05) are considered as statistically equivalent; see the horizontal bars joining the markers of the respective algorithm variants.

The graphic in Fig. 1(a) shows the CD plot for the uniform capacity instances, and the one in Fig. 1(b) for the variable capacity instances. In both graphics it can be seen that both algorithm variants with negative learning (ACO_{neg} and ACO_{neg}^+) significantly improve over the standard ACO approach. Moreover, ACO_{neg}^+ improves over ACO_{neg} with statistical significance. This is also the general picture given by the numerical results in Tables 3 and 4.

Interestingly, when separating the instances according to different graph densities, it can be noticed that negative learning is especially useful in the context of sparse graphs. In contrast, when moving towards dense graphs the efficacy of negative learning is reduced. In the context of graphs with uniform capacities,

Table 4. Results for general graphs with variable capacity.

n	m	Cap.	Best known	CMSA		ACO		ACO_{neg}		ACO_{neg}^+	
				Avg.	Time	Avg.	Time	Avg.	Time	Avg.	Time
800	1000	$(2,5)$	248.1	248.2	79.2	269.2	131.7	251.8	47.4	249.9	68.6
800	2000	$(2,5)$	181.2	181.5	341.7	195.0	98.7	180.8	73.1	**179.8***	79.7
800	5000	$(2,5)$	134.1	**134.1**	28.1	139.1	99.6	138.4	127.3	**134.1**	94.3
1000	1000	$(2,5)$	333.8	**333.8**	3.9	365.6	146.9	**333.8**	8.8	**333.8**	1.9
1000	5000	$(2,5)$	169.0	**169.0**	85.1	182.8	86.3	171.2	109.7	169.6	105.0
1000	10000	$(2,5)$	167.0	**167.0**	27.5	168.4	92.3	198.3	7.7	**167.0**	170.1
800	1000	$(\alpha/5, \alpha/2)$	400.0	**400.0**	2.6	409.3	112.5	400.2	66.7	**400.0**	0.8
800	2000	$(\alpha/5, \alpha/2)$	273.4	**273.4**	6.5	283.2	87.5	274.6	101.6	**273.4**	7.6
800	5000	$(\alpha/5, \alpha/2)$	115.0	115.1	178.6	123.0	83.7	116.5	77.1	**115.0**	85.7
1000	1000	$(\alpha/5, \alpha/2)$	500.0	**500.0**	8.6	517.7	122.2	**500.0**	11.2	**500.0**	1.0
1000	5000	$(\alpha/5, \alpha/2)$	168.1	**168.1**	77.4	181.3	128.7	170.9	105.4	168.8	92.2
1000	10000	$(\alpha/5, \alpha/2)$	104.7	107.1	247.9	104.8	97.1	108.6	131.0	**95.6***	121.3
800	1000	$[1, \alpha]$	300.2	**300.2**	4.0	316.0	144.9	**300.2**	6.8	**300.2**	0.5
800	2000	$[1, \alpha]$	186.2	186.2	442.6	204.9	105.3	187.3	61.5	**185.8***	63.9
800	5000	$[1, \alpha]$	98.1	98.1	683.7	101.7	63.3	96.8	84.4	**95.6***	80.2
1000	1000	$[1, \alpha]$	400.8	**400.8**	6.4	409.6	141.5	**400.8**	7.3	**400.8**	0.5
1000	5000	$[1, \alpha]$	143.8	143.8	866.9	151.9	95.4	141.4	101.1	**140.6***	98.9
1000	10000	$[1, \alpha]$	90.1	90.1	541.8	88.2	66.8	87.8	132.1	**85.6***	108.0

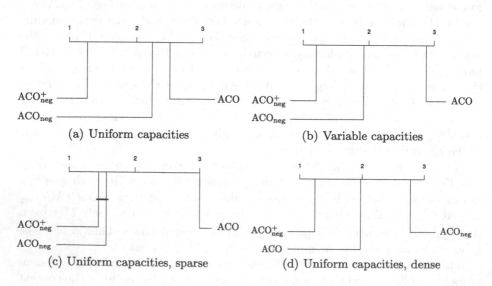

(a) Uniform capacities

(b) Variable capacities

(c) Uniform capacities, sparse

(d) Uniform capacities, dense

Fig. 1. Critical difference plots

it is even the case that standard ACO outperforms ACO_{neg} for dense graphs. This is shown in the context of uniform capacity graphs in Figs. 1(c) and (d).

5 Conclusions and Outlook

In this paper we introduced a new approach for making use of negative learning in ant colony optimization. This approach builds, at each iteration, a subinstance of the original problem instance by merging the solution components found in the solutions generated by the ant colony optimization algorithm in that iteration. Then it uses a different optimization technique—CPLEX was used here—for finding the best solution in this subinstance. The solution components from the subinstance that do not form part of this solution are penalized by means of increasing their negative pheromone values. The proposed approach is shown to be very beneficial for the capacitated minimum dominating set problem.

Future work will center along two lines. First, we plan to study why this new approach is more useful in sparse graphs. And second, we plan to apply this approach to a whole range of different combinatorial optimization problems.

Acknowledgements. This work was supported by project CI-SUSTAIN funded by the Spanish Ministry of Science and Innovation (PID2019-104156GB-I00).

References

1. Blum, C., Dorigo, M.: The hyper-cube framework for ant colony optimization. IEEE Trans. Syst. Man Cybern. Part B **34**(2), 1161–1172 (2004)
2. Blum, C., Roli, A.: Metaheuristics in combinatorial optimization: overview and conceptual comparison. ACM Comput. Surv. **35**(3), 268–308 (2003)
3. Calvo, B., Santafé, G.: scmamp: statistical comparison of multiple algorithms in multiple problems. R J. **8**(1) (2016)
4. Cordón, O., Fernández de Viana, I., Herrera, F., Moreno, L.: A new ACO model integrating evolutionary computation concepts: the best-worst ant system. In: Proceedings of ANTS 2000 - Second International Workshop on Ant Algorithms, pp. 22–29 (2000)
5. Dorigo, M., Stützle, T.: Ant Colony Optimization. MIT Press, Cambridge (2004)
6. Dorigo, M., Stützle, T.: Ant colony optimization: overview and recent advances. In: Gendreau, M., Potvin, J.Y. (eds.) Handbook of Metaheuristics. International Series in Operations Research & Management ScienceInternational Series in Operations Research & Management Science, vol. 272, pp. 311–351. Springer, Cham (2019). https://doi.org/10.1007/978-3-319-91086-4_10
7. García, S., Herrera, F.: An extension on "statistical comparisons of classifiers over multiple data sets" for all pairwise comparisons. J. Mach. Learn. Res. **9**, 2677–2694 (2008)
8. Iredi, S., Merkle, D., Middendorf, M.: Bi-criterion optimization with multi colony ant algorithms. In: Zitzler, E., Thiele, L., Deb, K., Coello Coello, C.A., Corne, D. (eds.) EMO 2001. LNCS, vol. 1993, pp. 359–372. Springer, Heidelberg (2001). https://doi.org/10.1007/3-540-44719-9_25

9. López-Ibáñez, M., Dubois-Lacoste, J., Cáceres, L.P., Birattari, M., Stützle, T.: The irace package: iterated racing for automatic algorithm configuration. Oper. Res. Perspect. **3**, 43–58 (2016)
10. Malisia, A.R., Tizhoosh, H.R.: Applying opposition-based ideas to the ant colony system. In: 2007 IEEE Swarm Intelligence Symposium, pp. 182–189. IEEE press (2007)
11. Maniezzo, V.: Exact and approximate nondeterministic tree-search procedures for the quadratic assignment problem. INFORMS J. Comput. **11**(4), 358–369 (1999)
12. Montgomery, J., Randall, M.: Anti-pheromone as a tool for better exploration of search space. In: Dorigo, M., Di Caro, G., Sampels, M. (eds.) ANTS 2002. LNCS, vol. 2463, pp. 100–110. Springer, Heidelberg (2002). https://doi.org/10.1007/3-540-45724-0_9
13. Pinacho-Davidson, P., Bouamama, S., Blum, C.: Application of CMSA to the minimum capacitated dominating set problem. In: Proceedings of GECCO 2019 - The Genetic and Evolutionary Computation Conference, pp. 321–328. ACM, New York (2019)
14. Potluri, A., Singh, A.: Metaheuristic algorithms for computing capacitated dominating set with uniform and variable capacities. Swarm Evol. Comput. **13**, 22–33 (2013)
15. Robinson, E.J.H., Jackson, D.E., Holcombe, M., Ratnieks, F.L.W.: 'No entry' signal in ant foraging. Nature **438**(7067), 442 (2005)
16. Rojas-Morales, N., Riff, M.-C., Coello Coello, C.A., Montero, E.: A cooperative opposite-inspired learning strategy for ant-based algorithms. In: Dorigo, M., Birattari, M., Blum, C., Christensen, A.L., Reina, A., Trianni, V. (eds.) ANTS 2018. LNCS, vol. 11172, pp. 317–324. Springer, Cham (2018). https://doi.org/10.1007/978-3-030-00533-7_25
17. Schlein, Y., Galun, R., Ben-Eliahu, M.N.: Abstinons - male-produced deterrents of mating in flies. J. Chem. Ecol. **7**(2), 285–290 (1981)
18. Schoonderwoerd, R., Holland, O., Bruten, J., Rothkrantz, L.: Ant-based load balancing in telecommunications networks. Adapt. Behav. **5**(2), 169–207 (1997)
19. Simons, C., Smith, J.: Exploiting antipheromone in ant colony optimisation for interactive search-based software design and refactoring. In: Proceedings of GECCO 2016 - Genetic and Evolutionary Computation Conference Companion, pp. 143–144. ACM (2016)

Ant Colony Optimization for Object-Oriented Unit Test Generation

Dan Bruce[1]([⊠]), Héctor D. Menéndez[2], Earl T. Barr[1], and David Clark[1]

[1] University College London, London, UK
{dan.bruce.17,e.barr,david.clark}@ucl.ac.uk
[2] Middlesex University London, London, UK
h.menendez@mdx.ac.uk

Abstract. Generating useful unit tests for object-oriented programs is difficult for traditional optimization methods. One not only needs to identify values to be used as inputs, but also synthesize a program which creates the required state in the program under test. Many existing Automated Test Generation (ATG) approaches combine search with performance-enhancing heuristics. We present Tiered Ant Colony Optimization (TACO) for generating unit tests for object-oriented programs. The algorithm is formed of three Tiers of ACO, each of which tackles a distinct task: goal prioritization, test program synthesis, and data generation for the synthesised program. Test program synthesis allows the creation of complex objects, and exploration of program state, which is the breakthrough that has allowed the successful application of ACO to object-oriented test generation. TACO brings the mature search ecosystem of ACO to bear on ATG for complex object-oriented programs, providing a viable alternative to current approaches. To demonstrate the effectiveness of TACO, we have developed a proof-of-concept tool which successfully generated tests for an average of 54% of the methods in 170 Java classes, a result competitive with industry standard RANDOOP.

1 Introduction

Generating unit tests for object-oriented programs is so difficult that the conventional wisdom in the ACO community is that Automated Test Generation (ATG) for complex object-oriented programs (OOP) is not currently possible for ACO. Indeed, in 2015 Mao et al. [23] said "for complex types such as String or Object, the current coding design in ACO cannot effectively handle them", and in 2018 Sharifipour et al. [27] identified generation of strings and objects as future work. Solving ATG for OOP requires calling methods in the correct order, with the correct inputs in order to explore the unit under test. This is a problem with a gigantic search space. Solving it automatically would be highly profitable, both in terms of time saved and potential increased coverage of a program.

ATG techniques can be broadly classified as static or dynamic, i.e. those that only observe the code or those that execute it. In recent years, many dynamic

© Springer Nature Switzerland AG 2020
M. Dorigo et al. (Eds.): ANTS 2020, LNCS 12421, pp. 29–41, 2020.
https://doi.org/10.1007/978-3-030-60376-2_3

approaches have used genetic algorithms (GA) [18]. GAs typically mutate and crossover candidate solutions, which, in the case of creating test programs, can lead to invalid states. Instead, following pheromone levels attributed to available methods produces legitimate test programs, guided by the fitness of previous tests. It is this observation that has motivated our exploration of ACO for object-oriented ATG. ACO has been applied to generating test cases for programs in the past. However, those applications were dominated by numerical programs, where the problem was simply finding the required values of primitive inputs [5,8,23,27]. Whilst these works show ACO's effectiveness at test data generation, they do not support its applicability to automated test generation for real world object-oriented software.

We introduce TACO, a Tiered Ant Colony Optimization algorithm that can generate complex test cases for complex object-oriented programs. TACO does so by following three tiers: 1) it selects a test coverage goal within the program under test, 2) it synthesizes test programs by creating sequences of methods, and 3) it generates numeric and string data values required as inputs by the test program. It is, to the best of our knowledge, the first complete ACO technique capable of generating valid Java test cases. TACO has been evaluated on 170 Java classes taken from SF110 [14], a well known Java testing benchmark, and successfully created tests for 54% of methods per class, covering an average of nearly 50% of lines of code. TACO achieves higher branch and line coverage than the industry standard tool, RANDOOP, not all of which overlaps, suggesting that additional engineering to cover further constructs would yield significant improvements. TACO demonstrates the potential of ACO for automated test generation for object-oriented code; further research and engineering effort may allow ACO to compete with the current state of the art in ATG.

Contributions

- TACO is the first complete ACO technique capable of creating real test cases for complex object-oriented programs.
- TACO's Tier II synthesizes test programs by building sequences of method calls, thereby creating complex objects required as inputs (Sect. 3.3).
- TACO has been realized as a tool and used to generate JUnit tests for real Java classes competitively with RANDOOP (Sect. 4).

2   Related Work

Relatively little research into automated test generation uses ant-based approaches [18,23]. Those that have applied ACO to software testing have focused on generating useful input values. The classical ACO-based test generation process typically follows 3 main steps: 1) partition the input space, 2) project each partition into each dimension or variable and, 3) decrease the partition granularity for those parts which are more interesting for the test purpose. In this way, each pair (partition, dimension) becomes a node for the graph that an ant can traverse to find inputs for a program [23]. This approach is usually applied to discreet domains, but there are several variations. Ayari et al. [5], for example, use $ACO_\mathbb{R}$, which is an ACO variant that was developed to optimize values from continuous domains [28]. They find numeric inputs required by specific methods. Similarly, Mao et al. [23] and Sharifipour et al. [27] independently compare ACO with other metaheuristics on small numerical programs. Both of these approaches generate values only for a specific method within the program under test, and they do not consider object-oriented programs. However, in both cases the ACO approach outperforms the other meta-heuristics. In previous work, we used an extended version of $ACO_\mathbb{R}$ for data generation, including strings, and then used heuristics to create objects and call methods [9]. In contrast, TACO uses ACO to synthesis a test program and build objects prior to data generation. Vats et al. [31] gather different ACO applications on software testing, where some of the work focus on OOP testing. The most relevant either generate simple inputs for programs following the classical method of partitioning the domain [12,22], or focus on other aspects of testing like iteration testing [11] and test suite minimization [29]. None of this work considers generating sequences of methods and input values as part of the same process, as TACO does.

Current, pure ACO test generation methodologies cannot deal with object-oriented programs, for which they need to create sequences of methods for each test case. Srivastava et al. proposed a technique related to our Tier II and generated test sequences of events, although the events considered did not require inputs [29]. Our technique has a stage after sequence synthesis which generates instances of all required inputs. ACO-based program synthesis is known as Ant Programming (AP) [20]. This field, inspired by genetic programming (GP), aims to create programs using ACO methods. Some examples are the work of Rojas and Bentley, which used ACO to synthesize programs which solve boolean functions [25], and Toffola et al. that compared A* and ACO for guiding program synthesis and found ACO to be effective at solving bottlenecks [30]. GP has been directly applied to the task of defining OOP test programs [16,32]. These techniques build trees of method calls and their required parameters. However, we contend that pheromone-guided synthesis is more suitable for generating method sequences than GP's mutation and crossover operators [32]. TACO creates test programs that are correct by construction, whereas mutation and crossover of existing test programs, as is done in GP, can lead to invalid programs that need to be fixed or discarded. Other approaches combine GP and ACO for program synthesis, such as the work of Hara et al. [17]. This combination is called Cartesian Ant Programming and is normally applied to circuits [17,21]. Although this work does solve some programming problems, to the best of our knowledge, there currently exists no pure ACO-based work that synthesizes complex Java programs.

3   TACO Algorithm

Solving ATG entails generating programs. Programs are discrete objects. Viewed as a combinatorial optimization problem, ATG is infeasible, because the search space is vast and under-constrained. Crucially, test programs require not only syntactic correctness but also semantic correctness in order to build a valid program state. Others have sought to constrain the problem via reformulation as multi-objective search; this tack constrains the fitness of solutions, but not the search space. Our key insight is that ATG naturally decomposes into a three deep hierarchy of successive subproblems – and each subproblem further constrains the overall problem. The three tiers are (1) goal selection, (2) test program synthesis, and (3) primitive and string types data generation.

Tier I prioritizes and selects goals within the program under test. As its performance metric is branch coverage, it targets a branch. In each iteration, Taco focuses on covering one branch at a time. Taco gains great leverage from its first tier: by selecting promising goals, Taco restricts the second two tiers to only the relevant subset of the search space. Taco's program synthesis (Tier II) is its core novelty. ATG of OOP is unique in that it often requires building a valid program state to reach a goal. This subtask requires calling a potentially large number of methods in sequence. A simple example is a goal guarded by an `if` that depends on a variable, which, in turn, can only be set by calling a method. Tier II searches over possible test programs, building objects and generating sequences of methods with data holes for primitives and strings. ACO shines at this subtask, quickly finding sequences of available methods that reach the goal. Finally, the last tier, data generation, is well-understood and well-solved by the ACO community [28]: here, ACO efficiently finds primitive and string values to fill the data holes in Tier II's method sequences.

Taco's three tiers operate independently. Therefore, one can easily replace or modify the specific algorithm each tier uses if a new state of the art emerges for its task—this could be another ACO variant or an entirely different algorithm. For example, TIII's algorithm could be modified to exploit research into numerical data generation when testing numerical programs.

Branch distance measures how far a conditional statement's control expression is from a different evaluation outcome in a particular program state. For example, the branch distance of numerical equality, a `==` b, is often computed as $|a - b|$. Given the conditional statement `x == 10`, `x := 8` creates a program state that is a distance of 2 from evaluating to true, while `x := 100` creates a state that is 90. Branch distance has long been used in ATG. All three of Taco's tiers use it as their fitness function to evaluate ants. For numerical conditional statements, Taco uses Korel's measures [20]; for string comparisons, it uses Levenshtein distance [2]. Often in ATG, approach level guides branch distance. Approach level is the number of control dependent statements, required to reach a target, that were not executed [6]. Tier I defines goals such that their approach level is always 0 (Sect. 3.2), so Taco does not use it.

Taco has two phases, shown in Fig. 1. The first phase generates ants using three-tiered ACO. The second phase converts the ants into test cases, which it executes, observes and then uses to calculate fitness. Taco uses these fitnesses to update its ant archives and pheromones lists for both goals and methods. Although Taco generates an ant for a specific goal, that ant may be effective,

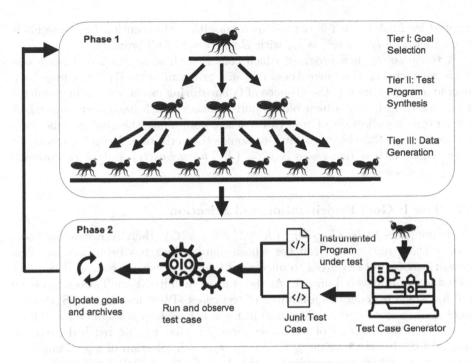

Fig. 1. An overview of the TACO approach. Phase 1 generates new ants. The distribution of work amongst the tiers is tunable; TACO chooses 1 goal per iteration, synthesizes 5 test programs per goal and generates 20 sets of data values per test program. Phase 2 converts each ant from phase 1 into a JUnit test case, then executes the instrumented program under test on that test case. Finally, TACO updates its goals and archives based on an ant's coverage and fitness.

or even cover, other goals. Therefore, once an ant's test case is executed, TACO checks for incidental coverage and passes the ant to every goal whose source unit it covered. TACO then checks the ant against a goal's best so far archive, and updates the pheromones and archives of each method in the ant's test program. If the ant is the new best for a goal, TACO resets the goal's counter. If an ant covers a new branch or block, TACO adds it to the test suite. TACO removes covered goals and updates the list of goals to include those that now meet the criteria (Sect. 3.2).

3.1 Problem Definition

A program, P, can be viewed as a Control Flow Graph (CFG) where the nodes are basic blocks, and edges are branches. A basic block (subsequently called block) is a sequence of statements which has one entry point, one exit point and, if the first statement is executed, all statements must be executed once and in order [1]. Every block has a unique label and the set of all blocks within P is denoted by L. A branch is defined as a transition of execution from one block to another $l_i \rightarrow l_j$ written as b_{ij} with B as the set of all branches.

A test case, t_k, is a program which calls P with some input and has a test oracle that checks the correctness of some program state [7]. This may be a specific output or merely the absence of failure during execution. From execution of t_k, one can observe which blocks and branches which have been covered. A test suite is a collection of test cases, where commonly the goal is some form of coverage, be that block, branch or some other criterion. Branch coverage of a test suite with n test cases is $C = \bigcup_{k=1}^{n} branches(t_k)$ and block coverage $U = \bigcup_{k=1}^{n} blocks(t_k)$.

3.2   Tier I: Goal Prioritization and Selection

We define goals to be $G = \{b_{i,j} \mid l_i \in U \wedge b_{i,j} \notin C\}$, thereby restricting goals to only those uncovered branches whose source block has been covered. This prevents allocating resources to uncovered branches that are control dependent upon another uncovered branch. At the start of each iteration, TACO selects one goal for which it generates a number of test cases. TACO uses an Ant System to select goals: a goal's probability, $p(g)$ in Eq. 1, is based upon its pheromone level τ_g, which is the number of uncovered branches that can be reached from the goal, and the heuristic value $\eta_g = (1 - c_g \cdot \delta)$. Each selection of a goal increases a counter, c_g, which is multiplied by the decay factor, δ (0.01 for TACO).

$$p(g) = \frac{\tau_g \cdot \eta_g}{\sum_{k \in G} \tau_k \cdot \eta_k} \tag{1}$$

Pheromone does not decay, instead the heuristic is used as a decay mechanism, using a counter to decay rather than reducing the pheromone at every timestep. TACO enforces a minimum pheromone level of 0.1. This process favours goals that lead to larger regions of uncovered code, and those for which TACO is regularly discovering new, best test cases. The counter helps to avoid wasting time on infeasible goals, as once TACO gets as close as is possible, the counter will not be reset again and the probability of selecting the target will only decrease. Previous ATG tools have used counters in this way [3].

3.3   Tier II: Test Program Synthesis

In Tier II of TACO, we take a non-traditional approach to program synthesis: holes in our program are considered as data holes and are missing primitive and string data values, not arbitrary code fragments [15]. Furthermore, to the best of our knowledge, we are the first to apply ACO to object-oriented program synthesis, which has been dominated by enumeration and constraint solving. For each goal selected by Tier I, many test programs (ants) are synthesized to allow optimization towards a covering test program (five in our implementation). Each goal has an archive of the best performing ants and pheromone levels for each method that has been called by ants considered the goal. Algorithm 1 and 2 show the pseudocode for Tier II.

Algorithm 1. Tier II: The **testSynth** algorithm builds a test program, represented as a sequence of method calls, $\langle m, i, o \rangle$. The $select_{1,2}$ and $getAvailableMethods$ functions are described in the text. $buildMethodSeq$ calls Algorithm 2.

Input: P, the program under test.
Input: g, the goal selected in Tier I.
Input: A_g, a list of previously generated Ms ordered by performance at g.
Input: s_g, a function that outputs the pheromone of a method at g.
Output: M, a sequence of method call tuples with holes for primitive or string data.
1: **if** $random(0.0, 1.0) > select\_threshold$ **then**
2:    **return** $select_1(A_g)$ {This helper function is described in text.}
3: $M := \langle \rangle$
4: **repeat**
5:    $M_a := getAvailableMethods(P, M)$ {methods in scope}
6:    $m := select_2(M_a, s_g)$ {This helper function is described in text.}
7:    $M := buildMethodSeq(M, method, s_g)$ {See Algorithm 2}
8: **until** $(resources\ exhausted \lor m = NULL)$
9: **return** M

When deciding which test program to execute next, TACO, can select an existing test program from the archive, or synthesize a new one. At line 1 of Algorithm 1, a global parameter dictates the probability of selecting versus generating a test program (in our implementation the probability of either is 50%). A test program is selected from the archive, $select_1$ on line 2, with probability proportional to its position within the archive. As the archive is sorted by branch distance, the test program with the lowest branch distance is the most likely.

At first, available methods are constructors or static methods of the program under test. Then, moving forwards, any method of an object that has been instantiated within M and in scope. The function s_g returns the current pheromone level of a method with respect to goal g. A method's pheromone starts at ρ_0; ants that perform well, and are added to a goal's archive, add pheromone to each method they visit. Pheromone change is shown in Eq. 2, where n is the number of ants added to the goal's archive that call method m_i, a is the number of ants generated that call m_i and γ_m, and δ_m are algorithm parameters that dictate amount of pheromone laid and removed[1]. Therefore, pheromone decays every time a method is added to a test program, rather than at each time-step. At time N, $\rho_{m_i}^N$ gives the pheromone of method m_i in Eq. 3.

$$\Delta\rho_{m_i} = (n \times \gamma_m) - (a \times \delta_m) \quad (2) \qquad \rho_{m_i}^N = \rho_{m_i}^{N-1} + \Delta\rho_{m_i} \quad (3)$$

At each step, TACO selects a method, line 6 of Algorithm 1 ($select_2$), probabilistically according to pheromone levels of available methods. TACO can choose to end the test program before reaching the max length at this point, by selecting

---

[1] For our implementation of TACO the following values were used: $\rho_0 = 50$, $\gamma_m = 0.5$, $\delta_m = 0.05$. With a minimum pheromone of 1 and maximum of 100.

Algorithm 2. Tier II: This **buildMethodSeq** adds a method call, m, its parameters, and a reference to its output, to the sequence of methods being generated, M. It is recursive, because some of m's parameters might be methods or be an abstract data type one of whose constructors we must call. **buildMethodSeq** leaves data holes in the sequences for primitive or string parameters. $select_2$ and $insert$ are described in the text.

Input: m, the method selected to be added to the test program.
Input: M, the method sequence (test program) which m should be added to.
Input: s_g, a function that outputs the pheromone of a method at g.
Output: M, a sequence of method call tuples with holes for primitive or string data.
1: $inputs := \langle \rangle, rid := NULL$
2: **if** $!isVoid(m)$ **then**
3:    $rid := getNonce()$
4: **for all** $p \in getParameters(m)$ **do**
5:   **if** $instanceof(p) \in primitives \cup \{String\}$ **then**
6:      $inputs += HOLE : instanceof(p)$
7:   **else**
8:      $C_a := getAvailableConstructors(p)$
9:      $c := select_2(C_a, s_g)$
10:     $M := buildMethodSeq(M, c, s_g)$ {Recursive call, returns M with c inserted}
11:      $inputs += getRid(M, c)$
12: $M := insert(M, \langle m, inputs, rid \rangle)$ {This helper function is described in text.}
13: **return** M

$NULL$ in place of a method. When adding the selected method to the sequence in Algorithm 2, any primitive or string values required are left as holes within the program (line 6). Tier III later searches over the input domain of these holes to find a set of instances that minimize branch distance to the goal. Object parameters are referenced using their rid, an output identifier which is independent of position within sequence. When the tuple defining a method call is added to a test program, line 12 of Algorithm 2 ($insert$), it is injected at the last position in the sequence where it still has an affect. For example, Fig. 2 shows the JUnit representation of a sequence of method calls. When `v1.methodUserObj()` (line 6) was selected, it had to be inserted after `v1` was defined (line 5), but before it was used (line 7).

Tier II's **testSynth** generates sequences, but a natural way to view them is as programs with data holes. The JUnit test case in Fig. 2 is obtained from the following sequence

$$M = \langle (\text{new ExClass}(), \langle \rangle, \text{v0}), (\text{new UserObj}, \langle HOLE: String \rangle, \text{v1}),$$
$$(\text{v1.methodUserObj}, \langle \rangle, NULL), (\text{v0.method1}(), \langle \text{v1} \rangle, \text{v2}),$$
$$(\text{v0.setValue}, \langle HOLE: int \rangle, NULL), (\text{v0.target}(), \langle \rangle, NULL) \rangle$$

To construct M, **testSynth** takes a goal, assumed to be within `target()`. Early in the search-process, **testSynth** generates nearly random method sequences, as pheromone levels initially provide no guidance. As TACO iterates, methods that create states that execute branches close to the goal will accumulate

```
public class ExampleClassTest() {
    @Test
    public void testMethod1() {
        ExClass v0 = new ExClass();
        UserObj v1 = new UserObj(<HOLE: String >);
        v1.methodUserObj();
        boolean v2 = v0.method1(v1);
        v0.setValue(<HOLE: int >)
        v0.target()
    }
}
```

Fig. 2. Output of Tier II: a method sequence realized as a JUnit test program with data holes.

pheromones. Pheromone levels will rapidly suggest selecting `target()`. The selection of **v0.method1()** triggers **addMethodSeq**, which processes **v0.method1()**'s parameter of type `UserObj`. **addMethodSeq** then probabilistically selects one of `UserObj`'s constructors, relying on the pheromone levels laid by ants in previous iterations. This constructor builds the v1 *rid*, which **addMethodSeq** passes to **v0.method1()**. While **addMethodSeq** does not directly change pheromone levels, it does indirectly affect them: its addition of methods to a method sequence means that ants will traverse and update those method's pheromone levels in subsequent iterations. **addMethodSeq**'s addition of `UserObj`'s constructor makes v1.methodUserObj() available to subsequent iterations, as the v1 *rid* is within the method sequence.

3.4   Tier III: Input Data Generation

Having progressed through the two previous tiers, the search space has been reduced from all valid test programs for the program under test to the input domain of the primitive and string holes. For Fig. 2, the input domain is one `String` and one `int`. For each test program, there are still a huge number of possibilities, which is why the optimization process samples many possible values for each (20 in the case of TACO's implementation).

A goal has an archive of primitive and string values for every method which has been called in a test program considered at the goal. Each of these archives operates in accordance with $ACO_{\mathbb{R}}$, allowing new values to be sampled based on the contents of the archive [28]. When values are needed for a method, a guide is selected from the method's archive based on position. For string values the method for sampling is as in Dorylus [9], mutating the guiding value by inserting, removing and swapping characters. For primitives, the guide value is used to define a Gaussian distribution, from which TACO samples a new value for each variable. Eq. 4 and 5 are the taken from directly from $ACO_{\mathbb{R}}$.

$$G_e^d(x) = \frac{1}{\sigma_e^d \sqrt{2\pi}} e^{-\frac{\left(x - v_e^d\right)^2}{2\sigma_e^{d2}}} \quad (4) \qquad\qquad \sigma_e^d = \zeta \sum_{l=1}^{k} \frac{abs\left(v_l^d - v_e^d\right)}{k - 1} \quad (5)$$

v_e^d is the value of variable d in the guide e. The standard deviation, σ_e^d, is calculated as the mean difference between the guide and all other values in the archive Eq. 5, its size controlled by ζ.

When an archive has spare capacity, TACO adds the input values of any ant that calls the method to it. Once capacity is reached, the ant must have a smaller branch distance than the current bottom of the archive.

This tier is where most related work on automated test generation operates, with the holes in a test program forming the vector of inputs for the ant algorithm. As such, the specific variant of ACO used could easily be swapped and experimented with, which we plan to do in future work.

4   Evaluation

To evaluate TACO, we implemented it in Java. It instruments the class under test and obtain control flow graphs for methods within classes. TACO handles arrays and lists, treating length as an integer hole and the contents as parameters. TACO does not currently handle other Java builtins, such as maps, sets, stacks etc. Our implementation used parameter values as given in Sect. 3; please note: these are not optimized values. Future work will study the effects of different values and search for optimal default settings.

We evaluated TACO's ability to automatically generate JUnit tests for 170 Java classes. These classes are part of the SF110 corpus [14]. SF110 contains 23,886 classes from 110 Java projects selected from SourceForge between 2012 and 2014. We selected these 170 classes uniformly at random. They came from 46 projects, and have an average of 21 branches, 66 lines of code and 16 methods each. When testing, we allowed two minutes of test case generation per class (each repeated ten times). The process of compiling, running and measuring coverage of the test suites was performed after, and not timed. Coverage data was obtained by running the output test suite on the original class with JUnit and Jacoco[2].

The state of the art in ACO applied to ATG does not handle object-oriented programs. Our central result is that TACO is the first ACO approach to ATG for object-oriented programs: TACO successfully generated test cases for an average of 54% of methods across the 170 classes, covering nearly 50% of lines of code. Java is a large language with huge industrial uptake. Generating test suites for the remaining 46% of methods would rely on further engineering to implement all of Java's many constructs. These include filesystem and network interactions, which TACO has no control over.

---

[2] JaCoCo is a free code coverage library for Java: https://www.eclemma.org/jacoco/.

Table 1. Average coverage of RANDOOP, TACO and EVOSUITE on the 170 classes selected from SF110, as reported by Jacoco; TACO's performance respectably falls between two state-of-the-art ATG tools that have enjoyed substantial, longterm, and ongoing engineering effort.

Tool	Coverage criterion				
	Branch	Line	Instruction	Complexity	Method
RANDOOP	19.0%	48.3%	44.3%	46.7%	56.0%
TACO	20.2%	48.7%	47.9%	47.5%	54.2%
EVOSUITE	47.5%	70.3%	69.1%	70.2%	78.4%

We ran the same experiments with two highly developed, industry standard, Java unit test generation tools; RANDOOP and EVOSUITE. RANDOOP has been under active development for over a decade, it uses feedback-directed random test generation to build a test suite for the class or program under test [24]. It has found previously unknown errors in widely used libraries and is currently used in industry[3]. It has been used as a baseline in the Search-Based Software Testing (SBST) tool competition, where it achieved the second highest score out of five tools in 2019 [19]. EVOSUITE is the state of the art in search-based unit test generation [13]. Similarly to RANDOOP, it has been actively developed for close to a decade. At its core it uses a genetic algorithm but has become a collection of state of the art techniques for generating unit tests for Java (including filesystem and network mocking [4]). The prowess of EVOSUITE is demonstrated by the fact it has won six of seven recent Search-Based Software Testing (SBST) tool competitions [10].

Despite the huge engineering advantage of RANDOOP, TACO's results are promising, beating RANDOOP in all measures except method coverage (Table 1). EVOSUITE's combination of advanced search techniques and enormous engineering effort allows it to generate tests for 78% of methods on average, covering close to 50% of branches. Unsurprisingly, it beats both TACO and RANDOOP in all measures. For future ACO ATG for OOP variants, EVOSUITE both defines a performance target to meet (or beat) and, given EVOSUITE's history of amalgamating best-of-class search techniques, a target to join and extend.

5   Conclusion

This paper has presented a novel Ant Colony Optimization algorithm, TACO, which applys ACO to object-oriented unit test generation. TACO combines a unique tiered structure with a new ACO technique for synthesising test programs for object-oriented code. We have developed a prototype tool which implements TACO and have run it on real Java programs, generating tests for more than 50% of methods, on average. ACO is a powerful meta-heuristic and we hope

---

[3] https://randoop.github.io/randoop/.

that this paper has served as a proof of concept that it can be used to generate complex test cases for complex object-oriented programs. Future work will close the engineering gap between TACO and the other tools to provide a framework for comparing ACO variants in the domain of object-oriented ATG.

References

1. Allen, F.E.: Control flow analysis. ACM SIGPLAN Not. **5**, 1–19 (1970)
2. Alshahwan, N., Harman, M.: Automated web application testing using search based software engineering. In: International Conference on Automated Software Engineering (ASE), pp. 3–12. IEEE/ACM (2011)
3. Arcuri, A.: Many independent objective (MIO) algorithm for test suite generation. In: Menzies, T., Petke, J. (eds.) SSBSE 2017. LNCS, vol. 10452, pp. 3–17. Springer, Cham (2017). https://doi.org/10.1007/978-3-319-66299-2_1
4. Arcuri, A., Fraser, G., Galeotti, J.P.: Generating TCP/UDP network data for automated unit test generation. In: Joint Meeting on Foundations of Software Engineering (ESEC/FSE), pp. 155–165. ACM (2015)
5. Ayari, K., Bouktif, S., Antoniol, G.: Automatic mutation test input data generation via ant colony. In: Annual Conference on Genetic and Evolutionary Computation (GECCO), pp. 1074–1081. ACM (2007)
6. Baars, A., et al.: Symbolic search-based testing. In: 2011 26th IEEE/ACM International Conference on Automated Software Engineering (ASE 2011), pp. 53–62. IEEE (2011)
7. Barr, E.T., Harman, M., McMinn, P., Shahbaz, M., Yoo, S.: The oracle problem in software testing: a survey. Trans. Softw. Eng. **41**(5), 507–525 (2014)
8. Bidgoli, A.M., Haghighi, H.: Augmenting ant colony optimization with adaptive random testing to cover prime paths. J. Syst. Softw. **161**, 110495 (2020)
9. Bruce, D., Menéndez, H.D., Clark, D.: Dorylus: an ant colony based tool for automated test case generation. In: Nejati, S., Gay, G. (eds.) SSBSE 2019. LNCS, vol. 11664, pp. 171–180. Springer, Cham (2019). https://doi.org/10.1007/978-3-030-27455-9_13
10. Campos, J., Panichella, A., Fraser, G.: EvoSuite at the SBST 2019 tool competition. In: International Workshop on Search-Based Software Testing (SBST), pp. 29–32. IEEE/ACM (2019)
11. Chen, X., Gu, Q., Zhang, X., Chen, D.: Building prioritized pairwise interaction test suites with ant colony optimization. In: International Conference on Quality Software, pp. 347–352. IEEE (2009)
12. Farah, R., Harmanani, H.M.: An ant colony optimization approach for test pattern generation. In: Canadian Conference on Electrical and Computer Engineering, pp. 001397–001402. IEEE (2008)
13. Fraser, G., Arcuri, A.: Evolutionary generation of whole test suites. In: International Conference On Quality Software (QSIC), pp. 31–40. IEEE (2011)
14. Fraser, G., Arcuri, A.: A large scale evaluation of automated unit test generation using EvoSuite. Trans. Softw. Eng. Methodol. (TOSEM) **24**(2), 8 (2014)
15. Gulwani, S., Polozov, O., Singh, R., et al.: Program synthesis. Found. Trends Program. Lang. **4**(1–2), 1–119 (2017)
16. Gupta, N.K., Rohil, M.K.: Using genetic algorithm for unit testing of object oriented software. In: International Conference on Emerging Trends in Engineering and Technology, pp. 308–313. IEEE (2008)

17. Hara, A., Watanabe, M., Takahama, T.: Cartesian ant programming. In: International Conference on Systems, Man, and Cybernetics, pp. 3161–3166. IEEE (2011)
18. Harman, M., Mansouri, S.A., Zhang, Y.: Search-based software engineering: trends, techniques and applications. Comput. Surv. (CSUR) 45(1), 11 (2012)
19. Kifetew, F., Devroey, X., Rueda, U.: Java unit testing tool competition-seventh round. In: International Workshop on Search-Based Software Testing (SBST), pp. 15–20. IEEE/ACM (2019)
20. Korel, B.: Automated software test data generation. Trans. Softw. Eng. 16(8), 870–879 (1990)
21. Kushida, J.i., Hara, A., Takahama, T., Mimura, N.: Cartesian ant programming introducing symbiotic relationship between ants and aphids. In: International Workshop on Computational Intelligence and Applications (IWCIA), pp. 115–120. IEEE (2017)
22. Li, K., Zhang, Z., Liu, W.: Automatic test data generation based on ant colony optimization. In: International Conference on Natural Computation, vol. 6, pp. 216–220. IEEE (2009)
23. Mao, C., Xiao, L., Yu, X., Chen, J.: Adapting ant colony optimization to generate test data for software structural testing. Swarm Evol. Comput. 20, 23–36 (2015)
24. Pacheco, C., Lahiri, S.K., Ernst, M.D., Ball, T.: Feedback-directed random test generation. In: International Conference on Software Engineering (ICSE), pp. 75–84. IEEE (2007)
25. Rojas, S.A., Bentley, P.J.: A grid-based ant colony system for automatic program synthesis. In: Late Breaking Papers at the Genetic and Evolutionary Computation Conference. Citeseer (2004)
26. Roux, O., Fonlupt, C.: Ant programming: or how to use ants for automatic programming. In: Proceedings of ANTS, vol. 2000, pp. 121–129. Springer, Berlin (2000)
27. Sharifipour, H., Shakeri, M., Haghighi, H.: Structural test data generation using a memetic ant colony optimization based on evolution strategies. Swarm Evol. Comput. 40, 76–91 (2018)
28. Socha, K., Dorigo, M.: Ant colony optimization for continuous domains. Eur. J. Oper. Res. 185(3), 1155–1173 (2008)
29. Srivastava, P.R., Baby, K.: Automated software testing using metahurestic technique based on an ant colony optimization. In: International Symposium on Electronic System Design, pp. 235–240. IEEE (2010)
30. Toffola, L.D., Pradel, M., Gross, T.R.: Synthesizing programs that expose performance bottlenecks. In: International Symposium on Code Generation and Optimization (CGO), pp. 314–326. ACM (2018)
31. Vats, P., Mandot, M., Gosain, A.: A comparative analysis of ant colony optimization for its applications into software testing. In: Innovative Applications of Computational Intelligence on Power, Energy and Controls with Their Impact on Humanity (CIPECH). pp. 476–481. IEEE (2014)
32. Wappler, S., Wegener, J.: Evolutionary unit testing of object-oriented software using strongly-typed genetic programming. In: Annual Conference on Genetic and Evolutionary Computation, pp. 1925–1932 (2006)

Branched Structure Formation in a Decentralized Flock of Wheeled Robots

Antoine Gaget[1], Jean-Marc Montanier[2], and René Doursat[3(✉)]

[1] Manchester Metropolitan University, Manchester, UK
antoine.gaget@gmail.com
[2] Tinyclues, Paris, France
[3] Complex Systems Institute Paris Ile-de-France (ISC-PIF), Paris, France
rene.doursat@iscpif.fr

Abstract. Swarm robotics studies how a large number of relatively simple robots can accomplish various functions collectively and dynamically. Modular robotics concentrates on the design of specialized connected parts to perform precise tasks, while other swarms exhibit more fluid flocking and group adaptation. Here we focus on the process of *morphogenesis* per se, i.e. the programmable and reliable bottom-up emergence of *shapes* at a higher level of organization. We show that simple abstract rules of behavior executed by each agent (their "genotype"), involving message passing, virtual link creation, and force-based motion, are sufficient to generate various *reproducible and scalable* multi-agent branched structures (the "phenotypes"). On this basis, we propose a model of collective robot dynamics based on "morphogenetic engineering" principles, in particular an algorithm of programmable network growth, and how it allows a flock of self-propelled wheeled robots on the ground to coordinate and function together. The model is implemented in simulation and demonstrated in physical experiments with the PsiSwarm platform.

1 Introduction

Swarm robotics studies how multiple relatively simple robots can accomplish various functions collectively and dynamically [2]. Boosted by recent advances in hardware technologies, the field is expected to bring many benefits as robot swarms can cover vast areas, decentralization is robust against individual failures (allowing replacements), and local communication with neighbors consumes less energy. Its core challenge, however, is to design or rather "meta-design" [4] the motion control and collective self-assembly of robots to make them operate as a single entity, whether physically connected or loosely aggregated. Meta-designing a complex system consists of finding the rules that each agent should individually follow to interact with others, decide, act, or even evolve new rules.

Background. Some swarm systems contain a large number of simple and cheap mobile robots, creating a dense "herd" such as the Kilobot platform [19]. On the

© Springer Nature Switzerland AG 2020
M. Dorigo et al. (Eds.): ANTS 2020, LNCS 12421, pp. 42–54, 2020.
https://doi.org/10.1007/978-3-030-60376-2_4

ground, units cluster together to maintain local communication and possibly display patterns, typically guided by "chemotaxis" based on virtual pheromones. Other works experiment with smaller flocks of unmanned aerial vehicles for indoor exploration [20], also examining self-reconfiguration in case of faulty rotors [9], or schools of (sub)marine robots performing synchronous encounters [23] and cooperative load transport [12]. At the other end of the spectrum, great effort can be spent on the design of sophisticated parts and actuators capable of physical attachment to achieve "modular robotics" [1]. In these cases, a limited number of units generally only permit sparse and precise formations, such as chains and T-junctions, typically by recursive attachment.

Historically, the Modular Transformer (M-TRAN) [15] was one of the first self-reconfigurable robotic kits. A group of M-TRANs can be placed in a certain initial state and go through a series of moves to achieve some target shape. Swarm-Bot [11] is a self-assembling system comprising smaller mobile robots called s-bots, which use mounted grippers and sensors/actuators (LEDs, camera) to cling to each other or static objects, following behavioral rules and local perception. Using the s-bot model, SwarmMorph [3,17] is a morphogenetic model based on a script language able to produce small 2D robot formations to achieve certain tasks. As for the SYMBRION project [13], it created an intricate piece of hardware in the form a cube that could dock precisely with its peers: the vision was to collectively form "symbiotic" robotic organisms that could move in 3D.

In recent years, modular robot systems have become more commonplace thanks to cheaper and faster hardware. For example, HyMod [18] is a set of cubic modules with full rotational freedom, which can combine to create 3D lattice structures such as snakes or wheeled vehicles. The Soldercube [16] in a similar building-block rotational unit equipped with magnetic ties and internal sensors to detect its orientation and occupied faces. Resting on top of a planar lattice of anchors, Soldercubes are able to transform their spatial arrangement by picking up and dropping off each other through attach/turn/detach combinations of moves. The Evo-bot [8] is another modular concept intended to physically implement the growth of artificial "creatures" as compounds of differentiated robotic modules, each one with a specific function such as resource harvesting or motion control. "Soft robotic" designs [22] also attempt to mimic cell aggregation with pneumatic and magnetic cubic elements that can shrink and inflate, giving rise to organisms capable of locomotion.

Motivation. The variety of systems briefly reviewed above is only a small sample of what is done in collective and modular robotics. Again, while some works focus on the design of specialized modules to generate super-robots that can perform specific tasks, others are rather inspired by artificial life swarms capable of fluid flocking and group adaptation. Few works, however, seem to pay much attention to the process of *morphogenesis* per se, i.e. the programmable and reliable bottom-up *development of shapes* at a higher level of organization. This will be our own focus: we want to show here that simple abstract rules of behavior executed by each agent (their "genotype"), involving message passing, link creation, and force-based motion, are sufficient to generate various *reproducible and scalable* multi-robot structures (the "phenotypes") by aggregation. Ideally, agent rules are independent of its physical embodiment—but of course we also present a proof of concept using real robots.

In summary, between swarm and modular robotics, the goal of the present work is to create flexible, yet at the same time highly specific spatial formations within a larger group of small wheeled robots, based on Morphogenetic Engineering (ME) principles. The field of ME [5,6] investigates the subclass of complex systems that self-assemble into nontrivial and reproducible structures, such as multicellular organisms built by cells, or the nests built by colonies of social insects. These natural examples can serve as a source of inspiration for the meta-design of self-organizing artificial and techno-social systems. In particular, we will follow here Doursat's abstract ME algorithm of "programmable network growth" [7]—which was later modified and hypothetically applied to the autonomous deployment of emergency response teams forming chains of agents using IoT devices [21].

The rest of the paper is as follows: In Sect. 2, we describe the abstract model of collective robot dynamics based on ME principles applied to network growth. We present the underlying mechanisms allowing a small swarm, or "flock", of self-propelled robots on the ground to coordinate and function together. Then, we show how the model unfolds in simulation (Sect. 3) and physical experiments (Sect. 4) to generate different structures.

2   Model

The "meta-design" methodology of this project consists of hand-made rules programmed in all agents to foster the development of a multi-robot structure. Given different rules, robots are able to form different target shapes by making local decisions based on what they detect and exchange with their neighbors.

Ambient Space: Neighborhoods and Forces. We consider a distributed, multi-agent system where each agent only relies on local perception of the environment to control its behavior and communicates with the agents that it detects in its vicinity. All the computational logic is embedded in the agents to obtain a fully decentralized system. At an abstract level, each agent represents a single robot potentially equipped with the following limited features and capabilities: a set of proximity sensors (infrared and/or ultrasonic) placed around the robot to detect nearby obstacles, evaluate distances and avoid collisions; a communication module to exchange messages with neighboring robots; a transportation module including wheels and motors to move around autonomously; and a small camera to identify the average position of the other agents (the center of mass of the flock).

The definition and computation of each agent's neighborhood is central to the cohesiveness of a collective robotic organism, as it ensures the proper coordinated propagation of information across the structure. To this aim, we use a hybrid "topological-metric" type of neighborhood implemented by a modified Delaunay triangulation, chosen for its accurate representation of physical contacts and robustness to change. The modification consists of pruning connections that are longer than a given threshold, set just below the average minimum distance of uniformly distributed agents in space in order to accommodate real-world constraints (Fig. 1a). Since the Delaunay triangulation is not metric-based, far away robots may also be connected and this is why a cutoff length was introduced to prevent unrealistic long-range communication.

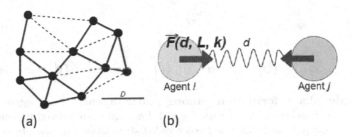

(a)   (b)

Fig. 1. Neighborhoods and forces. (a) Example of a Delaunay graph among a dozen agents where connections are trimmed (dashed lines) above a given cutoff distance D. (b) Two connected agents i and j at distance d exert virtual and opposite elastic forces of magnitude $F = k|d - L|$ onto each other (implemented by wheel-based movements, see later sections), where L denotes a free length and k a rigidity coefficient. If $d > L$, i and j move toward each other; otherwise, they pull apart.

Each neighborhood connection also carries a virtual spring creating elastic forces (Fig. 1b), which translates into wheel-based movements by each agent to stay at a certain optimal distance from its neighbors, neither too close to avoid collisions, nor too far to remain within signal range.

Network Components: Ports, Links and Gradients. The morphogenetic core of the model is derived from Doursat's original algorithm of programmable network growth [7]. It involves input/output *ports* on the nodes, *links* between nodes (on top of their neighborhood connections), and *gradients* (integer values) sent and received by the nodes over the links through the ports. All agents are endowed with the same set of pairs of input/output ports, denoted by (X_{in}, X_{out}), (Y_{in}, Y_{out}), etc. A port can be in one of three states: "open", where it accepts (in input) or creates (in output) links with neighbors; "busy", where it is already linked to a maximum number of agents (generally one) and cannot accept or create new links; and "closed", where it is disabled and devoid of links. An open input port on agent i can accept link requests originating only from its mirror output port located on a neighboring agent j, for example: $X_{\text{in}}^i \leftarrow X_{\text{out}}^j$ (but not $X_{\text{in}}^i \leftarrow Y_{\text{out}}^j$ or $X_{\text{in}}^i \leftarrow X_{\text{in}}^j$).

Each type of pair of ports is associated with a gradient field across the network, composed of integer values representing important positional information about the nodes relative to each other within the topology (essentially their hop distance), and denoted by x_g^i, y_g^i, etc. When a link $i \leftarrow j$ is created between two agents, the gradient associated with the ports is propagated through this link from j to i via an increment, i.e. $x_g^i = x_g^j + 1$ if it concerns the X ports. Then both agents switch the corresponding ports to the busy state.

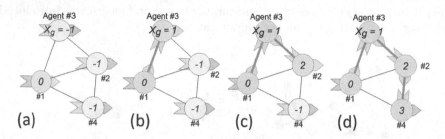

Fig. 2. Simple chain formation among static agents. Each agent executes Algorithm 1 (non-bracketed parts) with $x_N = 4$. Ports are symbolized by thick arrows (inputs as tails, outputs as heads) and color-coded states: open in blue, busy in green, closed in red. Nodes are also colored: recruiting in blue, accepting in gray, integrated in green. Edges can be of two types: neighborhood connections in black, structural links in green. (a) Initial state: gradient x_g^1 is set to 0 in a seed Agent 1 and undefined everywhere else. (b) Agents 1 and 3 agree on creating a chain link and propagate the gradient, i.e. $x_g^3 = 1$. (c,d) The chain continues to grow, with Agent 3 recruiting 2, and 2 recruiting 4 in turn. This results in $x_g^2 = 3$, which reaches the given threshold x_N (the maximum length) prescribed in the genome, therefore shuts the output port X_{out}^2 and ends the chain. (Color figure online)

In the context of collective robotics, this abstract port-link-gradient framework translates into the self-organization of branched structures made of chains of robots (Figs. 2 and 3). These structures are a subset of the background communication mesh described above. Therefore, at every time step each agent may have two types of neighbors: ones that are simply within signal range, or "connected" (thin black edges), and ones that are formally and durably "linked" to it (thick green edges)—albeit not physically for lack of hooks or magnets.

Within a static connectivity graph, network growth proceeds by peer-to-peer recruitment and aggregation of agents as follows: if agent j is already part of the growing structure and has an open output port X_{out}^j, it will look if one of its neighbors i has a corresponding open input port X_{in}^i (i.e. is not yet in the structure) to request a link creation—which it does by sending requests to each neighbor in turn.

The specifics of the growth process (which ports to open or update, how many links to create in a chain, etc.) are prescribed by an identical set of rules, or "genome", executed by each agent. The genome dictates how an agent should behave, i.e. the local decisions it should make at every time step, which will vary depending on its current neighborhood configuration and the gradient values it carries. In essence, a genomic ruleset is composed of a list of *condition→action*

Algorithm 1: Genome of a simple chain/[[branched structure]] growth

---

x_N = prescribed chain length
[[y_N = branch length]]

---

if $t = 0$ **then**
    if *is seed* **then** {close X_{in}, open X_{out}; $x_g = 0$; [[close Y_{in}, Y_{out}; $y_g = -1$]]}
    else {open X_{in}, close X_{out}; $x_g = -1$; [[open Y_{in}, close Y_{out}; $y_g = -1$]]}
    return

---

if $x_g = x_N - 1$ **then** close X_{out}
else if $x_g \geq 0$ *and* X_{out} *is closed* **then** open X_{out}
if [[x_g *is odd* **then** {open Y_{out}; $y_g = 0$}]]
if [[$y_g = y_N - 1$ **then** close Y_{out}]]

---

clauses, where conditions are based on gradients and port states, and actions update the ports. Examples of genomes and structures developed from them are shown in the next subsection.

In the beginning, agents are scattered at random across the arena. One agent is chosen to be the seed of the structure and is initialized differently from the others. Typically its input port is closed, its output port open, and its gradient value set to 0. Conversely, all other agents start with open inputs, closed outputs, and undefined gradients at -1. Then, each agent repeatedly executes four main steps in a loop: (a) port states are changed according to the genomic rules; (b) links are created where possible; (c) gradient values are propagated and updated; and (d) the robot moves by applying spring forces and/or a search behavior. The latter step is explained below in the subsection about mobile network growth.

Examples of Genomes and Structures. In this section, we give two examples of abstract network growth among static agents on top of their background communication graph, omitting spring forces and motion. The first system involves four agents forming a simple chain based on one pair of X ports (Fig. 2). The genome is described in Algorithm 1 (non-bracketed parts only), where x_N is set to 4. As explained above, at first ($t = 0$) the unique seed agent is initialized differently from the other agents. Then, as soon as an agent is recruited into the structure, its gradient x_g becomes positive by propagation and triggers the opening of the output port X_{out}, unless $x_g = x_N - 1$, which means that it found itself at the end of the chain and should close X_{out}.

The second example shows a slightly more complicated branched structure, or "creature" composed of a "body" chain of five agents and two short "leg" chains of two agents each, sprouting from the even-positioned body agents (Fig. 3). Ports X are used to form the body, while different ports Y support the legs. The genomic rules are described in Algorithm 1, with $x_N = 5$ and $y_N = 3$. Compared to the previous example, the added complication consists of managing ports Y and their associated gradient y_g depending on certain values of x_g. Here, if $x_g = 1$ or 3 it means that the agent is second or fourth along the main chain, therefore it should open its other output port Y_{out} and set $y_g = 0$ to start a branch by recruiting free agents via Y_{in}. For branch termination, the same condition is used in Y, i.e. closing Y_{out} when $y_g = y_N - 1$.

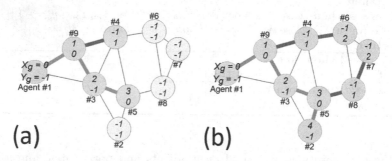

Fig. 3. Branched structure formation among static agents. Each agent possesses two pairs of ports, X and Y (not represented), and executes Algorithm 1 with $x_N = 5, y_N = 2$. Color coding is the same as Fig. 2, with red lines symbolizing Y links (leg branches). (a) Seed is Agent 1, which recruited Agent 9 via an X link, thus starting the body chain. Then, Agent 9 applying both X_{out} and Y_{out} opening rules recruited Agent 3 on X (extending the body) and Agent 4 on Y (starting a leg branch). The main chain then continued with Agent 5, which was also preparing to grow a second le.g. (b) Finished structure, where all agents have satisfied the rules and no more recruitment is attempted. (Color figure online)

Mobile Network Growth in Space. In reality, as robots move around, the background mesh is not static but continually updated (as per Fig. 1a) so that new connections may appear and existing ones disappear. In spite of this, already created structural links will persist: if communication between linked robots is accidentally interrupted, they keep tabs on each other and resume regular gradient exchange whenever possible. This should rarely happen, however, as elastic forces tend to keep them close to each other, as if physically attached.

To maximize matching opportunities, agents not yet recruited navigate toward, and stay close to, the existing structure. If an agent finds itself isolated far away without neighborhood connections, it uses the camera to search for the bulk of the flock and head over there. When its front proximity sensors detect a close obstacle, then two scenarios can happen: (α) in *simulation* (Sect. 3), it initiates a clockwise exploration behavior by turning left and keeping its right-side sensors active until it receives a link request; (β) in the *physical experiments* (Sect. 4), it just sticks near the first encountered neighbor(s) by applying default elastic forces. In this last case, an added condition is to receive a "connected-component" flag propagated from the seed agent over the graph connections: if it does not get it, then it moves again toward the flock's center.

To be more precise, different types of springs are used or not depending on the local state of neighboring nodes. Three cases can be distinguished: (i) if both nodes are integrated into the structure and linked to each other, then a strong attractive elastic force is applied between them with a coefficient k_{att} and a length L_{att} significantly smaller than the cutoff communication distance D to keep them close; (ii) if both nodes belong to the structure but are not directly linked (yet spatially close, e.g. if the chain is folded), then a weak repulsive force is used to pull them apart, with a coefficient k_{rep} and a length L_{rep} greater than or about equal to D; (iii) if one or both nodes are outside of the structure, then two variants happen: (iii. γ) in *simulation*, no spring force is applied and the free agents rely on proximity sensing for their search behavior (the linked agents ignore them when calculating their forces); (iii. δ) in the *physical experiments*, the repulsion force $L_{\text{rep}}, k_{\text{rep}}$ is used to keep them at an optimal distance.

Altogether, this combination of attractive and repulsive elastic forces leads the robot flock to form a tight chain-like structure visible to the naked eye (although without physical links) while at the same time making this structure unfold in space.

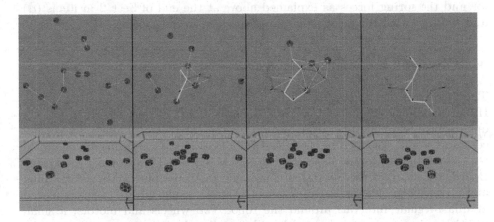

Fig. 4. Branched structure formation in simulation. Bottom: screenshots of the MORSE display at time steps $t = 0, 13, 94, 385$. Top: custom 2D visualization tool based on log files at same time steps with color code of Fig. 3. Each virtual robot executes Algorithm 1 with $x_N = 7, y_N = 3$. The 13 robots self-organize into a 7-robot chain body with three 2-robot legs at odd-numbered positions. This network structure also unfolds in space under the influence of the spring forces with parameters $D = 3.56d, L_{\text{att}} = 1.8d, k_{\text{att}} = 1, L_{\text{rep}} = 5.4d, k_{\text{rep}} = 0.5$, where d is a robot's diameter and $d = 0.5$ MORSE unit. The simulation stops at $t = 427$ when robots cannot form new links and elastic forces have reached equilibrium. Videos available at https://tinyurl.com/gaget20.

3   Simulations

Before trying our model with real robots (see Sect. 4), we implemented it in a realistic simulation. This allowed us to test and adjust the model more flexibly, in order to prepare the ground toward bridging the reality gap with the physical experiments. To this aim, we chose the MORSE simulator environment[1], a platform written in Python and powered by the Blender physics engine. MORSE offers accurate representations, physics simulation and detailed graphic display of robotic components and external objects. In our case, each agent is instantiated by an autonomous low-height cylindrical robot endowed with its own virtual control and sensorimotor abilities: two wheels and motors, eight infrared proximity sensors on its periphery, and a communication module for short-range broadcast. For simplicity, the turret camera is not simulated but replaced by global information about the center of mass of the flock sent to the robots that needed it. A Python script encoding the behavior of the simulated robots, including their genomic rules of self-assembly, is running in parallel with the MORSE engine.

The simulated experiment shown here is a flock of 13 robots forming a branched structure based on the complete genome of Algorithm 1 (Fig. 4). In addition to the networking rules, spatial motion relied on the exploration behavior and the spring forces as explained above at the end of Sect. 2 in items (α) and (i–iii. γ) with the parameter values specified in the caption.

4   Physical Experiments

The PsiSwarm platform, a disc-shaped robot on wheels, was designed by James Hilder and Jon Timmis at the York Robotics Lab[2]. It runs on Mbed OS, an open-source real-time operating system for the Internet of Things. Its control code in C++ is uploaded to the board via a USB link. PsiSwarms are equipped with the following components (Fig. 5a): an Mbed LPC1768 Rapid Prototyping Board, the heart of the operation containing the code, plus a Micro-USB plug and a Bluetooth emitter/receptor; an LCD screen; eight infrared sensors placed at quasi-regular intervals around the robot; two wheels and motors; a small joystick to input commands into the robot; and a single battery.

To centrally monitor the PsiSwarms in real time, whether to read out their trajectories or intervene in the experiment, we relied on the ARDebug software [14], an augmented-reality tool that can track the robots with ArUco square markers pasted on top of them [10] (each one carrying a binary pixel matrix that encodes a unique ID number; Fig. 5g–i), and can exchange information with them via Bluetooth. The software uses a top-down bird's-eye camera (mounted on the ceiling) to evaluate the location of every PsiSwarm and link this information with the right Bluetooth sockets to communicate with them individually. This allowed us to send to/receive from each robot data, both in the beginning (e.g., whether it is the seed) and during the experiment. However, although it is in principle possible to transfer the genomic code Algorithm 1 in that manner, the executable was instead input into each robot by hand using a USB cable.

---

[1] http://www.openrobots.org/morse.
[2] https://www.york.ac.uk/robot-lab/psiswarm/.

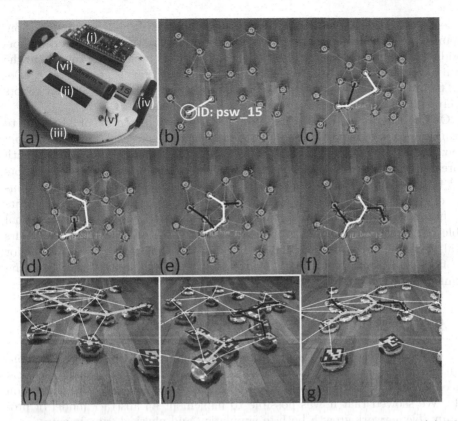

Fig. 5. Formation of linked structures in a flock of wheeled robots. (a) The PsiSwarm platform: *i.* mother board, *ii.* LCD screen, *iii.* infrared sensors, *iv.* wheels & motors, *v.* joystick, and *vi.* battery. (b–g) 20 PsiSwarm robots execute Algorithm 1 (in full) with $x_N = 5, y_N = 3$ inside a 170×180 cm arena. (b–f) Top views from the ceiling camera at times $t = 4, 13, 26, 45, 100$ s. Structural links (thick green & red) and graph connections (thin white) are automatically visualized by ARDebug in real time. (b) Shortly after initialization, the seed robot *psw_15* created a first link with a neighbor. (c) Search behavior and spring forces bring robots closer, while two more body links (green) and one leg link (red) appeared. (d–f) The branched structure continued growing until it was complete and stabilized (robots stopped moving) under the attractive/repulsive forces, with parameters $D = 43.2$ cm, $L_{att} = 11$ cm, $k_{att} = 0.6$, $L_{rep} = 32.4$ cm, $k_{rep} = 0.42$. (g) Same final state in perspective view. (h, i) Other examples of "phenotypes" based on Algorithm 1: (h) a simple 9-robot chain and (i) a 3 + 6-robot T-shape. Videos of all three experiments available at https://tinyurl.com/gaget20. (Color figure online)

Toward our goal of flock formation, we faced three technical issues: (1) the IR sensors are not powerful enough to detect the positions of neighboring robots beyond a few cm; (2) PsiSwarms are not equipped to communicate locally with each other; (3) they also lack a turret camera to spot the flock from afar. To compensate for these shortcomings, we had to infringe somewhat the principle of decentralization by resorting to proxy-mediated detection & communication. Thanks to the ceiling camera and ArUco markers, (1) the Delaunay neighborhoods were computed centrally by ARDebug and fed back to the robots (in the form of relative angle-distance pairs); (2) this information also served for ARDebug to broker peer-to-peer requests via the Bluetooth links; and (3) stray robots received from ARDebug the direction back toward the flock.

On the other hand, we also made sure to keep the intervention of ARDebug to a minimum, i.e. only provide the robots with the raw, low-level information from their surroundings that they could have otherwise gathered by themselves with more hardware. In no instance was ARDebug actually controlling the robots and telling them what messages to send and how to move; these calculations and decisions were made by each of them. Based on the relative positions of its neighbors (obtained from its fictive detectors via ARDebug), and its internal table of structural links and graph connections, each robot could compute its total vectorial force and next move, as per items (i–iii. δ) above—or head for the flock and apply protocol (β) if it was stranded far away. Three resulting formations are shown in Fig. 5b–i.

5   Conclusion

In conclusion, we proposed a morphogenetic engineering model and a demonstration of self-organized branched structure formation among small identical wheeled robots, based on local neighborhood perception and communication only. We showed that it was possible to implement an abstract model of programmable network growth both in simulation and physical experiments.

The technical problems encountered in the experiments were essentially due to limitations in the PsiSwarm's capabilities. Its lack of hardware for mid-range peer-to-peer detection & communication, and flock recognition, had to be remedied by the central monitoring system ARDebug, which tracked robots and brokered information and message-passing among them. ARDebug's role, however, remained minimal in the sense that it only emulated the neighborhood data that would otherwise be handled by extra sensors and emitters, while the core computation and decision-making modules remained on board.

Beyond these workarounds, the experiments presented so far are encouraging, although at this point they only constitute a proof of concept. To complete this study, an extended statistical analysis over many trials, whether exploring different genotypes or variable random conditions on the same genotype, should be conducted to adjust parameters and establish the resilience of the model in real-world settings. In addition, simulations and experiments with more robots must be conducted to insure the scalability of our model. For future work, more complex branched chains or loops involving other port types and a larger swarm could be attempted. Last but not least, the loose flocking structures thus created should demonstrate their usefulness by moving across space and behaving as single cohesive "creatures". Even deprived of physical hooks, they should be able e.g.. to encircle and push bulky objects—or interact in any other way with their environment via specialized differentiated robots and division of labor, similar to multicellular organisms.

Acknowledgements. AG's thesis, supervised by RD and JMM, is funded by the Dept of Computing & Mathematics at MMU. AG thanks James Hilder at the YRL for his support with the PsiSwarms that he built, and ARDebug. RD thanks MMU for funding, and Jon Timmis (YRL Head) for managing, the purchase of 40 PsiSwarms. RD's former intern, Philip Boser, did the first experiments with 3 prototypes.

References

1. Ahmadzadeh, H., Masehian, E., Asadpour, M.: Modular robotic systems: characteristics and applications. J. Intell. Robot. Syst. **81**(3–4), 317–357 (2016)
2. Brambilla, M., Ferrante, E., Birattari, M., Dorigo, M.: Swarm robotics: a review from the swarm engineering perspective. Swarm Intell. **7**(1), 1–41 (2013)
3. Christensen, A.L., O'Grady, R., Dorigo, M.: SWARMORPH-script: a language for arbitrary morphology generation in self-assembling robots. Swarm Intell. **2**(2–4), 143–165 (2008)
4. Doursat, R.: Organically grown architectures: creating decentralized, autonomous systems by embryomorphic engineering. In: Würtz, R.P. (ed.) Organic Computing. Understanding Complex SystemsUnderstanding Complex Systems, pp. 167–199. Springer, Heidelberg (2009). https://doi.org/10.1007/978-3-540-77657-4_8
5. Doursat, R., Sánchez, C., Dordea, R., Fourquet, D., Kowaliw, T.: Embryomorphic engineering: emergent innovation through evolutionary development. In: Doursat, R., Sayama, H., Michel, O. (eds.) Morphogenetic Engineering. Understanding Complex Systems, pp. 275–311. Springer, Heidelberg (2012). https://doi.org/10.1007/978-3-642-33902-8_11
6. Doursat, R., Sayama, H., Michel, O.: A review of morphogenetic engineering. Nat. Comput. **12**(4), 517–535 (2013)
7. Doursat, R., Ulieru, M.: Emergent engineering for the management of complex situations. In: Proceedings of the 2nd International Conference on Autonomic Computing and Communication Systems, pp. 1–10 (2008)
8. Escalera, J.A., Doyle, M.J., Mondada, F., Groß, R.: Evo-bots: a simple, stochastic approach to self-assembling artificial organisms. In: Groß, R., et al. (eds.) Distributed Autonomous Robotic Systems. SPAR, vol. 6, pp. 373–385. Springer, Cham (2018). https://doi.org/10.1007/978-3-319-73008-0_26
9. Gandhi, N., Saldaña, D., Kumar, V., Phan, L.T.X.: Self-reconfiguration in response to faults in modular aerial systems. IEEE Robot. Autom. Lett. **5**(2), 2522–2529 (2020)
10. Garrido-Jurado, S., Munoz-Salinas, R., Madrid-Cuevas, F.J., Marin-Jimenez, M.J.: Automatic generation and detection of highly reliable fiducial markers under occlusion. Pattern Recognit. **47**(6), 2280–2292 (2014)
11. Groß, R., Bonani, M., Mondada, F., Dorigo, M.: Autonomous self-assembly in swarm-bots. IEEE Trans. Robot. **22**(6), 1115–1130 (2006)
12. Hajieghrary, H., Kularatne, D., Hsieh, M.A.: Differential geometric approach to trajectory planning: cooperative transport by a team of autonomous marine vehicles. In: Annual American Control Conference, pp. 858–863. IEEE (2018)
13. Kernbach, S., et al.: Symbiotic robot organisms: REPLICATOR and SYMBRION projects. In: Proceedings of the 8th Workshop on Performance Metrics for Intelligent Systems, pp. 62–69 (2008)

14. Millard, A.G., et al.: ARDebug: an augmented reality tool for analysing and debugging swarm robotic systems. Front. Robot. AI **5**, 87 (2018)
15. Murata, S., Yoshida, E., Kamimura, A., Kurokawa, H., Tomita, K., Kokaji, S.: M-TRAN: self-reconfigurable modular robotic system. IEEE/ASME Trans. Mechatron. **7**(4), 431–441 (2002)
16. Neubert, J., Lipson, H.: Soldercubes: a self-soldering self-reconfiguring modular robot system. Auton. Robot. **40**(1), 139–158 (2016)
17. O'Grady, R., Christensen, A.L., Dorigo, M.: SWARMORPH: multirobot morphogenesis using directional self-assembly. IEEE Trans. Robot. **25**(3), 738–743 (2009)
18. Parrott, C., Dodd, T.J., Groß, R.: HyMod: a 3-DOF hybrid mobile and self-reconfigurable modular robot and its extensions. In: Groß, R., et al. (eds.) Distributed Autonomous Robotic Systems. Springer Proceedings in Advanced Robotics, vol. 6, pp. 401–414. Springer, Cham (2018). https://doi.org/10.1007/978-3-319-73008-0_28
19. Rubenstein, M., Ahler, C., Nagpal, R.: Kilobot: a low cost scalable robot system for collective behaviors. In: IEEE Conference on Robotics and Automation, pp. 3293–3298 (2012)
20. Stirling, T., Wischmann, S., Floreano, D.: Energy-efficient indoor search by swarms of simulated flying robots without global information. Swarm Intell. **4**(2), 117–143 (2010)
21. Toussaint, N., Norling, E., Doursat, R.: Toward the self-organisation of emergency response teams based on morphogenetic network growth. In: Proceedings of the Artificial Life Conference, pp. 284–291. MIT Press (2019)
22. Vergara, A., Lau, Y.s., Mendoza-Garcia, R.F., Zagal, J.C.: Soft modular robotic cubes: toward replicating morphogenetic movements of the embryo. PLoS ONE **12**(1), e0169179 (2017)
23. Yu, X., Hsieh, M.A., Wei, C., Tanner, H.G.: Synchronous rendezvous for networks of marine robots in large scale ocean monitoring. Front. Robot. AI **6**, 76 (2019)

Collective Decision Making in Swarm Robotics with Distributed Bayesian Hypothesis Testing

Qihao Shan$^{(\boxtimes)}$ (iD) and Sanaz Mostaghim (iD)

Faculty of Computer Science,
Otto von Guericke University Magdeburg, Magdeburg, Germany
{qihao.shan,sanaz.mostaghim}@ovgu.de

Abstract. In this paper, we propose Distributed Bayesian Hypothesis Testing (DBHT) as a novel collective decision-making strategy to solve the collective perception problem. We experimented with different sampling and dissemination intervals for DBHT and concluded that the selection of both intervals presents a trade-off between speed and accuracy. After that, we compare the performance of DBHT in simulation with that of 3 other commonly used collective decision-making strategies, DVMD, DMMD and DC. We tested them on collective perception problems with different difficulties and feature patterns. We have concluded that DBHT outperforms considered existing algorithms significantly in collective perception tasks with high difficulty, namely close proportion of features and clustered feature distribution.

1 Introduction

Collective decision making has been a longstanding topic of study within swarm intelligence. The aim of this research area is to explain how groups of natural intelligent agents make decisions together, as well as to construct decision-making strategies that enable groups of artificial intelligent agents to come to a decision. The problems being investigated usually require the agents to form a collective decision using only their individual information and local interaction with their peers. There are two categories of problems that are primarily investigated within collective decision making, consensus achievement and task allocation. In the former category, agents need to form a singular opinion, while in the latter category, agents need to be allocated to different tasks.

In this paper, we address the problem of collective perception, which is a discrete consensus achievement problem. Morlino et al. [8] introduced this problem in 2010. Many collective decision-making strategies have been adopted to address this problem. Valentini et al. proposed Direct Modulation of Voter-based Decisions (DMVD) in [16]. Valentini et al. also proposed Direct Modulation of Majority-based Decisions (DMMD) in [15], and further analyzed it in [14]. Direct Comparison (DC) of option quality was proposed by Parker et al. in [9].

© Springer Nature Switzerland AG 2020
M. Dorigo et al. (Eds.): ANTS 2020, LNCS 12421, pp. 55–67, 2020.
https://doi.org/10.1007/978-3-030-60376-2_5

These strategies usually draw inspirations from natural systems and focus on consensus forming among robots of different opinions.

In this paper, we are proposing Distributed Bayesian Hypothesis Testing (DBHT) as a novel method to perform collective perception. Our method takes inspiration from sensor fusion techniques used in sensor networks, where Bayesian reasoning has been widely used. Hoballah et al. [6] and Varshney et al. [17] have designed various decentralized detection algorithms based on Bayesian hypothesis testing in sensor networks. Alanyali et al. explored how to make multiple connected noisy sensors reach consensus using message passing based on belief propagation [1]. Such algorithms usually require strong and fixed communication between sensor nodes and are thus rarely used for swarm robotics. We have improved upon this characteristics by adopting a self-organizing communication topology. In our method, individual robots first form their estimation of the likelihood of various hypotheses by observing their immediate surrounding environment. Then a leader periodically collects opinions from other robots, and forms the final estimate of the whole swarm. In this paper, we will evaluate the performance of our proposed decision-making strategy and compare it to that of DMVD, DMMD and DC.

This paper is structured as follows. Sect. 2 presents previous works related to this paper and details about benchmark decision-making strategies. Sect. 3 provides the mathematical derivation and detailed description of our proposed algorithm. In Sect. 4, we show our experiments and evaluations of the results. Finally, Sect. 5 contains the conclusion and future work.

2   Problem Statement and Related Works

We follow Valentini et al. [13], Strobel et al. [11] and Bartashevich et al.'s [2] definition of the collective perception problem. There is an arena with a number of tiles that can either be black or white. The goal is to determine whether black or white tiles are in the majority using N mobile robots. The robots are assumed to have rudimentary low level control and perform random walk to sample the environment. They can only observe the color of the tile directly beneath themselves. We also assume that the robots have a maximum communication radius and can only perceive and communicate with their peers within the radius. The collective perception problem was first proposed by Morlino et al. [8]. The current form of the problem was established by Valentini et al., who explored the performance of several widely used collective decision-making algorithms, DMVD, DMMD and DC, in solving the collective perception problem [13]. Strobel et al., in their exploration of collective decision-making performance with Byzantine robots, tested the performance of DMVD, DMMD and DC on environments with different proportion of black tiles [11]. Ebert et al. extended collective decision making to environments with more than 2 colors [4]. Bartashevich et al. proposed novel benchmarks for collective decision-making task. They showed that apart from the proportion of black and white tiles, the pattern of them also have a great impact on the performance of collective decision-making strategies [2]. Ebert et al. then also proposed a novel collective decision-making strategy based on Bayesian statistics [3]. In this paper, we closely examine DMVD, DMMD and DC and test how their performances compare to that of our proposed algorithm.

DMVD and DMMD implement decision making by individual robot as a probabilistic finite state machine [15,16]. When applied to the collective perception problem, each robot can be in one of 4 states, exploration $E_i, i \in \{B, W\}$ and dissemination $D_i, i \in \{B, W\}$. B and W indicates the color that the robot thinks is in the majority. Both strategies start by randomly assigning half of the robots to state E_B and the other half E_W. In the exploration states, a robot samples the environment at every control loop on its own and computes the quality of its current opinion $\rho_i \in (0, 1]$. The durations of exploration states are random and exponentially distributed with a mean of σ sec. In the dissemination states, a robot broadcasts its opinion to its neighbors. The duration of dissemination states is exponentially distributed with a mean of $\rho_i g$, which is proportional to the quality of the robot's opinion. Here we follow [11,13] and set σ and g to 10 sec. Random walk routine as used in [13] is executed in all states. The process will continue until all robots' opinion become the same. DMVD and DMMD differ in their behaviors during the dissemination states. When using DMMD strategy, a robot takes the opinion favored by a majority of its neighbors plus itself. When using DMVD strategy, a robot takes the opinion of a random neighbor.

DC uses similar probabilistic finite state machines on an individual level to achieve collective decision making, except that the mean length of dissemination states is no longer $\rho_i g$ but g [9]. Thus all states have a mean duration of 10 s in our case. During dissemination, a robot will broadcast both its opinion and its quality estimate of the opinion ρ_i. At the end of dissemination states, a robot will compare its own quality estimate ρ_i with that of a random neighbor ρ_j. If $\rho_i < \rho_j$, the robot will switch its decision to j.

3   Distributed Bayesian Hypothesis Testing

In this section, we introduce our proposed approach Distributed Bayesian Hypothesis Testing (DBHT). In order to obtain an estimation of the proportion of white tiles, we model the environment as a discrete random variable with 2 possible states $V = White/Black$. $P(V = White) = P_W$ is the probability that a random place in the arena is white, thus the proportion of white tiles in the arena. At a given point of time, N robots each have made S observations of the tile colors beneath themselves. The observations can be either black or white, and we label $N \times S$ observations as $ob1, 1...ob_{N,S}$.

Given a hypothesis of the proportion of white tiles, $P_W = h$, we compute its likelihood given the past observations made, i.e., $P(P_W = h|ob_{1,1}...ob_{N,S})$. We first use Bayes rule:

$$P(P_W = h|ob_{1,1}...ob_{N,S}) = \frac{P(ob_{1,1}...ob_{N,S}|P_W = h)P(P_W = h)}{P(ob_{1,1}...ob_{N,S})} \quad (1)$$

Here $P(P_W = h)$ is the prior and $P(ob_{1,1}...ob_{N,S})$ is the marginal likelihood, both of which we assume to be the same for all hypotheses. In this case, we can apply the chain rule to $P(ob_{1,1}...ob_{N,S}|P_W = h)$

$$= P(ob_{1,1}|P_W = h)P(ob_{1,2}|P_W = h, ob_{1,1})...P(ob_{N,S}|P_W = h, ob_{1,1}, ..., ob_{N,S-1}) \quad (2)$$

Assuming the observations are all independent from each other, we have:

$$= P(ob_{1,1}|P_W = h)P(ob_{1,2}|P_W = h)...P(ob_{N,S}|P_W = h) \quad (3)$$

Thus the likelihood of a hypothesis can be computed by multiplying the conditional probability of each observation given the hypothesis. An individual robot's estimate based on its observations and hence its opinion can be computed as follows:

$$Op_1 = P(P_W = h|ob_{1,1}...ob_{1,S}) \quad (4)$$

$$= P(ob_{1,1}|P_W = h)...P(ob_{1,S}|P_W = h)P(P_W = h) \quad (5)$$

Therefore, the estimate of the whole swarm can be computed by calculating the product of the opinion of individual robots:

$$P(P_W = h|ob_{1,1}...ob_{N,S}) = Op_1 Op_2 ... Op_N \quad (6)$$

Numerically, we set 10 hypotheses expressed in the following matrix. The first column is the proportion of white tiles (P_W) and the second column is the proportion of black tiles ($1 - P_W$).

$$H = \begin{bmatrix} 0.05 & 0.95 \\ 0.15 & 0.85 \\ ... & \\ 0.95 & 0.05 \end{bmatrix} \quad (7)$$

An observation can be either $ob = \begin{bmatrix} 1 \\ 0 \end{bmatrix}$ for white or $ob = \begin{bmatrix} 0 \\ 1 \end{bmatrix}$ for black tiles. Therefore, the opinion of individual n can be calculated:

$$Op_n = \Pi_{s=1..S} H \cdot ob_s \quad (8)$$

Algorithm 1: Distributed Bayesian Hypothesis Testing

Result: $MaxIndex(Op^*)$
$n \in [1..N]$ Index of the robot itself, Leader has $n = 1$
$Op_n = [0.1, 0.1, 0.1, 0.1, 0.1, 0.1, 0.1, 0.1, 0.1, 0.1]$
$Neigh(n) = []$ List of Neighbors of robot n
$State_n = 0$, t=0, Randomly Place Robots in the Arena
Define T_S, T_D, $MaxCommDistance$, M_{max}= Max Number of Neighbors
while $Max(Op^*) < 0.99$ **do**

    if $State_n = 0$ **then**
        Random Walk
        if $t\%T_S = 0$ **then**
            Collect Observation ob
            $Op_n = Op_n \circ (H \cdot ob)$
        if $t\%T_D = 0$ & *is Leader* **then**
            $State_n = 1$
    else if $State_n = 1$ **then**
        for *m=1..N* **do**
            if $Distance(n, m) < MaxCommDistance$ & $State_m = 0$ & $m \neq n$
            & $length(Neigh(n) < M_{max})$ **then**
                Append $Neigh(n)$ with m; Append $Neigh(m)$ with n
                $State_m = 1$
        $State_n = 2$
    else if $State_n = 2$ **then**
        if *is Leader* **then**
            if *Messages from all Neighbors Received* **then**
                $Op^* = Normalize(Op_1 \circ \Pi_{m \in \text{Neigh}(1)} Message_{m \to 1})$
                $State_n = 3$
        else
            if *Messages from all Neighbors except m Received* **then**
                $Message_{n \to m} =$
                $Normalize(Op_n \circ \Pi_{x \in \text{Neigh}(n) \setminus \{m\}} Message_{x \to n})$
                $State_n = 3$
    else
        if *is Leader or Any Neighbor is in State 0* **then**
            $State_n = 0$; $Neigh(n) = []$
    $t = t + 1$

Fusion of the opinions of the individual is done through a self-organizing communication network with a tree topology. To construct such topology, each robot's behavior is designed as a finite state machine with 4 states as shown in Algorithm 1. One robot is designated as the leader, which is tasked with both observing its immediate vicinity and collecting the opinion of other robots. The leader can be chosen by the user or elected in a self-organizing way, e.g. as in [12], before the decision making process. In our experiments, the robot with index 1 is designated as the leader. All other robots are also assigned an unique index. More in details, all robots start in state 0, where they perform random walk routine the same as in [13] and modify their own opinion by sampling the color of the ground beneath themselves as shown below. Sampling is done periodically with the interval T_S. Individual robot's opinion is computed iteratively, thus only opinion after the last sampling need to be stored, making the memory and computation complexity of our method $O(N_H M_{max})$, where N_H is the number of hypotheses and M_{max} is the maximum number of neighbors a robot can have.

$$Op_{n,s} = Op_{n,s\text{-}1} \circ (H \cdot ob_{n,s}) \tag{9}$$

$$Op_{n,0} = \begin{bmatrix} 0.1\ 0.1\ 0.1\ 0.1\ 0.1\ 0.1\ 0.1\ 0.1\ 0.1\ 0.1 \end{bmatrix}^T \tag{10}$$

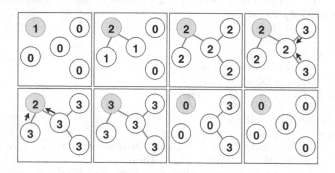

Fig. 1. Illustration of states (numbers), communication links and message passing during dissemination.

At every dissemination interval T_D, the leader will start a dissemination session. It switches to state 1, stops moving and sends out signals to look for robots nearby. Only robots within communication distance that are in state 0 will respond, to ensure that the final communication topology has a tree structure. They will establish connection with the leader and switch to state 1 too. They will also stop moving and send out signals to look for neighbors themselves, and the process continues. After searching for neighbors, a robot will go into state 2 and prepare for the transmission of messages. It is likely that some robots cannot be reached by the network due to maximum communication distance and number of neighbors. Robots in state 2 will perform message passing. It is done similarly to belief propagation algorithm. Message from robot n_1 to n_2 is an array of 10 numbers and defined as the follows.

$$Message_{n \to m} = Op_n \circ \Pi_{x \in \text{Neigh}(n) \backslash \{m\}} Message_x \to n \quad (n \neq 1) \tag{11}$$

In practice, messages are normalized before being sent to avoid underflow. Once a robot sends its message, it switches to state 3 and rest. Message passing starts from the leave nodes and gradually converges towards the leader. The leader will compute the estimate of the whole swarm as follows.

$$Op^* = Op_1 \circ \Pi_{m \in \text{ Neigh}(1)} Message_m \to 1 \tag{12}$$

Once Op^* is computed, the leader will switch back to state 0 and send out signals to its neighbors. The neighbors will switch to state 0 and send out signals as well. The process continues until all robots are in state 0. Communications thus stop and will start again at the next dissemination session. There is therefore no need for the robots to be in constant communication with each other. A new topology of the communication network will be constructed at every new dissemination session. The illustration of robot states during a dissemination session is shown in Fig. 1. Op^* will converge to one of the 10 hypotheses eventually. The algorithm stops when one of the hypotheses has a normalized likelihood of at least 0.99. The leader can then report the result to the user or direct the swarm to perform other tasks dependent on the decision.

4   Experiments

In this section, we describe our experiments and analyze the results. The setting of our experiments are largely the same as [2,11,13]. The arena is $2\,m * 2\,m$ with 400 tiles. We use 20 simulated e-puck robots [7] to perform the designated task. As for the maximum communication radius, we follow the settings proposed in [2,11] and set it 50 cm. E-pucks can only perceive up to 5 neighbors simultaneously [5] and can communicate with up to 7 neighbors [7], therefore we can set M_{max} to 5. In addition, they cannot receive multiple messages simultaneously [10], therefore setting M_{max} to 2 greatly reduces the probability of communication failure. We have tested both limits and the results are shown in Sect. 4.2 and 4.3. The low level random walk routine we used is the same as described in [13]. We use the same Matlab simulation environment as [2] to simulate the robots and arena.

Table 1. Average error using different sampling and dissemination intervals

/10%	Sampling interval T_S/s					
Dissem interval T_D /s	0.1	0.2	0.5	1	2	5
1	0.3	0.216	0.072	0.032	0.026	0.036
2	0.278	0.204	0.064	0.038	0.022	0.028
5	0.208	0.124	0.044	0.028	0.018	0.024
10	0.17	0.128	0.052	0.01	0.018	0.024
20	0.094	0.076	0.026	0.026	0.016	0.006

4.1   Finding the Optimal Sampling and Dissemination Interval

The first set of experiments aims at determining the optimal sampling and dissemination interval, T_S and T_D. It is expected that reducing the sampling interval could provide more samples per unit time. However, these samples will be collected closer to each other, thus highly correlated. It then means that the independence assumption used in Sect. 3 will not closely reflect reality. Therefore the algorithm will produce less accurate result. On the other hand, increasing the

Table 2. Average decision time using different sampling and dissemination intervals

/s	Sampling interval T_S /s					
Dissem interval T_D/s	0.1	0.2	0.5	1	2	5
1	3.09	4.75	9.53	15.80	29.66	65.82
2	3.84	6.13	11.08	18.17	32.72	69.62
5	7.36	9.22	14.98	23.11	39.08	80.05
10	12.43	14.59	21.06	30.06	45.12	87.06
20	22.23	23.87	30.07	39.95	57.40	103.64

dissemination interval means it takes longer on average for the leader to connect to a large number of robots. However, the robots can travel for a longer distance and collect more samples without violating the independence assumption during the time. Therefore the result will be more accurate. To determine the optimal values for T_S and T_D, we run 100 tests each for scenarios with white tile proportions of 0.55, 0.65, 0.75, 0.85, 0.95, and random distribution of white tiles. We compare the average error and consensus time across different T_S and T_D. Error is defined as the difference between the true proportion of white tiles and the computed proportion by the swarm.

Our results (Tables 1 and 2) show that as T_S increase from 0.1 to 1 s, there is a significant decrease in error. This is because the speed of our robots is 0.16 m/s, thus for 1 s, a robot would have traveled for 0.16 m. This is a bit bigger than the width (0.1 m) and diagonal length (0.141 m) of an individual tile, meaning that when collecting the next sample, the robot would have moved to the neighboring tile. Since the distribution of tiles is random, there is a very weak correlation between the colors of adjacent tiles, thus moving to the neighboring tile is enough to reduce the correlation to near zero. This is why when T_S is beyond 1 s, increasing it no longer reduces the error much but still increases the decision time. We thus choose 1 s as the optimal T_S. This also agrees with the findings of Ebert et al. [3], that collecting less correlated samples sparsely can improve the accuracy of decision making. However, when T_S is 5 s, there is an increase in error compared to 2 s, since too high a T_S means too few sample would be collected, thus impeding accurate decision making. At the same time, increasing T_D provides moderate improvements on accuracy and in exchange moderate increase in decision time. A trade-off need to be considered when applying our method to a real problem. Here we choose a T_D of 5 s to be used in later experiments.

4.2   Comparison with Other Collective Perception Algorithms

To determine the effectiveness of our algorithm and how it compares to other state-of-the-art algorithms, we follow the experimental framework of Bartashevich et al. [2] and perform simulations of all considered algorithms for different difficulty ($\rho_b^* = number_{\text{black tiles}}/number_{\text{white tiles}}$) as well as different patterns of black and white tiles. For difficulty, we use the same test cases as in [2,11], $\rho_b^* = [0.52, 0.56, 0.61, 0.67, 0.72, 0.79, 0.85, 0.92]$. White color is always kept in majority. The patterns are selected from the patterns used for matrix visualization and classified according to entropy (E_c) and Moran index (MI). E_c describes the densities of clusters in the pattern and MI describes the level of connectivity between clusters. See [2] for detailed definition. For every ρ_b^* and pattern, the simulation is run 100 times. The performance of the proposed algorithms is measured by the exit probability. It is the probability that the swarm will come to the correct decision that white color is in majority. For DBHT, a result is considered correct when the computed hypothesis for white tile proportion is among 0.55, 0.65 ... 0.95. The test results are shown in Fig. 2. The first row shows an example of each pattern tested. The second row shows the exit probability for every algorithm considered. The shaded area indicates the standard deviation of our measurement of exit probability. It is computed theoretically by treating the exit probability as the mean of 100 Bernoulli trials. Thus the standard deviation is $\sqrt{p(1-p)/100}$, where p is the measured exit probability. The third row shows the mean decision time for every algorithm considered. The shaded area indicates the standard deviation of all the samples.

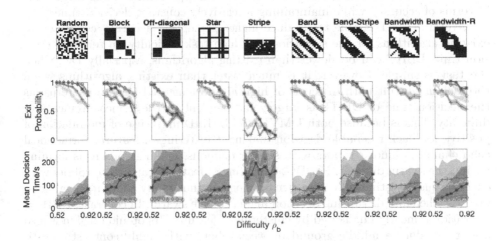

Fig. 2. Exit probability and decision time for all algorithms in 9 patterns over various difficulties. Red+:DMVD, Greeno:DMMD, Blue*:DC, Cyan□:DBHT $M_{\max} = 5$, Magenta◇:DBHT $M_{\max} = 2$ (Color figure online)

It can be observed that maximum neighbor limit only has a small impact on the performance of DBHT algorithm. Exit probability when $M_{\max}$ is 2 and 5 are usually very close. Decision time is slightly longer when the limit is 2.

Random $(E_c \approx 0.5, MI \approx 0)$ pattern is the most studied pattern in previous works. All considered algorithms have comparable performance when ρ_b^* is low, with DC having the lowest decision time. As ρ_b^* increase, DMVD and DMMD have a significant drop in accuracy and a rise in decision time. The accuracy of

DBHT and DC is more resilient to the difficulty increase, however, DC also has a significant increase in decision time.

Star ($E_c \approx 0.8, MI \approx 0.4$) and **Band** ($E_c \approx 0.7, MI \approx 0.3$) pattern are observed to be more challenging than random pattern for collective perception. DMVD and DMMD have lower accuracy and higher decision time compared to DC and DBHT. DC and DBHT are comparable in accuracy. They are also comparable in decision time when the difficulty is low, but DC's decision time increases significantly when the difficulty is high.

Bandwidth ($E_c \approx 0.9, MI \approx 0.6$) and **Bandwidth-R** ($E_c \approx 1, MI \approx 0.7$) see DBHT outperforming DMVD and DMMD in both accuracy and decision time. However its accuracy is not as high as DC especially at the highest tested difficulty of 0.92. In terms of decision time, DBHT and DC have similar performance at low difficulty, but the decision time of DC quickly rises as difficulty increases.

Block ($E_c \approx 0.9, MI \approx 0.8$), **Off-diagonal** ($E_c \approx 0.9, MI \approx 0.8$), **Stripe** ($E_c \approx 1, MI \approx 0.8$) and **Band-Stripe** ($E_c \approx 0.9, MI \approx 0.6$) are observed to be the most difficult collective perception scenarios, with highly clustered black tiles. In these scenarios, existing algorithms often becomes very inaccurate with long and volatile decision time. DBHT is able to outperform existing algorithms in terms of accuracy while maintaining a relatively constant decision time.

Overall, DBHT is able to produce higher perception accuracy compared to existing algorithms for the scenarios with high task difficulty. In terms of decision time, DBHT can be slower than existing algorithms, especially DC, when the task is simple. However, it is much faster than existing algorithms when the task is difficult both in terms of high ρ_b^* and clustered feature patterns, as the decision time of DBHT is largely independent of the 2 measurements of difficulty. This is because both DMVD and DMMD make use of modulation of positive feedback to enable decision making. Clustering of features forms local echo chambers among robots with similar opinions, making consensus forming within the swarm difficult. For DC, clustering of features can create robots with extreme quality estimation of its own opinion and thus rarely adopt its neighbors' opinion, therefore disrupting decision making of the swarm. In contrast, DBHT's operation is mostly unaffected by clustering of features. Its opinion fusion is also able to produce a middle ground between robots with highly contrasting estimates. In addition, if collective perception is to be applied in the real world, ρ_b^* and feature patterns are usually unknown. Thus the volatile decision time of the three existing algorithms causes a halting problem. When the algorithm is running for a long time with no clear decision, there is a dilemma whether to keep the algorithm running for even longer time, or to recognize a failed run and restart the process, wasting past progress.

4.3   Estimation Accuracy and Effects of Limiting Maximum Neighbors

The performance of DBHT can also be measured by the difference between the most likely hypothesis computed by the swarm and the correct hypothesis that is closest to the true P_W. Among our test cases, we classify ρ_b^* value of [0.52, 0.56, 0.61] to hypothesis $P_W = 0.65$ and the rest to $P_W = 0.55$. The average errors for all tested scenarios and different M_{max} are shown in the top row in Fig. 3.

Across all patterns, the error usually spikes at ρ_b^* value of 0.61 or 0.67. These ρ_b^* are in the middle of 2 classes which can cause some error during classification. In addition, the 2 plotted curves are very close to each other and thus a change in M_{max} in DBHT does not significantly impact the accuracy of collective perception. This is because although a low M_{max} does reduce the average number of opinions that can be collected as shown in the bottom row in Fig. 3, to meet the likelihood threshold on the final chosen hypothesis of 0.99, the robots have to collect more samples over longer periods of time, as shown in Fig. 3 (middle row). Therefore, the limit on maximum number of neighbors gives a trade-off between decision time, design complexity and robustness of the system.

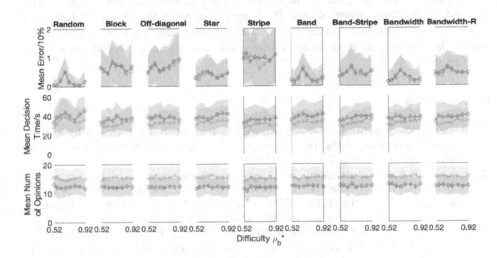

Fig. 3. Mean error, mean decision time, and mean number of opinions collected for DBHT with different M_{max}, Cyan □:DBHT $M_{max} = 5$, Magenta ◇:DBHT $M_{max} = 2$ (Color figure online)

5   Conclusion

In this paper, we have proposed DBHT as a novel collective perception strategy. We have argued that a distributed perception but centralized opinion fusion strategy for decision making is easier to use in the swarm robotics than a fully decentralized decision-making strategy used in most state-of-the-art strategies. After that, we tested the performance of DBHT with different sampling and dissemination intervals. We have concluded that, up to a limit, collecting sparse and uncorrelated samples could increase perception accuracy but also increase the decision time. Changing the dissemination interval presents a similar trade-off. We compared DBHT's performance with that of 3 other state-of-the-art collective decision-making strategies, DMVD, DMMD and DC, in how well they determine which color is in the majority. We have shown that DBHT often has superior performance due to its resilience in high ρ_b^* and feature patterns with large clusters, as well as its stable decision time regardless of the difficulty of the environment. Finally, we examined DBHT's ability to accurately estimate the proportion of colors as well as the effect of limiting the maximum number of neighbors during dissemination. We have found out that the error not only generally increase with ρ_b^* and pattern difficulty, but also tend to spike around boundary cases between classes. Also, the limit on maximum number of neighbors does not have a significant impact on estimation accuracy. The decrease in number of opinions that can be collected is made up by a longer decision time, which means more uncorrelated samples.

In future work, we plan to utilize Bayesian reasoning further to model the correlation between observations. Here we circumvent this issue by having a large sampling interval. Taking correlation between samples into consideration can make decision making more robust when a decision is needed on a short time frame.

References

1. Alanyali, M., Venkatesh, S., Savas, O., Aeron, S.: Distributed Bayesian hypothesis testing in sensor networks. In: Proceedings of the 2004 American Control Conference, vol. 6, pp. 5369–5374. IEEE (2004)
2. Bartashevich, P., Mostaghim, S.: Benchmarking collective perception: new task difficulty metrics for collective decision-making. In: Moura Oliveira, P., Novais, P., Reis, L.P. (eds.) EPIA 2019. LNCS (LNAI), vol. 11804, pp. 699–711. Springer, Cham (2019). https://doi.org/10.1007/978-3-030-30241-2_58
3. Ebert, J.T., Gauci, M., Mallmann-Trenn, F., Nagpal, R.: Bayes bots: collective Bayesian decision-making in decentralized robot swarms. In: International Conference on Robotics and Automation (ICRA 2020) (2020)
4. Ebert, J.T., Gauci, M., Nagpal, R.: Multi-feature collective decision making in robot swarms. In: Proceedings of the 17th International Conference on Autonomous Agents and MultiAgent Systems, pp. 1711–1719. International Foundation for Autonomous Agents and Multiagent Systems (2018)
5. Francesca, G., Brambilla, M., Trianni, V., Dorigo, M., Birattari, M.: Analysing an evolved robotic behaviour using a biological model of collegial decision making. In: Ziemke, T., Balkenius, C., Hallam, J. (eds.) SAB 2012. LNCS, vol. 7426, pp. 381–390. Springer, Heidelberg (2012). https://doi.org/10.1007/978-3-642-33093-3_38
6. Hoballah, I.Y., Varshney, P.K.: Distributed Bayesian signal detection. IEEE Trans. Inf. Theory **35**(5), 995–1000 (1989)
7. Mondada, F., et al.: The e-puck, a robot designed for education in engineering. In: Proceedings of the 9th Conference on Autonomous Robot Systems and Competitions, vol. 1, pp. 59–65. IPCB: Instituto Politécnico de Castelo Branco (2009)
8. Morlino, G., Trianni, V., Tuci, E., et al.: Collective perception in a swarm of autonomous robots. In: IJCCI (ICEC), pp. 51–59 (2010)

9. Parker, C.A., Zhang, H.: Biologically inspired collective comparisons by robotic swarms. Int. J. Robot. Res. **30**(5), 524–535 (2011)
10. Salomons, N., Kapellmann-Zafra, G., Groß, R.: Human management of a robotic swarm. In: Alboul, L., Damian, D., Aitken, J. (eds.) TAROS 2016. LNCS, vol. 9716, pp. 282–287. Springer, Cham (2016). https://doi.org/10.1007/978-3-319-40379-3_29
11. Strobel, V., Castelló Ferrer, E., Dorigo, M.: Managing Byzantine robots via blockchain technology in a swarm robotics collective decision making scenario. In: Proceedings of the 17th International Conference on Autonomous Agents and MultiAgent Systems, pp. 541–549. International Foundation for Autonomous Agents and Multiagent Systems (2018)
12. Trabattoni, M., Valentini, G., Dorigo, M.: Hybrid control of swarms for resource selection. In: Dorigo, M., Birattari, M., Blum, C., Christensen, A., Reina, A., Trianni, V. (eds.) ANTS 2018. LNCS, vol. 11172, pp. 57–70. Springer, Cham (2018). https://doi.org/10.1007/078_3_030_00533_7_5
13. Valentini, G., Brambilla, D., Hamann, H., Dorigo, M.: Collective perception of environmental features in a robot swarm. In: Dorigo, M., et al. (eds.) ANTS 2016. LNCS, vol. 9882, pp. 65–76. Springer, Cham (2016). https://doi.org/10.1007/978-3-319-44427-7_6
14. Valentini, G., Ferrante, E., Hamann, H., Dorigo, M.: Collective decision with 100 kilobots: speed versus accuracy in binary discrimination problems. Auton. Agents Multi-Agent Syst. **30**(3), 553–580 (2016)
15. Valentini, G., Hamann, H., Dorigo, M.: Efficient decision-making in a self-organizing robot swarm: on the speed versus accuracy trade-off. In: Proceedings of the 2015 International Conference on Autonomous Agents and Multiagent Systems, pp. 1305–1314 (2015)
16. Valentini, G., Hamann, H., Dorigo, M., et al.: Self-organized collective decision making: the weighted voter model. In: AAMAS, pp. 45–52 (2014)
17. Varshney, P.K., Al-Hakeem, S.: Algorithms for sensor fusion, decentralized Bayesian hypothesis testing with feedback, vol. 1. Technical report, Kaman Sciences Corp, Colorado Springs, CO (1991)

Constrained Scheduling of Step-Controlled Buffering Energy Resources with Ant Colony Optimization

Jörg Bremer[1(✉)] and Sebastian Lehnhoff[1,2]

[1] Division of Energy Informatics, Department of Computing Science,
University of Oldenburg,Oldenburg, Germany
joerg.bremer@uol.de
[2] OFFIS Institute for Information Technology, Oldenburg, Germany
lehnhoff@offis.de

Abstract. The rapidly changing paradigm in energy supply with a shift of operational responsibility towards distributed and highly fluctuating renewables demands for proper integration and coordination of a broad variety of small generation and consumption units. Many use cased demand for optimized coordination of electricity production or consumption schedules. In the discrete case, this is an NP-hard problem for step-controlled devices if some sort of intermediate energy buffer is involved. Systematically constructing feasible solutions during optimization degenerates to a difficult task. We present a model-integrated approach based on ant colony optimization. By using a simulation model for deciding on feasible branches (follow-up power operation levels), ants construct the feasible search graph on demand, thus avoiding exponential growth in this combinatorial problem. Applicability and competitiveness are demonstrated in several simulation studies using a model for a co-generation plant as typical small sized smart grid generation unit.

1 Introduction

A dwindling share of traditional, large, mostly fossil-fueled power plants necessitates a transfer of responsibility for the safe operation of the power grid to small and volatile renewable energy resources. Such tasks comprise predictive/day-ahead scheduling (planning production based on forecasts zeg for the next day) as well as ancillary services for power conditioning and supply security (e.g. voltage or frequency control) [28,36].

Day-ahead scheduling tasks in applications for the future smart grid have already been investigated for years and many solutions are meanwhile available. Most of these solutions are suitable for continuously (in power control, not in time domain) controllable appliances or are limited in number of time frames that may be considered. A problem arises when devices with some (probably thermal) storage are considered for time ahead scheduling that may only be controlled by step control. Step-control allows altering power input (or output

© Springer Nature Switzerland AG 2020
M. Dorigo et al. (Eds.): ANTS 2020, LNCS 12421, pp. 68–81, 2020.
https://doi.org/10.1007/978-3-030-60376-2_6

respectively) only in few discrete steps. Often devices may just be switched on or off. An attached thermal buffer store with naturally limited capacity constrains the choice of possible power levels in each time interval. The limitation in each time interval results from power level decisions for (at least a recent subset of) previous time intervals. Thus, the decision for a specific time interval cannot be made independently from the other decisions. This discrete scheduling problem has been proven to be NP-complete (number of planned time intervals) already for a single device [8].

We apply ant colony optimization (ACO) [13–15] to solve the problem of finding an optimal schedule for a given energy resource under model-predicted technical operation constraints and given objectives defining optimality. Optimality of an operation schedule can be given for example by the product of generated electricity and some given energy prices or by the similarity to (correlation with) some wanted generation profile. In this paper, we use time varying energy tariffs and load balancing as examples.

The rest of the paper is organized as follows. The paper starts with a brief overview on related work on the similar continuous problem. After formalizing the discrete problem, the proposed solution based on ant colony optimization with integrated simulation model is described. The paper concludes by demonstrating the applicability of the proposed approach by several simulation studies.

2 Related Work

Within the future smart grid many use cases will require the adaption of operation of even small sized energy resources to some given electric schedule [2, 34, 35]. Schedule usually refers to a real-valued vector $x \in \mathbb{R}^n$ with each element x_i denoting mean active power (or equivalently amount of energy) generated or consumed during the i-th time interval. Negative values denote consumption. In the discrete step-control case a schedule $x \in \mathbb{Z}^n$ will denote the number of the respective power level (operation mode) – with zero denoting off. Example use cases are demand-side management [16], microgrid control [31], demand response [33], or virtual power plants [36]. Such behavior will be required from generation units like small co-generation plants, controllable consumers like cool storages, as well as from prosumers like batteries to provide some flexibility for grid operation. Applications comprise decentralized production planning, ancillary services like frequency or voltage control, virtual power plants, or microgrid control. Coordination of decentralized energy resources might be achieved by direct control or by indirect control via incentives like varying energy prices [11, 23, 39]. We will consider both cases with the same algorithm.

The number of algorithms applied to (in-house) energy management problems is numerous, but solutions to devices with step-control are scarce [1]. Many approaches consider only continuous power level variations; examples can be found in [3, 24]. In [9], an example for binary decision variables and a mixed integer non-linear problem can be found, but without including appliances with storage constraint characteristics. In contrast, for the continuous case, several solutions are available that have been designed to handle constraints entailed by some integrated buffer store for example by decoders as constraint-handling technique [7]. Examples are given in [4,6,26,40]. We will now consider the step-control case that constitutes the discrete optimization problem. A good overview can also be found in [1].

In [12] an ant colony based solution to the related combinatorial problem of finding starting times of shiftable loads in demand-side management has been presented. One possible solution to the step-control case with discrete power levels is given by [21,22]. In this agent-based system, the flexibility of each energy resource is (in the discrete case) represented by a fixed set of a-priori model-predicted schedules. The problem is treated as a (distributed) constraint optimization problem. This approach does not consider large portions of the real flexibility of the energy resources as it uses just a small example subset. Thus, we strive for an approach that integrates the prediction-model directly into the algorithm and hence harnesses the full potential and flexibility of an energy resource.

3   Problem Statement

We consider the following planning problem. For a given device, a schedule has to be found for the operation of the device during some future time horizon such that individual operation constraints are satisfied and some given goal is met. A typical schedule comprises 96 dimensions for all 15 min intervals of a day. For the real valued case several solutions exist. We extend the planning problem to the case of step-control devices.

When starting the planning procedure, the following is known: starting time t_0, the initial operational state of the device at t_0, and some cost $c_1, \ldots, c_n$ for n discrete time intervals $\mathcal{T} = \{T_1, \ldots, T_n\}$ (not necessarily equidistant). For the period of one time interval cost are considered constant. Examples for possible cost are monetary cost for electricity given by varying prices, or set points given by some external smart grid control strategy [37]. In the latter case, cost are actually given by the deviation of the actual from the desired value.

A step-control may choose among a limited number of j discrete power levels $\mathcal{P} = \{P_1, P_2, \ldots, P_j\}$, for the duration of each time interval. The relation between power levels (denoting operated power) and the discrete number of operation mode is given by a device specific mapping

$$\Pi : \mathbb{Z} \to \mathbb{R}, \; \Pi(x_i) \mapsto P_i. \tag{1}$$

During each interval we assume a constant power output or feed-in. Equivalently, the amount of energy consumed or produced during T_i may be used.

Controlling devices with some integrated (energy) storage like cooling devices, batteries, co-generation plants or similar impose a dependency of possible choices at later planning intervals on previous choices, but allow on the other hand for additional flexibility. Let R be the controlled variable, e.g. room temperature controlled by heating through a co-generation plant. Even if the temperature has to be kept on a constant level, this might be achieved with different operation schedules if heat can be intermediately stored in an attached thermal buffer store. In this case, the temperature within the thermal store may vary within a given range. In general, operation modes in every time interval have to be chosen in a way that for every point in time: $R \in \mathcal{R}$, with $\mathcal{R} = [R_{\min}, R_{\max}]$.

Which operation modes $\mathcal{P}_i^{T_i} \subseteq \mathcal{P}$ specifically are possible depends on the operational state of the device at the beginning of an interval T_i and thus depend on the previous operations. $\mathcal{P}_i^{T_i}$ denotes the set $\mathcal{P}_i^{T_i} = \{P \in \mathcal{P} | (R_{i-1} \oplus P) \in \mathcal{R}\}$, where $\oplus$ denotes the operation that determines R_i in interval i from R_{i-1} (if not given by the initial state) and from the chosen operation mode.

When looking at a single device, the problem can be defined as follows. Assuming constant power input (or feed-in) the amount of energy consumed or generated during time interval T_i is given by $E_i = \delta_{T_i} P_i$, with $\delta_{T_i} = t_i - t_{i-1}$. Thus, the cost for the whole planning horizon are given by:

$$C = \sum_{k=1}^{n} \delta_{T_k} P_k c_k. \tag{2}$$

In case all intervals have the same width, the objective function can be simplified and expressed with operation mode schedules x:

$$\arg\min_{x \in \mathcal{F}} \delta_T \sum_{k=1}^{n} \Pi(x) \circ c = \arg\min_{x \in \mathcal{F}} \sum_{k=1}^{n} \Pi(x) \circ c, \tag{3}$$

with $\circ$ denoting the Hadamard product. $\mathcal{F}$ denotes the feasible region specific to the energy resource that is scheduled.

In the following, we consider only cases with equidistant time intervals. Constraints are given by the controlled variable R that has to stay within the given range:

$$P_k \in \mathcal{P}_k^{T_k} \wedge R_k \in \mathcal{R} \quad \forall\, 1 \le k \le n, \tag{4}$$

And a set of device specific operational constraints that further restrict the possible operations. Examples for such constraints are allowed power range, buffer restriction, min./max. on/off times, or ramping [5, 10, 30].

For each time interval an assignment of electrical power has to be found such that the controlled variable R (buffer store temperature in case of a co-generation plant) stays within the allowed range and technical operation constraints are fulfilled. Due to the interconnection (induced by the buffer) of possible power levels in each time interval, they cannot be chosen independently. Each subset of possible feasible power levels for time interval T_i depends (in the worst case) on all choices for $T_1, \ldots, T_{i-1}$. For this reason, the difficult part of this optimization is due to this constraint, not due to the (rather simple) objectives. This problem has shown to be NP-complete with growing planning horizon already for a single energy resource in case of step-control resources [8].

4   Algorithm for Step-Control

When optimizing the schedule for an energy resource (or an ensemble of devices), the recursive dependencies on prior decisions turns out to be problematic for the definition of a neighborhood relation between different solutions. The domain of x_i depends on the assignments of $(x_1, \ldots, x_{i-1})$, because the operation of the sub-schedule $(x_1, \ldots, x_{i-1})$ has an impact on the feasible operational phase-space for x_i of the device. It is an NP-complete problem in itself to construct the whole feasible search graph in advance.

Thus, we used an ACO approach with each ant equipped with a simulation model of the respective device as a means to ask for possible branches at each node. In this way, the search graph can be generated locally on demand. Figure 1(a) shows the general idea of the used search graph.

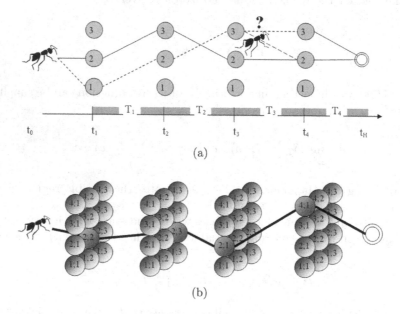

(a)

(b)

Fig. 1. Path construction for an operation schedule (top: single resource; bottom: multiple constraint-coupled resources, 2-dimensional example)

Nodes represent different power levels during each planning interval and are organized in layers at the beginning of each time period. Each ant has to make its way from the beginning of the planning horizon to the end. The layers represent the set of all existing power levels (independent of the current operational state); the power level for the next time interval within the schedule is chosen at the beginning of the interval and fixed during the interval. Costs for a time interval – cf. Eq. 3 – are represented by the chosen power level and some global cost information; e.g. on energy prices during the interval. Edges are allowed only in between neighboring layers and are existent just virtually.

For path – and thus solution construction – each ant starts at time t_0 with the power level given be the initial operation state of the energy resource. At each layer the path taken so far is passed to the simulation model to calculate the feasible choices for the current decision for the next edge to a new power level. Each ant walks from layer to layer and knows the so far covered path; the simulation model calculates based on initial state and path a selection of feasible power levels (subset of feasible edges) for the next time interval; the ant choses from this selection and moves on the next layer. Edges materialize real only after the simulation model acknowledges feasibility based on the ants previous path. In this way, feasibility of constructed paths (representing an operation schedule for the energy resource) is ensured.

The graph of feasible schedules is obviously not complete as often the case in ACO algorithms. The graph of feasible schedules is a k-partite graph

$$V = V_1, \ldots, V_{|\mathcal{T}|} \quad , \text{with} \tag{5}$$

$$V_i = \{((\mathcal{P}^{T_i}_{G_1})_1 \ldots (\mathcal{P}^{T_i}_{G_m})_1), \ \ldots, \ ((\mathcal{P}^{T_i}_{G_1})_k \ldots (\mathcal{P}^{T_i}_{G_m})_\ell)\}, \tag{6}$$

whose edges all point towards immediate future time intervals:

$$\forall i < |\mathcal{T}| \bullet \{(v, v') \mid v \in V_{i+1} \wedge v' \in V_i\} = \emptyset. \tag{7}$$

Each node set V_i consists of different compositions of power levels that are operable during interval T_i, where $V_i \subseteq \mathcal{M}$ is a subset of the set of all theoretically compositions $\mathcal{M}$:

$$\mathcal{M} = \{(P^{G_1}_{i_1} \ldots P^{G_m}_{i_m}) \mid 0 < i_j \le \mathcal{P}^{G_i} \, , \, 0 < j \le |\mathcal{G}|\}. \tag{8}$$

$\mathcal{M}$ corresponds to the set of virtual nodes. Which nodes V_i become existent for ant A and are thus reachable must always be evaluated based on current operational state and so far covered path (course of previously operated power levels). Each time an ant has decided on a power level for a planning interval T_i it moves along this edge and then involves the simulation model to discover possible branches for time interval T_{i+1}; cf. Fig. 1(a). The simulation model (already parametrized with the initial state at T_0) is passed the previous path of the ant $(P_1, \ldots, P_i)$ and returns a set of feasible power modes (or respectively a set of possible absolute power values). This set constitutes possible new edges x_{ij} pointing from the i-th power level in time interval T_i to the j-th power level in T_{i+1}. The weight of each of these new edges results from the electric power and the given cost for this time interval. A decision for an edge from the set of power levels is made by calculating a probability for each edge x_{ij}:

$$P(x_{ij}) = \frac{\tau_{ij}^\alpha \cdot \nu_{ij}^\beta}{\sum_{k \in \mathcal{I}^{V_i}} \tau_{ik}^\alpha \cdot \nu_{ik}^\beta} \tag{9}$$

By taking into account the amount of pheromone associated to x_{ij} and a priority rule (e.g. use the highest power level as in a greedy approach this would promise the highest profit), a local search is induced. A weighting of both rules is given by α and β. According to these probabilities the next edge is chosen by wheel selection. With a given probability an alternative rule is used [13]. This rule uses the maximum product of priority and pheromone $(\max_{j \in \mathcal{I}^{V_i}} (\tau_{ij}^{\alpha} \cdot \nu_{ij}^{\beta}))$ as criterion and allows for searching the immediate neighborhood.

It may occur that an ant finds itself in a dead-end and the simulation model is not able to find any feasible power level. In this case, something in the previous path has to be changed to go further. For this reason we integrated a classical back tracking as rollback. In case of a dead end, the last edge is removed for the ant's path and from the previous set of feasible power levels. Then, another decision is made with the removed set but with the same mechanisms. If the set is empty this is again treated as a dead-end and triggers another step backward.

After each ant has constructed a path (representing an operation schedule denoting electric power for each time interval), all paths are evaluated. The κ best ants are selected to deposit pheromone on the trail:

$$
\tau_{ij} = \tau_{ij} \cdot (1 - \rho) + \rho \cdot \begin{cases} \frac{1}{F(x)} & \forall (i,j) \in x \\ 0 & \forall (i,j) \notin x \end{cases}. \tag{10}
$$

Parameter ρ controls evaporation. As the search graph is not known in advance, we cannot use a static data structure for deposition. Instead we used maps for each time interval containing the edges as key-value-pairs, with a default value of zero if an edge has so far not been present in the map. The key for edge identification is composed of both power-levels ij.

In case the given problem instance comprises more than on energy resource and each energy resource cannot be optimized individually due to a joint constraint, the approach can easily be extended to a higher-dimensional graph for an ensemble of more than just one energy resource, as shown in Fig. 1(b).

5   Results

For testing the algorithm, we used the simulation model of a co-generation plant that has already been evaluated and proved useful in several projects [20,27,29]. This model comprises a micro CHP with 4.7 kW of rated electrical power (12.6 kW thermal power) bundled with a thermal buffer store. Constraints restrict power band, buffer charging, gradients, min. on and off times, and satisfaction of thermal demand. Thermal demand is determined by simulating losses of a detached house (including hot water drawing) according to given weather profiles. Electric power feed-in is either zero (off) or between 1.3 and 4.7 kW. This operational range is always divided into m equidistant discrete power levels resulting in $m + 1$ operation modes in total.

The parameters of our algorithm (weighting of pheromone and priority rule α, β; priority rule share γ, share of depositing ants κ, evaporation ρ, min. pheromone μ) have been tuned in advance by using a Halton sequence [17,25] for random search. If not otherwise stated, 10 ants were used in each run.

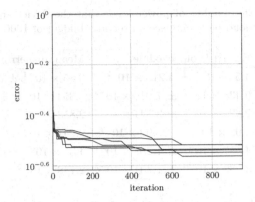

Fig. 2. Convergence of different runs for a 96-dimensional load-balancing problem for a 4-modes co-generation plant (known optimum: zero).

We use two scenarios: (1) load balancing and (2) variable energy pricing. In scenario one the objective is to minimize the distance between the operation schedule $x \in \mathbb{Z}^n$ and a wanted (probably market given) target schedule $\zeta \in \mathbb{Z}^d$. As $\Pi(x)$ and ζ denote the generated and the wanted amount of energy (or equivalently the mean electrical power) for n succeeding time intervals, any distance measure would do. We used $| \cdot |_2$ during optimization. In scenario two, a time varying tariff $c \subset \mathbb{R}^n$ is given denoting the energy price for each of the n time intervals. Thus, the objective is to minimize the overall cost

$$\Pi(x) \circ c \to \min \tag{11}$$

or for generators (in order to maximize profit)

$$\sum_{i=1}^{n} (c_{\max} - c_i) \cdot \Pi(x_i) \to \min. \tag{12}$$

For all simulations, models of the co-generation plant with random initial state have been integrated into the algorithm. Figure 2 shows as a first example the convergence of different runs for a 96-dimensional load balancing scenario. The given budget was 1000 iterations resulting in max. 10.000 objective evaluations. The residual error denotes the mean absolute error in power feed-in.

Co-generation plants with a controller that allows a more fine-grained power control are easier to plan although the search space grows significantly with the number of power levels per time interval. On the other hand, the relation of feasible to infeasible space becomes advantageous with more power levels. Thus, as the tricky part is ensuring feasibility of solutions, planning becomes easier with fine-grained control. In fact, with a continuous controller, the problem is no longer NP-hard. Table 1 shows the impact of the number of power levels on the solution quality with a fixed budget of evaluations. Table 2 shows the impact of the number of used ants (with a fixed number of iterations).

Table 1. Impact of the number of power levels that the device controller may pilot (for a 32-dimensional load balancing scenario and a budget of 1000 iterations).

# of op. modes	Error (d_1 on schedule)	Mean abs. error/ kW
2	$4.5 \times 10^{-1} \pm 4.218 \times 10^{-2}$	$5.85 \times 10^{-1} \pm 5.484 \times 10^{-2}$
3	$9.438 \times 10^{-2} \pm 7.016 \times 10^{-2}$	$1.354 \times 10^{-1} \pm 9.757 \times 10^{-2}$
4	$2.423 \times 10^{-2} \pm 3.841 \times 10^{-2}$	$2.853 \times 10^{-2} \pm 4.521 \times 10^{-2}$
5	$1.02 \times 10^{-2} \pm 2.502 \times 10^{-2}$	$9.247 \times 10^{-3} \pm 2.282 \times 10^{-2}$
6	$4.557 \times 10^{-3} \pm 1.288 \times 10^{-2}$	$4.31 \times 10^{-3} \pm 1.109 \times 10^{-2}$
7	$3.989 \times 10^{-3} \pm 1.401 \times 10^{-2}$	$2.748 \times 10^{-3} \pm 9.461 \times 10^{-3}$

Table 2. Impact of the number of ants (for a 96-dimensional load balancing scenario, 4 power levels fixed, and a budget of 1000 iterations).

# of ants	Error (d_1 on schedule)	Mean abs. error/kW
2	$4.23 \times 10^{-2} \pm 3.63 \times 10^{-2}$	$4.715 \times 10^{-2} \pm 4.05 \times 10^{-2}$
5	$2.958 \times 10^{-2} \pm 2.85 \times 10^{-2}$	$3.349 \times 10^{-2} \pm 3.323 \times 10^{-2}$
10	$3.827 \times 10^{-2} \pm 3.374 \times 10^{-2}$	$4.394 \times 10^{-2} \pm 3.943 \times 10^{-2}$
15	$3.827 \times 10^{-2} \pm 2.59 \times 10^{-2}$	$4.476 \times 10^{-2} \pm 3.206 \times 10^{-2}$
20	$3.168 \times 10^{-2} \pm 2.675 \times 10^{-2}$	$3.572 \times 10^{-2} \pm 3.121 \times 10^{-2}$
50	$3.419 \times 10^{-2} \pm 2.266 \times 10^{-2}$	$3.928 \times 10^{-2} \pm 2.836 \times 10^{-2}$

Next, we compared the ACO approach with the covariance matrix adaption evolution strategy (CMA-ES) [18,32]. CMA-ES is a well-known evolution strategy for solving multi modal black box problems and has demonstrated excellent performance [19] especially for non-linear, non-convex black-box problems. CMA-ES improves its operations by harnessing lessons learned from previously successful evolution steps for future search directions. A new population of solution candidates is sampled from a multi variate normal distribution $\mathcal{N}(0, C)$ with covariance matrix C which is adapted such that it that it maximizes the occurrence of improving steps according to previously seen distributions for good steps. Sampling offspring is weighted by a selection of solutions of the parent generation. In a way, the method learns a second order model of the objective function and exploits it for structure information and for reducing calls of objective evaluations. In order to use CMA-ES for this non-linear discrete problem, we relaxed it to an equidimensional continuous search space and rounded the results back to $\mathbb{Z}$. The constraints for co-generation plant operation were integrated using a classical penalty approach [38], by adding a penalty term to the objective reflecting the inverse length of the feasible part of a solution schedule. A second penalty reflecting infeasible power levels improved the result. The weighting for the scalarization of the objectives has again been tuned using Halton sequences. For algorithm parametrization we relied on the recommendations from [38] giving recommendations for a wide range of applications.

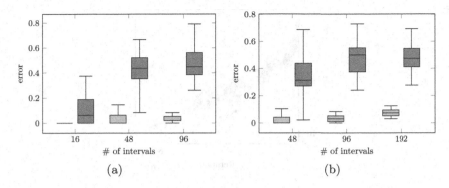

Fig. 3. Comparison of ACO (light gray) and CMA-ES (dark gray) for problems with different dimensionality. Figure 3(a) shows examples with a budgt of 20000 objective evaluations, Fig. 3(b) shows examples with a budget of 10^6 evaluations.

Table 3. Resulting share of feasible results (and intervals) from the experiments in Fig. 3.

Budget	# of intervals	Feasible schedules	Feasible intervals
20000	16	84%	95%
	48	12%	61.2%
	96	0%	30.5%
10^6	48	24%	68.42%
	96	0%	33.04%
	192	0%	13.54%

Figure 3 shows two results with budgets of 20000 Fig. 3(a) and 10^6 Fig. 3(b) objective evaluations. Both algorithms have been tested on load balancing problems with different dimensions. As can be seen, the ACO performs better probably because the ACO may construct feasible solutions in a systematic manner whereas CMA-ES is just guided by the degree of feasibility calculated afterwards. This is also reflected in the results from Table 3 showing the respective feasibility of the found results. The ACO approach yields a share of 100% feasible results regardless of the dimension of the problem. CMA-ES achieves an acceptable share of feasible results only for low-dimensional problems. Significantly increasing the budget of objective evaluations improves the result but still fails for higher dimensional problems. For the quite usual problem of planning one day with 15-min. resolution (96 dimensions), CMA-ES fails completely. We have tested other standard heuristics with even worst results.

Figure 4 shows some example runs for a multi-plant scenario. Here, 10 cogeneration plants are controlled by ants at the same time. Finally, two example results on time varying energy prices are shown in Fig. 5; 5(a) shows an example with two peak prices in the morning and the evening, Fig. 5(b) shows the example

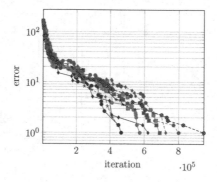

Fig. 4. Example runs for scenarios with 10 concurrently controlled co-generation plants.

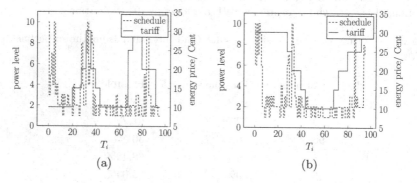

Fig. 5. Time varying energy prices and resulting altered energy feed-in.

of a photovoltaics dominated regime with the aim of shifting other feed-in to the dark hours. In both cases feed-in is successfully altered to the more profitable hours except the early hours of the day where the empty buffer store had to be charged to gain enough flexibility for the rest of the day.

6   Conclusions

The control paradigm shift towards smartly connected small and distributed energy resources within the electricity grid entails new challenges to algorithmic control. Optimizing and orchestrating step-controlled devices in day-ahead scheduling is NP-hard already for single devices with growing number of time intervals within the planning horizon. Providing ancillary services for power conditioning (like voltage or frequency control) require even shorter periods and thus induce higher-dimensional optimization problems in the future. New types of generation units and controllable consumers entail new difficulties in constraint-handling especially if some buffer technology for intermediate energy storage does not allow for an independent consideration of different time intervals.

We proposed and ant colony based approach with model integration for constraint-handling. By integrating a model of the planned energy resource, ants may decide while already constructing their path in the search graph on feasible further directions. The feasible graph – whose construction would also be NP-hard – does not have to be known in advance. Simulation result rendered this approach valid, competitive and sufficiently fast.

References

1. Beaudin, M., Zareipour, H.: Home energy management systems: a review of modelling and complexity. Renew. Sustain. Energy Rev. **45**, 318–335 (2015). https://doi.org/10.1016/j.rser.2015.01.046
2. Behrangrad, M.: A review of demand side management business models in the electricity market. Renew. Sustain. Energy Rev. **47**, 270–283 (2015). https://doi.org/10.1016/j.rser.2015.03.033
3. Boynuegri, A.R., Yagcitekin, B., Baysal, M., Karakas, A., Uzunoglu, M.: Energy management algorithm for smart home with renewable energy sources. In: 4th International Conference on Power Engineering, Energy and Electrical Drives, pp. 1753–1758 (2013)
4. Bremer, J., Lehnhoff, S.: A decentralized PSO with decoder for scheduling distributed electricity generation. In: Squillero, G., Burelli, P. (eds.) EvoApplications 2016. LNCS, vol. 9597, pp. 427–442. Springer, Cham (2016). https://doi.org/10.1007/978-3-319-31204-0_28
5. Bremer, J., Lehnhoff, S.: Hybridizing S-metric selection and support vector decoder for constrained multi-objective energy management. In: Madureira, A.M., Abraham, A., Gandhi, N., Varela, M.L. (eds.) HIS 2018. AISC, vol. 923, pp. 249–259. Springer, Cham (2020). https://doi.org/10.1007/978-3-030-14347-3_24
6. Bremer, J., Rapp, B., Jellinghaus, F., Sonnenschein, M.: Tools for teaching demand-side management. In: EnviroInfo (1), pp. 475–483. Shaker Verlag, Aachen (2009)
7. Bremer, J., Sonnenschein, M.: Constraint-handling for optimization with support vector surrogate models-a novel decoder approach. In: International Conference on Agents and Artificial Intelligence, vol. 2, pp. 91–100. SciTePress (2013)
8. Bremer, J.: Agenten-basierte simulation des planungsverhaltens adaptiver verbraucher in stromversorgungssystemen mit real-time-pricing. Diploma thesis, C.v.O. Universität Oldenburg, Department für Informatik (Abteilung Umweltinformatik), March 2006
9. Capone, A., Barbato, A., Martignon, F., Chen, L., Paris, S.: A power scheduling game for reducing the peak demand of residential users, October 2013. https://doi.org/10.1109/OnlineGreenCom.2013.6731042
10. De Angelis, F., Boaro, M., Fuselli, D., Squartini, S., Piazza, F., Wei, Q.: Optimal home energy management under dynamic electrical and thermal constraints. IEEE Trans. Ind. Inf. **9**(3), 1518–1527 (2013)
11. Deng, R., Yang, Z., Chow, M.Y., Chen, J.: A survey on demand response in smart grids: mathematical models and approaches. IEEE Trans. Ind. Inf. **11**(3), 570–582 (2015)
12. Dethlefs, T., Preisler, T., Renz, W.: Ant-colony based self-optimization for demand-side-management. In: Weber, C., Derksen, C. (eds.) Proceedings SmartER Europe Conference. Essen (2015)

13. Dorigo, M., Caro, G.D., Gambardella, L.M.: Ant algorithms for discreteoptimization. Artif. Life **5**(2), 137–172 (1999). https://doi.org/10.1162/106454699568728
14. Dorigo, M., Di Caro, G.: Ant colony optimization: a new meta-heuristic. In: Proceedings of the 1999 congress on evolutionary computation-CEC99 (Cat. No. 99TH8406), vol. 2, pp. 1470–1477. IEEE (1999)
15. Dorigo, M., Stützle, T.: The ant colony optimization metaheuristic: algorithms, applications, and advances. In: Glover, F., Kochenberger, G.A. (eds.) Handbook of Metaheuristics. International Series in Operations Research & Management Science, vol. 57, pp. 250–285. Springer, Boston (2003). https://doi.org/10.1007/0-306-48056-5_9
16. Gellings, C.W., Parmenter, K.E.: Demand-side management. In: Energy Management and Conservation Handbook, pp. 399–420. CRC Press (2016)
17. Halton, J., Smith, G.: Radical inverse quasi-random point sequence, algorithm 247. Commun. ACM **7**, 701 (1964)
18. Hansen, N.: The CMA evolution strategy: a comparing review. In: Lozano, J., Larranaga, P., Inza, I., Bengoetxea, E. (eds.) Towards a New Evolutionary Computation. Advances on Estimation of Distribution Algorithms, vol. 192, pp. 75–102. Springer, Heidelberg (2006). https://doi.org/10.1007/3-540-32494-1_4
19. Hansen, N.: The CMA evolution strategy: a tutorial. Technical report (2011)
20. Hinrichs, C., Bremer, J., Sonnenschein, M.: Distributed hybrid constraint handling in large scale virtual power plants. In: IEEE PES Conference on Innovative Smart Grid Technologies Europe (ISGT Europe 2013). IEEE Power & Energy Society (2013). https://doi.org/10.1109/ISGTEurope.2013.6695312
21. Hinrichs, C., Lehnhoff, S., Sonnenschein, M.: A decentralized heuristic for multiple-choice combinatorial optimization problems. In: Stefan, H., et al. (eds.) Operations Research Proceedings 2012. ORP, pp. 297–302. Springer, Cham (2014). https://doi.org/10.1007/978-3-319-00795-3_43
22. Hinrichs, C., Sonnenschein, M.: A distributed combinatorial optimisation heuristic for the scheduling of energy resources represented by self-interested agents. Int. J. Bio-Inspired Comput. **10**, 69 (2017). https://doi.org/10.1504/IJBIC.2017.085895
23. Khan, A.R., Mahmood, A., Safdar, A., Khan, Z.A., Khan, N.A.: Load forecasting, dynamic pricing and DSM in smart grid: a review. Renew. Sustain. Energy Rev. **54**, 1311–1322 (2016)
24. Koch, S., Zima, M., Andersson, G.: Potentials and applications of coordinated groups of thermal household appliances for power system control purposes. In: 2009 IEEE PES/IAS Conference on Sustainable Alternative Energy (SAE), pp. 1–8 (2009)
25. Kuipers, L., Niederreiter, H.: Uniform Distribution of Sequences. Dover Books on Mathematics, Dover Publications (2006)
26. Li, Y., Rezgui, Y., Zhu, H.: District heating and cooling optimization and enhancement - towards integration of renewables, storage and smart grid. Renew. Sustain. Energy Rev. **72**, 281–294 (2017). https://doi.org/10.1016/j.rser.2017.01.061
27. Neugebauer, J., Kramer, O., Sonnenschein, M.: Classification cascades of overlapping feature ensembles for energy time series data. In: Woon, W.L., Aung, Z., Madnick, S. (eds.) DARE 2015. LNCS (LNAI), vol. 9518, pp. 76–93. Springer, Cham (2015). https://doi.org/10.1007/978-3-319-27430-0_6
28. Nieße, A., et al.: Market-based self-organized provision of active power and ancillary services: an agent-based approach for smart distribution grids. In: Proceedings on Complexity in Engineering (COMPENG), pp. 1–5. IEEE (2012)

29. Nieße, A., Sonnenschein, M.: A fully distributed continuous planning approach for decentralized energy units. In: Cunningham, D.W., Hofstedt, P., Meer, K., Schmitt, I. (eds.) Informatik 2015. GI-Edition - Lecture Notes in Informatics (LNI), vol. 246, pp. 151–165. Bonner Köllen Verlag (2015)

30. Nieße, A., Sonnenschein, M., Hinrichs, C., Bremer, J.: Local soft constraints in distributed energy scheduling. In: Ganzha, M., Maciaszek, L., Paprzycki, M. (eds.) Proceedings of the 2016 Federated Conference on Computer Science and Information Systems. Annals of Computer Science and Information Systems, vol. 8, pp. 1517–1525. IEEE (2016). https://doi.org/10.15439/2016F76

31. Nosratabadi, S.M., Hooshmand, R.A., Gholipour, E.: A comprehensive review on microgrid and virtual power plant concepts employed for distributed energy resources scheduling in power systems. Renew. Sustain. Energy Rev. **67**, 341–363 (2017)

32. Ostermeier, A., Gawelczyk, A., Hansen, N.: A derandomized approach to self-adaptation of evolution strategies. Evol. Comput. **2**(4), 369–380 (1994)

33. Palensky, P., Dietrich, D.: Demand side management: demand response, intelligent energy systems, and smart loads. IEEE Trans. Industr. Inf. **7**(3), 381–388 (2011)

34. Ramchurn, S.D., Vytelingum, P., Rogers, A., Jennings, N.R.: Putting the 'smarts' into the smart grid: a grand challenge for artificial intelligence. Commun. ACM **55**(4), 8697 (2012). https://doi.org/10.1145/2133806.2133825

35. Ruiz-Romero, S., Colmenar-Santos, A., Mur-Pérez, F.: Putting the 'smarts' into the smart grid: a grand challenge for artificial intelligence. Renew. Sustain. Energy Rev. **38**, 223–234 (2014). https://doi.org/10.1016/j.rser.2014.05.082

36. Saboori, H., Mohammadi, M., Taghe, R.: Virtual power plant (VPP), definition, concept, components and types. In: Asia-Pacific Power and Energy Engineering Conference, pp. 1–4. IEEE (2011)

37. Sarstedt, M., et al.: Standardized evaluation of multi-level grid control strategies for future converter-dominated electric energy systems. In: at-Automatisierungstechnik, vol. 67 (2019)

38. Smith, A., Coit, D.: Handbook of Evolutionary Computation, chap. Penalty Functions, p. Section C5.2. Department of Industrial Engineering, University of Pittsburgh, USA. Oxford University Press and IOP Publishing (1997)

39. Sonnenschein, M., Stadler, M., Rapp, B., Bremer, J., Brunhorn, S.: A modelling and simulation environment for real-time pricing scenarios in energy markets. In: Managing Environmental Knowledge (2006)

40. Yu, T., Kim, D.S., Son, S.Y.: Home appliance scheduling optimization with time-varying electricity price and peak load limitation. In: The 2nd International Conference on Information Science and Technology, IST, pp. 196–199 (2013)

Construction Task Allocation Through the Collective Perception of a Dynamic Environment

Yara Khaluf[1]([✉]) [iD], Michael Allwright[2] [iD], Ilja Rausch[1] [iD], Pieter Simoens[1] [iD], and Marco Dorigo[2] [iD]

[1] Department of Information Technology, Ghent University - imec, Gent, Belgium
{yara.khaluf,ilja.rausch,pieter.simoens}@ugent.be
[2] IRIDIA, Université Libre de Bruxelles, Brussels, Belgium
{michael.allwright,mdorigo}@ulb.ac.be

Abstract. Building structures is a remarkable collective process but its automation remains an open challenge. Robot swarms provide a promising solution to this challenge. However, collective construction involves a number of difficulties regarding efficient robots allocation to the different activities, particularly if the goal is to reach an optimal construction rate. In this paper, we study an abstract construction scenario, where a swarm of robots is engaged in a collective perception process to estimate the density of building blocks around a construction site. The goal of this perception process is to maintain a minimum density of blocks available to the robots for construction. To maintain this density, the allocation of robots to the foraging task needs to be adjusted such that enough blocks are retrieved. Our results show a robust collective perception that enables the swarm to maintain a minimum block density under different rates of construction and foraging. Our approach leads the system to stabilize around a state in which the robots allocation allows the swarm to maintain a tile density that is close to or above the target minimum.

1 Introduction

Building structures are among the most remarkable production processes we humans undertake. However, it is both costly and time-consuming. Consequently, integrating robots into construction processes can be of great benefit. Social insects provide important examples of collective behaviors such as creating large complex structures. These stem from simple behaviors of agents without centralized control or pre-planning. One prominent example is termites: millions of small insects successfully self-organize to build massive, complex mounds that sometimes exceed 12 m in height.

Inspired by natural swarms, researchers have started looking into swarm robotics systems that are able to construct increasingly complex structures [1, 2,38]. Similar to social insects, construction by a robot swarm involves agents arranging building materials in an environment to form structures. To do so, usually robots coordinate through stigmergy. In contrast to direct communication,

© Springer Nature Switzerland AG 2020
M. Dorigo et al. (Eds.): ANTS 2020, LNCS 12421, pp. 82–95, 2020.
https://doi.org/10.1007/978-3-030-60376-2_7

stigmergy enables the coordination of the agents' activities through changing their shared environment [5,8,32]. Upon sensing such environmental changes, the agents conclude their next activity. For example, the availability of building material provides a cue for robots, triggering their decision-making process. Nevertheless, in some situations, stigmergy may not be a sufficient or appropriate means of communication; in these cases, direct communication becomes necessary to exchange particular pieces of information for successful task completion.

We base our study on the Swarm Robotics Construction System (SRoCS) [3], a simulation of which is shown in Fig. 1(a). In this system, robots use computer vision to monitor other robots' actions, and based on these observations, they perform predefined construction actions that advance to complete a partially built structure. In this paper, we consider a scenario where a swarm is divided into two groups of robots. The first group is responsible for exploring the environment, finding building material in form of building blocks, and transporting it to the construction site. This behavior is referred to as the foraging task. The second group is responsible for assembling the foraged blocks into a structure. This behavior is referred to as the construction task. We consider two groups of robots since some robots must remain in the cache area over extended periods of time to estimate the tile density. This estimation is paramount to effectively allocating the robots between the foraging and construction tasks.

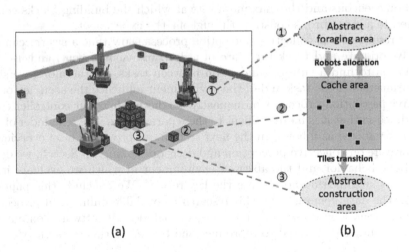

(a)   (b)

Fig. 1. A structure being built using the swarm robotics construction system: (a) in the ARGoS simulator, and (b) an abstraction of the environment with the cache area (pink), and building blocks (black). (Color figure online)

In this study, we focus on the region surrounding the construction site where blocks can be temporarily placed for construction. We call this the cache area (pink in Fig. 1). In order for construction robots to perform their task efficiently, there must be enough blocks in the cache area. Such blocks are discovered and transported to the cache by the foraging robots. The goal in this study is to maintain an amount of blocks that maximizes the rate of construction. To achieve this, the swarm must find a suitable allocation between the robots adding blocks to the cache area (i.e., robots in the foraging task) and the robots removing blocks from the cache area (i.e., robots in the construction task) such that a certain amount of blocks is always preserved in the cache. To achieve this proper allocation, the construction robots need to collectively perceive the number of blocks in the cache area. This estimate helps the individual robots switch to foraging when the estimate drops below a given threshold, or to continue estimating otherwise. This collective perception process is highly challenging as it is performed in a dynamic environment, where the number of blocks in the cache area changes continuously. To address this challenge, we consider an abstraction of the construction scenario, where we can focus on the collective perception and decision-making dynamics involved in finding a suitable robots allocation between the two tasks. We use a homogeneous swarm, in which all robots are capable of performing either the construction task or the foraging task. The robots performing the construction task can communicate with each other if they are within range. Differently, robots performing the foraging task are unable to interact with the robots in the cache and vice versa. Once a robot switches to the foraging task, it always returns to the construction task with a new block after a period of time spent on foraging. Finally, the building material is assumed to be homogeneous and the maximum rate at which the building blocks can be attached to a structure is constant throughout the experiment.

The results of our collective perception process show that a swarm can collectively estimate and track the state of a dynamic environment, and use this information to find a suitable allocation between tasks. Furthermore, we show how parameters of the task and of the environment influence the accuracy of the collective perception process. Consequently, we discuss how our controller could be modified so that it is more resilient to these parameters. The remainder of this paper is organized as follows. In the next section, we provide a brief overview of the literature on collective perception and decision-making. In Sect. 3, we detail the robots' behavior and the abstract environment in which we evaluate it. In Sect. 4, we summarize and discuss the key results. We conclude this paper in Sect. 5. The data generated by this research is available online as a project on the Open Science Framework.[1] This project includes all software components and documentation required to reproduce and to extend this research [15].

2   Related Work

In robot swarms, collective decision-making has been extensively researched [10, 12, 21, 22, 25, 27–29, 33, 34, 37], due to its essential role in a wide range of tasks including foraging, flocking, and construction. In general, two types of system outputs are associated with a collective decision-making process: (i) a consensus in which the swarm agrees on a single option [35], and (ii) a division of the swarm

---

[1] Project on the Open Science Framework: https://osf.io/n7kr3/.

among different tasks—i.e., task allocation—[6,16,19,24]. The goal of the swarm in the consensus-achievement is to converge as quickly as possible on one option, referred to as a symmetry-breaking decision [11,18], or to converge on the best option as quickly as possible, referred to as a value-sensitive decision [9,23]. In the case of task allocation, the goal of the swarm is to maximize the system performance by minimizing the idle time of individual robots.

In this study, we tackle a specific process of collective decision-making that is referred to as collective perception. Collective perception is a process, in which the robots are engaged in perceiving an environmental stimulus collaboratively [14,30,31,36]. In general, robots need to perceive particular signals in their environments to make decisions based on the perceived values. We assume that the environments where robot swarms are deployed are large so that the perception of a single robot is far from sufficient to sense a system-wide stimulus—i.e., a stimulus that spreads across a large space of the environment. This motivates the need for collective perception, where robots combine their perceptions in a distributed manner, and self-organize to act as a single unit.

Collective perception is observed in social insects such as honeybees, where individuals tend to evaluate, for instance, queuing delays to optimize their task allocation [26]. Also bee foragers transfer their nectar load to multiple receivers suggesting the use of this behavior to estimate the environmental nectar flow [13]. These studies inspired collective perception in artificial systems such as robot swarms. For example, authors in [30] developed a bio-inspired algorithm enabling a robot swarm to aggregate at two locations, where the size of each group corresponds to the size of the selected location. In [36], the authors use a robot swarm to collectively decide which of two colors is the most represented in a pattern drawn on arena ground. The algorithm developed in [36] was tested for benchmarking and generalization in [4] across a larger number of patterns (nine). Contrary to [36], the authors in [4] find that the difficulty of the collective perception process doesn't depend mainly on the ratio of one color to the other, but on the distribution of each color in the environment. The authors in [7] proposed a distributed Bayesian algorithm to solve the collective perception task of a similar two-color environment. They define the speed vs. accuracy trade-off of the collective perception as a multi-objective optimization problem. Additionally, the authors have shown that it is possible to guarantee the accuracy of the collective perception, at the cost of decision time.

None of the aforementioned algorithms, however, support collective perception in a dynamic environment, where the perceived features change over time. Dynamic environments impose a serious challenge to collective perception algorithms, i.e. the rate at which the swarm reaches a consensus vs. the rate at which the environment changes. Also, contrary to other algorithms, our collective perception algorithm attempts to estimate absolute values of the perceived feature (e.g., the percentage of a particular color), instead of merely providing a decision on its relative properties (e.g., color x is represented more than color y).

3 The Model

We approach the collective construction problem using an abstract model, as depicted in Fig. 1(b). In the abstract model, we focus on the cache area, in which building blocks are modeled as 2D tiles. These tiles are moved from the foraging area to the cache area by robots performing the foraging task. Robots that are allocated to the construction task, explore the cache area and pick up tiles (when encountered) and use these to build a structure. We omit the details of the foraging process, and instead we replace it by a stochastic process that characterizes the retrieval of tiles. We define a lower-bound density of tiles in the cache area that we call the target density Γ. This density enables us to minimize the idle time of constructing robots—i.e., to maximize the construction rate. We assume the density Γ to be known and provided to the swarm. The goal of the swarm is to allocate the robots to the foraging and construction tasks so that Γ is satisfied and maintained in the cache area.

3.1 The Retrieval Process of Tiles

We model the output of the foraging process—i.e., the retrieval of tiles—using a renewal process. This is a sequence of random variables, which are referred to as the arrival times, at which a repeating event occurs—i.e., retrieving a tile. The inter-arrival time is the period between two consecutive events, these in our study are two consecutive retrieval of tiles. We model these inter-arrival times by sampling from an exponential distribution with the density function:

$$f_T = \frac{1}{\lambda_f} e^{-\frac{1}{\lambda_f} T}, \tag{1}$$

where the parameter λ_f is the average time a robot spent foraging before returning to the cache area with a new tile. Modeling the inter-arrival times to be exponentially distributed results in the renewal process to be a Poisson process, a common way to model arrival events [17,20,39]; therefore, the average number of tiles retrieved by foraging robots within the time period δt is given by:

$$\langle M_i(\delta t) \rangle = \lfloor \frac{\delta t}{\lambda_f} \rfloor N_{f \to c}(\delta t), \tag{2}$$

where $N_{f \to c}(\delta t)$ is the number of robots switching from foraging to construction in the time interval δt. The value of this variable changes over time as a function of the number of robots in the cache and the estimated and target tile density.

3.2 The Simulated Environment

To evaluate our collective perception process, we use the ARGoS simulator to create a $4 \times 4\,\text{m}^2$ arena that is divided into 2500 tiles. We run experiments with 80 robots that can drive around the arena, avoid obstacles such as walls and each other, sense whether or not they are driving over a shaded tile, and communicate with each other over wifi. We restrict the communication distance of the robot to be a maximum of one meter to prevent global communication in the swarm.

We simulate a robot switching to the foraging task by removing it from the simulation and setting up a countdown timer that is initialized following Eq. (1). The mean of this distribution λ_f in Eq. (1) is one of the key parameters which we vary in our experiments. Once the timer reaches zero, we simulate the robot switching back to the construction task by adding it back into the simulation. Using this strategy, the foraging robots are unable to communicate with the robots performing the construction task, which is consistent with our scenario.

When a robot returns from the foraging task, a cell in the arena is shaded to represent the tile this robot retrieved. We follow a probabilistic approach to place the retrieved tile in the cache. Our approach has the effect of creating clusters of tiles in the cache area (see Fig. 2). This is achieved by increasing the probability to select a candidate location x by a factor of $\kappa = 5$ for each tile that is adjacent (max. 8 tiles) to the location x. Hence candidate locations with more tiles in the neighborhood have a higher probability to be selected. Performing construction—i.e. removing a block from the cache and attaching it to a hypothetical structure—is simulated by a tile being unshaded. This transition occurs whenever a robot moves off a tile and onto another tile. The maximum number of tiles that can be unshaded per second is the construction limit ξ_c, which is another key parameter that we vary in our experiments. A full simulation run with default parameters is hosted on Open Science Framework.[2]

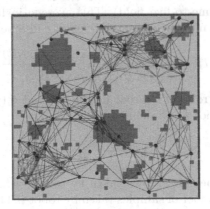

Fig. 2. Screenshot of a simulation in the cache that shows the tiles clustering effect. The magenta lines represent the communication links between the robots.

3.3   The Robot Behavior

A robot moves through the cache in a straight line unless it encounters an obstacle. In case of an obstacle, the robot turns on the spot until its heading is

---

[2] Complete run of a simulation (video): https://osf.io/6mgys/.

clear. When the robot is not avoiding obstacles, it samples the ground beneath it to determine whether it is on top of a tile. The robot keeps track of the total number of samples it has taken and the number of times a sample was taken on a tile. In addition to these counts, the robot's memory also contains a table that records these counts received from the robot's neighbors. An entry in this table contains three fields: (i) the neighbor's total number of samples, (ii) the neighbor's number of samples taken over a tile, and (iii) a time-to-live value that is used to drop the neighbor's entry from the table when its value reaches zero.

At each time step, a robot adjusts the table in its memory by decrementing the time-to-live field for each entry. It then sends this table with an additional entry, representing its sample counts, to all of the robot's direct neighbors. The time-to-live value in this additional entry is initialized to its maximum value. When a direct neighbor receives this information, it updates its table by replacing the contents of each neighbor's entry with the one with the highest time-to-live value found among the existing entries in its table and the entries from the received messages. In this way, each robot always has the most up-to-date entry for each robot that it has recently communicated with. This time-to-live value is also used to avoid loops and to prevent duplicate entries in a robot's table.

After a robot has updated its table, it uses both its local sample counts and the sample counts from its neighbors to estimate the tile density. To compute this estimate $\langle \gamma_i(t) \rangle$ the robot i constructs a weighted average, in which the contribution of each entry (neighbor j's information) is weighted by (i) the number of samples that neighbor has taken and (ii) the hop distance of that neighbor (calculated from the time-to-live field). This average includes i's own sample counts weighted by the number of samples i took and a hop distance of one.

$$\langle \gamma_i(t) \rangle = \frac{\omega_i(t)\gamma_i(t) + \sum_{j \in N_i} \omega_j(t)\gamma_j(t)}{\omega_i(t) + \sum_{j \in N_i} \omega_j(t)}, \tag{3}$$

where N_i is the set of robot i's direct neighbors, and $\gamma_i(t)$ is the tile density measured locally by robot i and defined as:

$$\gamma_i(t) = \frac{c_i(t)}{s_i(t)} \tag{4}$$

where $c_i(t)$ is the number of samples taken by robot i while driving over tiles, and $s_i(t)$ is the total number of samples taken by robot i. The weight ω_i assigned to the locally-measured density by robot i in Eq. (3) is defined as:

$$\omega_i(t) = s_i(t)h_{ij}(t) \tag{5}$$

where $h_{ij}(t)$ is the hop distance between robot i and robot j. The weighted average increases the influence of robots that have sampled larger areas of the arena and that are further away. The latter weighting makes the swarm more resilient against over-estimating the tile density which would otherwise occur due to the clustering of tiles.

After estimating the density of tiles, each robot i decides probabilistically to switch from the construction to the foraging task, as long as $\langle \gamma_i(t) \rangle$ is lower than the target density Γ. The switching probability $Pr_i^{c \to f}(t)$ of robot i at time step t is proportional to the difference between robot i's estimate and Γ:

$$Pr_i^{c \to f}(t) = \begin{cases} \eta |\Gamma - \langle \gamma_i(t) \rangle| & \text{if} \langle \gamma_i(t) \rangle < \Gamma \\ 0 & \text{otherwise,} \end{cases} \tag{6}$$

where η is a design parameter used to keep $Pr_i^{c \to f}(t)$ in the interval $[0, 1]$.

4   Results and Discussion

We have investigated the performance of the collective perception process as well as the robots' task allocation for different experiment configurations. Specifically, our results were obtained over the following set of parameters: (i) the mean foraging time λ_f (tested values $\lambda_f \in \{5, 10, 20\}$), (ii) the lower-bound of the target density Γ (tested values $\Gamma \in \{0.1, 0.3, 0.4\}$), (iii) the construction limit ξ_c (tested values $\xi \in \{5, 10, 20\}$), and (iv) the constant η in the switching probability as defined in Eq. (6) (tested values $\eta \in \{0.1, 0.15, 0.2\}$). We have published the data from these experiments online [15]. The evaluation of our approach spans over four metrics. The first is the time trajectory of the density of tiles $\rho_{gt}(t)$, which is the ground truth; the second is the time trajectory of the swarm estimate $\rho_s(t)$; the third is the time evolution of the individual deviation from the swarm estimate of the tile density: $\Delta_i(t) = |\langle \gamma_i(t) \rangle - \rho_s(t)|$; and the fourth illustrates how the robots allocation to construction and foraging evolves over time. We run all experiments for 2 500 s (12 500 time steps) with a swarm size of 80 robots and average the results of each experiment across 30 runs. In the following we discuss our findings over a subset of the parameters' tested values.

Let us start with the first metric, tile density. Figure 3 shows that the swarm was able to increase the tile density in the cache area and keep it above the target density Γ for $\Gamma \in \{0.3, 0.4\}$. The swarm estimate is initially in full agreement with the ground truth as both start at 0 tiles. Over time, the swarm estimate $\rho_s(t)$ stabilizes around the target density Γ, with a minority of robots (see the standard deviation) estimating the tile density to be higher than Γ. This minority acts to reduce the number of robots sent to retrieve tiles, while the majority acts to increase this number, leading the ground truth to a value higher than the target density. The interplay of these two groups in the swarm causes both the swarm estimate and ground truth to stabilize with the ground truth higher than the swarm estimate. We also notice that increasing the mean foraging time ($\lambda_f \in \{10, 20\}$) leads to a slower increase in $\rho_{gt}(t)$. This is because higher values of λ_f imply longer periods between the retrieval of tiles, on average. Furthermore, we see that for a specific target density Γ, $\rho_s(t)$ seems to be maintained across different values of λ_f and ξ_c. This suggests that the collective perception process is relatively robust to both parameters. This robustness implies that our algorithm is suitable for real-world construction tasks, where complicated foraging that takes longer to find building materials, or prolonged assembly of building material does not affect the collective perception performance.

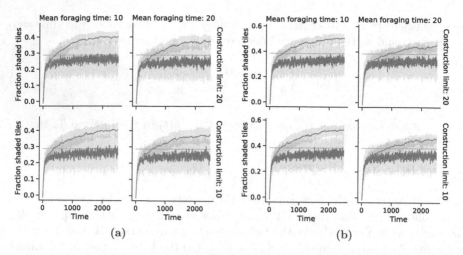

Fig. 3. Tile density for $\eta = 0.2$ and target density (a) $\Gamma = 0.3$ and (b) $\Gamma = 0.4$: ■ swarm estimate $\rho_s(t)$, ■ ground truth $\rho_{gt}(t)$ (averaged across 30 runs).

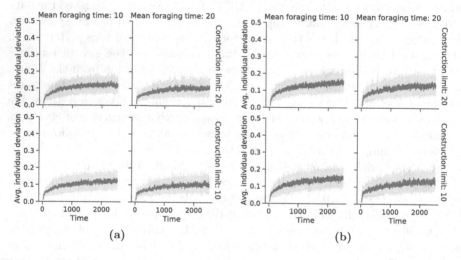

Fig. 4. Average individual deviation $\langle \Delta_i(t) \rangle$ from swarm estimate $\rho_s(t)$ for $\eta = 0.2$ and target density (a) $\Gamma = 0.3$ and (b) $\Gamma = 0.4$ (averaged across 30 runs).

Figure 4 illustrates the mean individual deviation $\langle \Delta_i(t) \rangle$ from the swarm estimate for different target densities $\Gamma \in \{0.3, 0.4\}$, mean foraging times $\lambda_f \in \{10, 20\}$, and construction limits $\xi_c \in \{10, 20\}$. Our results show robustness of the individual deviation with regard to changes in these parameters, with a maximum average deviation of 0.15 ($\langle \Delta_i(t) \rangle \leq 0.15$). Such a small variance $\langle \Delta_i(t) \rangle$ indicates a strong agreement between the individual and the group estimate.

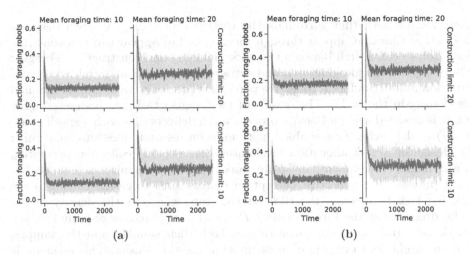

Fig. 5. Fraction of foraging robots for $\eta = 0.2$ and target density (a) $\Gamma = 0.3$ and (b) $\Gamma = 0.4$ (averaged across 30 runs).

Finally, Fig. 5 shows the fraction of robots allocated to foraging over time. At the beginning, all 80 robots are in the cache area for all experiment configurations. Thus, initially, there is a jump in the number of robots leaving from the construction task to the foraging task. This jump is due to the low swarm estimate $\rho_s(t)$ of the tile density during the first 100 s of the experiments (see Fig. 3). During this initial period, the estimate $\rho_s(t)$ of the swarm remains consistently below the target density. Hence the condition $\langle \gamma_i(t) \rangle < \Gamma$ is true for a large majority of the robots, that then switch to the foraging task following Eq. (6) with a relatively high probability. The magnitude of the spike increases with the mean foraging time λ_f. This is due to the longer time it takes for the robots to arrive back from the foraging area, and thus for the tile density and the swarm estimate to rise. Nevertheless, as soon as the swarm estimate stabilizes and the ground truth of tiles in the cache reaches or exceeds its target $\rho_{gt}(t) \geq \Gamma$, the fraction of the foraging robots starts to drop until it stabilizes, leading the system into an equilibrium state with respect to the robots allocation. The drop takes longer time for larger λ_f. Furthermore, the fraction of robots that continue foraging is higher for higher Γ values. This is because tiles need to be retrieved at a faster rate. Additionally, the fraction of foraging robots is higher for larger λ_f given the same Γ. This is due to the slower rate of tile retrieval when λ_f is larger. Thus, this slower rate pushes more robots to foraging while the robots that remain in the cache are performing construction and estimating the tile density $\langle \gamma_i(t) \rangle$ for a longer time.

5   Conclusions

We studied an abstract scenario of a collective construction process by a robot swarm in a dynamic environment. In this scenario, we focused on the cache area—i.e., an area that surrounds the construction area—, where blocks are modeled as tiles that appear through foraging and disappear through construction. Robots can switch between two tasks: foraging and construction. Foraging robots explore the foraging area and retrieve the building material, while construction robots conduct a collective perception process to maintain a minimum tile density in the cache. The details of the foraging task are abstracted away and it is modeled using a Poisson process which delivers tiles with a specific rate $(1/\lambda_f)$ to the cache. This enabled us to focus on research questions concerning the design of a task allocation mechanism that exploits collective perception in a dynamic environment. In future work, we plan to simulate a detailed foraging process. The collective perception process aims to assign robots to the foraging task to increase the tiles retrieval rate whenever the density in the cache drops below the target density Γ—i.e., the required lower-bound on the tile density. Robots in the cache rely on both their samples and the samples of their neighbors to compute an estimate of the tile density. This estimate is computed as a weighted average that assigns higher importance to (i) robots that are further away, making the swarm more resilient to over-estimation or under-estimation due to the clustering of tiles; (ii) robots with larger samples, as these contributions are more representative. Robots use their estimate from Eq. (3) to probabilistically decide whether to switch to the foraging task or to continue estimating/performing the construction task.

Our results show that the proposed collective perception process leads to a proper robots allocation, which in turn guarantees a minimum tile density in the cache. This allocation changes as a function of the average time it takes a robot to find and retrieve a tile λ_f, and the target density Γ. Furthermore, our results show a strong agreement between the individual estimate $\langle \gamma_i(t) \rangle$ and the swarm estimate $\rho_s(t)$, with a maximum variance of 0.15. This study is a first step towards designing collective perception processes in dynamic environments, in which the perceived feature (e.g., tile density) changes over time and the goal is to estimate its absolute value. In future work, we plan to study the influence of the size and update rate of the robots' memorized samples on the performance of the perception process. We also intend to extend the proposed algorithm to enable the swarm to maintain a tile density close to its target.

Acknowledgements. This work is partially supported by the European Union's Horizon 2020 research and innovation programme under the Marie Skłodowska-Curie grant agreement No. 846009. Marco Dorigo acknowledges support from the Belgian F.R.S.-FNRS, of which he is a Research Director.

References

1. Allwright, M., Bhalla, N., Dorigo, M.: Structure and markings as stimuli for autonomous construction. In: Eighteenth International Conference on Advanced Robotics - ICAR 2017, pp. 296–302. IEEE Press, Piscataway (2017)

2. Allwright, M., Bhalla, N., Pinciroli, C., Dorigo, M.: Simulating multi-robot construction in ARGoS. In: Dorigo, M., Birattari, M., Blum, C., Christensen, A.L., Reina, A., Trianni, V. (eds.) ANTS 2018. LNCS, vol. 11172, pp. 188–200. Springer, Cham (2018). https://doi.org/10.1007/978-3-030-00533-7_15
3. Allwright, M., Zhu, W., Dorigo, M.: An open-source multi-robot construction system. HardwareX **5**, e00049 (2019)
4. Bartashevich, P., Mostaghim, S.: Benchmarking collective perception: new task difficulty metrics for collective decision-making. In: Moura Oliveira, P., Novais, P., Reis, L.P. (eds.) EPIA 2019. LNCS (LNAI), vol. 11804, pp. 699–711. Springer, Cham (2019). https://doi.org/10.1007/978-3-030-30241-2_58
5. Bonabeau, E., Dorigo, M., Theraulaz, G.: Swarm Intelligence: From Natural to Artificial Systems. Oxford University Press, New York (1999)
6. Brutschy, A., Pini, G., Pinciroli, C., Birattari, M., Dorigo, M.: Self-organized task allocation to sequentially interdependent tasks in swarm robotics. Auton. Agent. Multi-Agent Syst. **28**(1), 101–125 (2012). https://doi.org/10.1007/s10458-012-9212-y
7. Ebert, J., Gauci, M., Mallmann-Trenn, F., Nagpal, R.: Bayes bots: collective Bayesian decision-making in decentralized robot swarms. In: International Conference on Robotics and Automation (ICRA) (2020)
8. Garnier, S., Gautrais, J., Theraulaz, G.: The biological principles of swarm intelligence. Swarm Intell. **1**(1), 3–31 (2007). https://doi.org/10.1007/s11721-007-0004-y
9. Gray, R., Franci, A., Srivastava, V., Leonard, N.E.: Multiagent decision-making dynamics inspired by honeybees. IEEE Trans. Control Netw. Syst. **5**(2), 793–806 (2018)
10. Gutiérrez, A., Campo, A., Monasterio-Huelin, F., Magdalena, L., Dorigo, M.: Collective decision-making based on social odometry. Neural Comput. Appl. **19**(6), 807–823 (2010). https://doi.org/10.1007/s00521-010-0380-x
11. Hamann, H., Schmickl, T., Wörn, H., Crailsheim, K.: Analysis of emergent symmetry breaking in collective decision making. Neural Comput. Appl. **21**(2), 207–218 (2012). https://doi.org/10.1007/s00521-010-0368-6
12. Hamann, H., Valentini, G., Khaluf, Y., Dorigo, M.: Derivation of a micro-macro link for collective decision-making systems. In: Bartz-Beielstein, T., Branke, J., Filipič, B., Smith, J. (eds.) PPSN 2014. LNCS, vol. 8672, pp. 181–190. Springer, Cham (2014). https://doi.org/10.1007/978-3-319-10762-2_18
13. Huang, M., Seeley, T.: Multiple unloadings by nectar foragers in honey bees: a matter of information improvement or crop fullness? Insectes Sociaux **50**(4), 330–339 (2003). https://doi.org/10.1007/s00040-003-0682-4
14. Khaluf, Y.: Edge detection in static and dynamic environments using robot swarms. In: IEEE 11th International Conference on Self-Adaptive and Self-Organizing Systems (SASO), pp. 81–90. IEEE (2017)
15. Khaluf, Y., Allwright, M., Rausch, I., Simoens, P., Dorigo, M.: Construction task allocation through the collective perception of a dynamic environment (2020). https://osf.io/n7kr3/
16. Khaluf, Y., Birattari, M., Hamann, H.: A swarm robotics approach to task allocation under soft deadlines and negligible switching costs. In: del Pobil, A.P., Chinellato, E., Martinez-Martin, E., Hallam, J., Cervera, E., Morales, A. (eds.) SAB 2014. LNCS (LNAI), vol. 8575, pp. 270–279. Springer, Cham (2014). https://doi.org/10.1007/978-3-319-08864-8_26
17. Khaluf, Y., Dorigo, M.: Modeling robot swarms using integrals of birth-death processes. ACM Trans. Auton. Adapt. Syst. (TAAS) **11**(2), 1–16 (2016)

18. Khaluf, Y., Pinciroli, C., Valentini, G., Hamann, H.: The impact of agent density on scalability in collective systems: noise-induced versus majority-based bistability. Swarm Intell. **11**(2), 155–179 (2017). https://doi.org/10.1007/s11721-017-0137-6
19. Khaluf, Y., Rammig, F.: Task allocation strategy for time-constrained tasks in robots swarms. In: Artificial Life Conference Proceedings, vol. 13, pp. 737–744. MIT Press (2013)
20. Kim, S.H., Whitt, W.: Choosing arrival process models for service systems: tests of a nonhomogeneous poisson process. Naval Res. Logist. (NRL) **61**(1), 66–90 (2014)
21. Meyer, B.: Optimal information transfer and stochastic resonance in collective decision making. Swarm Intell. **11**(2), 131–154 (2017). https://doi.org/10.1007/s11721-017-0136-7
22. Montes de Oca, M.A., Ferrante, E., Scheidler, A., Pinciroli, C., Birattari, M., Dorigo, M.: Majority-rule opinion dynamics with differential latency: a mechanism for self-organized collective decision-making. Swarm Intell. **5**(3–4), 305–327 (2011)
23. Pais, D., Hogan, P.M., Schlegel, T., Franks, N.R., Leonard, N.E., Marshall, J.A.: A mechanism for value-sensitive decision-making. PloS One **8**(9), e73216 (2013)
24. Pini, G., Brutschy, A., Frison, M., Roli, A., Dorigo, M., Birattari, M.: Task partitioning in swarms of robots: an adaptive method for strategy selection. Swarm Intell. **5**(3–4), 283–304 (2011). https://doi.org/10.1007/s11721-011-0060-1
25. Prasetyo, J., De Masi, G., Ferrante, E.: Collective decision making in dynamic environments. Swarm Intell. **13**(3–4), 217–243 (2019). https://doi.org/10.1007/s11721-019-00169-8
26. Ratnieks, F.L., Anderson, C.: Task partitioning in insect societies. ii. use of queueing delay information in recruitment. Am. Nat. **154**(5), 536–548 (1999)
27. Rausch, I., Khaluf, Y., Simoens, P.: Collective decision-making on triadic graphs. In: Barbosa, H., Gomez-Gardenes, J., Gonçalves, B., Mangioni, G., Menezes, R., Oliveira, M. (eds.) Complex Networks XI. SPC, pp. 119–130. Springer, Cham (2020). https://doi.org/10.1007/978-3-030-40943-2_11
28. Reina, A., Miletitch, R., Dorigo, M., Trianni, V.: A quantitative micro-macro link for collective decisions: the shortest path discovery/selection example. Swarm Intell. **9**(2–3), 75–102 (2015). https://doi.org/10.1007/s11721-015-0105-y
29. Reina, A., Valentini, G., Fernández-Oto, C., Dorigo, M., Trianni, V.: A design pattern for decentralised decision making. PloS One **10**(10), e0140950 (2015)
30. Schmickl, T., Möslinger, C., Crailsheim, K.: Collective perception in a robot swarm. In: Şahin, E., Spears, W.M., Winfield, A.F.T. (eds.) SR 2006. LNCS, vol. 4433, pp. 144–157. Springer, Heidelberg (2007). https://doi.org/10.1007/978-3-540-71541-2_10
31. Strobel, V., Castelló Ferrer, E., Dorigo, M.: Managing byzantine robots via blockchain technology in a swarm robotics collective decision making scenario. In: Dastani, M., Sukthankar, G., André, E., Koenig, S. (eds.) Proceedings of the 17th International Conference on Autonomous Agents and Multiagent Systems, International Foundation for Autonomous Agents and Multiagent Systems, pp. 541–549 (2018)
32. Theraulaz, G., Bonabeau, E.: A brief history of stigmergy. Artif. Life **5**(2), 97–116 (1999)
33. Trianni, V., De Simone, D., Reina, A., Baronchelli, A.: Emergence of consensus in a multi-robot network: from abstract models to empirical validation. IEEE Robot. Autom. Lett. **1**(1), 348–353 (2016)

34. Valentini, G., Hamann, H.: Time-variant feedback processes in collective decision-making systems: influence and effect of dynamic neighborhood sizes. Swarm Intell. 9(2–3), 153–176 (2015). https://doi.org/10.1007/s11721-015-0108-8
35. Valentini, G.: Achieving Consensus in Robot Swarms. SCI, vol. 706. Springer, Cham (2017). https://doi.org/10.1007/978-3-319-53609-5
36. Valentini, G., Brambilla, D., Hamann, H., Dorigo, M.: Collective perception of environmental features in a robot swarm. In: Dorigo, M., et al. (eds.) ANTS 2016. LNCS, vol. 9882, pp. 65–76. Springer, Cham (2016). https://doi.org/10.1007/978-3-319-44427-7_6
37. Valentini, G., Ferrante, E., Hamann, H., Dorigo, M.: Collective decision with 100 kilobots: speed versus accuracy in binary discrimination problems. Auton. Agent. Multi-Agent Syst. 30(3), 553–580 (2016)
38. Werfel, J., Petersen, K., Nagpal, R.: Designing collective behavior in a termite-inspired robot construction team. Science 343(6172), 754–758 (2014)
39. Wolff, R.W.: Poisson arrivals see time averages. Oper. Res. 30(2), 223–231 (1982)

Control Parameter Importance and Sensitivity Analysis of the Multi-Guide Particle Swarm Optimization Algorithm

Timothy G. Carolus[1]([✉]) [iD] and Andries P. Engelbrecht[1,2] [iD]

[1] Department of Industrial Engineering, Stellenbosch University,
Stellenbosch, South Africa
t.g.carolus@gmail.com
[2] Computer Science Division, Stellenbosch University, Stellenbosch, South Africa
engel@sun.ac.za

Abstract. The multi-guide particle swarm optimization (MGPSO) algorithm is a multi-objective optimization algorithm that uses multiple swarms, each swarm focusing on an individual objective. This paper conducts an importance and sensitivity analysis on the MGPSO control parameters using functional analysis of variance (fANOVA). The fANOVA process quantifies the control parameter importance through analysing variance in the objective function values associated with a change in control parameter values. The results indicate that the inertia component value has the greatest sensitivity and is the most important control parameter to tune when optimizing the MGPSO.

1 Introduction

Particle swarm optimization (PSO) [11] performance can be greatly affected by the apt selection of control parameters [1–4]. Regardless of establishing standardized values for the control parameters of the PSO, control parameter values need problem specific tuning to achieve optimal performance for each problem [2,10]. The tuning process requires a large number of possible control parameter configurations to be analyzed, which is a computationally expensive task. When tuning the control parameters of multi-objective optimization algorithms (MOAs), the cost of tuning increases.

The multi-guide PSO (MGPSO) is a highly competitive MOA [13]. The MGPSO assigns a swarm per objective, with each swarm optimizing only that objective. Additionally, the MGPSO uses an archive shared across the swarms, which stores previously found non-dominated solutions. An archive guide is added to the velocity update equation to facilitate knowledge exchange about previously found non-dominated solutions. The MGPSO has five control parameters to tune. Scheepers et al. have shown that performance of the MGPSO is sensitive to control parameter values, and that the optimal values are problem dependent [14]. Scheepers et al. derived, via stability analysis, a guideline for

© Springer Nature Switzerland AG 2020
M. Dorigo et al. (Eds.): ANTS 2020, LNCS 12421, pp. 96–106, 2020.
https://doi.org/10.1007/978-3-030-60376-2_8

setting control parameter values such that an equilibrium is guaranteed. This guideline helps to reduce computational expensive tuning.

An alternative to tuning control parameters is to identify which control parameter(s) is (are) most influential to the performance of the MGPSO. Knowledge of control parameter importance allows for the relocation of parameter tuning resources to control parameters that best influence the overall performance of the algorithm. The main objective of this paper is to study the relative importance of each of the MGPSO control parameters. Functional analysis of variance (fANOVA) [8,16], which is the analysis of the variance of a response relative to each of the inputs (control parameter configurations), is used to determine the importance of each of the MGPSO control parameters.

The remainder of the paper is structured as follows: Sect. 2 provides background concepts of multi-objective optimization. Section 3 outlines the experimental procedure, followed by a discussion of the results in Sect. 4. Section 5 provides concluding remarks.

2   Background

2.1   Multi-Objective Optimization

A multi-objective optimization problem (MOOP) has two or three sub-objectives [18]. MOAs aim to find solutions as close to the true Pareto-optimal front (POF) as possible and with as many non-dominated solutions as possible, whilst obtaining an even spread of these solutions. The POF are objective vectors of non-dominated particle positions. A decision vector $\mathbf{x}_1$ dominates another decision vector $\mathbf{x}_2$ if and only if $f_i(\mathbf{x}_1) \leq f_i(\mathbf{x}_2)$ $\forall i \in \{1, \cdots, m\}$ and $\exists i \in \{1, \cdots, m\}$ such that $f_i(\mathbf{x}_1) < f_i(\mathbf{x}_2)$, assuming minimization of the sub-objectives.

Assuming minimization, a boundary constrained MOOP is defined as

$$\min_{\mathbf{x}} \left(f_1(\mathbf{x}), f_2(\mathbf{x}), \cdots, f_m(\mathbf{x})\right) \text{ s.t. } x_j \in [x_{j,\min}, x_{j,\max}], \forall j = 1, \cdots, n \quad (1)$$

where m is the number of objectives and $\mathbf{x}$ is a particular solution within the boundaries of the solution space of dimension n.

2.2   Multi-Guide Particle Swarm Optimization

PSO is a population-based search algorithm based on the dynamics of a flock of birds [11]. Each particle within the swarm retain knowledge of its personal best and neighbourhood best positions throughout the duration of the search period. This information is used to update each particle's position in the search space through the addition of a velocity vector to its current position vector:

$$\mathbf{x}_i(t+1) = \mathbf{x}_i(t) + \mathbf{v}_i(t+1) \quad (2)$$

where the velocity is defined for the basic inertia PSO [15] as

$$\mathbf{v}_i(t+1) = w\mathbf{v}_i(t) + c_1\mathbf{r}_1(t)(\mathbf{y}_i(t) - \mathbf{x}_i(t)) + c_2\mathbf{r}_2(t)(\hat{\mathbf{y}}_i(t) - \mathbf{x}_i(t)) \quad (3)$$

where $\mathbf{x}_i(t)$ and $\mathbf{v}_i(t)$ are the position and velocity vectors of particle i at time t, respectively. The PSO control parameters are the inertia weight, w, and the cognitive and social acceleration coefficients given by c_1 and c_2 respectively. The personal and neighbourhood best position vectors at time t are given by $\mathbf{y}_i(t)$ and $\hat{\mathbf{y}}_i(t)$, respectively.

The MGPSO algorithm is a PSO algorithm that uses multiple swarms, as well as a bounded archive to share non-dominated solutions between swarms. The velocity update equation of the MGPSO is defined as

$$\begin{aligned}
\mathbf{v}_i(t+1) = w\mathbf{v}_i(t) + c_1\mathbf{r}_1(t)(\mathbf{y}_i(t) - \mathbf{x}_i(t)) + \lambda_i c_2 \mathbf{r}_2(t)(\hat{\mathbf{y}}_i(t) - \mathbf{x}_i(t)) \\
+ (1 - \lambda_i)c_3\mathbf{r}_3(t)(\hat{\mathbf{a}}_i(t) - \mathbf{x}_i(t))
\end{aligned} \tag{4}$$

where $\mathbf{r}_1, \mathbf{r}_2$ and $\mathbf{r}_3$ are vectors of random values sampled from a uniform distribution between 0 and 1; c_1, c_2 and c_3 are the cognitive, social and archive acceleration coefficients respectively and λ_i is the archive balance coefficient for particle i; λ_i is selected randomly from a uniform distribution between 0 and 1.

The MGPSO archive stores new non-dominated solutions, given that they are not dominated by any other solutions in the archive and that the archive is not full. When the archive has reached its capacity, crowding distance is used to remove the most crowded non-dominated solution from the archive. The crowding distance of a particular solution is defined as an estimate of the density of solutions surrounding that solution [12]. The archive guide, $\hat{\mathbf{a}}_i(t)$, is selected from the archive using tournament selection, usually with a tournament size of 2 or 3 [14]. The solution with the largest crowding distance is selected.

2.3   Stability Analysis

Extensive theoretical analysis of the stability of the standard, single-objective PSO has been done [6]. An order-1 and order-2 stability analysis of the MGPSO was done by Scheepers et al. [14], considering the limits

$$\lim_{t \to \infty} E[s_t] = s_E \text{ and } \lim_{t \to \infty} V[s_t] = s_V \tag{5}$$

where $E[s_t]$ is the expectation value and $V[s_t]$ is the variance of a sequence, s_t. The stability region of the MGPSO was derived as

$$0 < c_1 + \lambda c_2 + (1-\lambda)c_3 < \frac{4(w+1)}{1 - w + \frac{(c_1^2 + \lambda^2 c_2^2 + (1-\lambda)^2 c_3^2)(1+w)}{3(c_1 + \lambda c_2 + (1-\lambda)c_3)^2}}, \quad |w| < 1. \tag{6}$$

2.4   Functional Analysis of Variance

fANOVA [16] is a statistical approach used to decompose the variance of response (objective function) values into additive components associated with each subset of control parameter values [8].

For a given algorithm A with n control parameters, each with domain $\Theta_c, c = 1, \cdots, n$, the control parameter configuration space of algorithm A is defined as $\Theta = \Theta_1 \times \cdots \times \Theta_n$. A complete instantiation of the control parameters for algorithm A is a vector, $\boldsymbol{\theta} = (\theta_1, \cdots, \theta_n)$. Control parameter tuning aims to find a $\boldsymbol{\theta}_i \in \Theta$ that optimizes the performance metric $m(\boldsymbol{\theta}_i, \pi_j)$ for a given problem π_j. The performance metric can be any performance aspect relevant to the context of the algorithm in question. This optimization procedure is applied over all possible problems $(\pi_1, \cdots, \pi_k)$ such that the control parameter configuration, $\boldsymbol{\theta}_i$, optimizes the overall performance given by

$$f(\boldsymbol{\theta}_i) := \sum_{j=1}^{k} m(\boldsymbol{\theta}_i, \pi_j). \tag{7}$$

The goal of fANOVA is to quantify the variance of the performance measure that can be attributed to each subset of control parameters. The specific performance measure calculated is the marginal predicted performance of algorithm A for all partial instantiation $\theta_\phi \in \Theta$ of a given model $\hat{y} : \Theta \to \mathbb{R}$, i.e.

$$\hat{m}(\theta_\phi) = \frac{1}{||\Theta_T||} \int \hat{y}(\theta_T) d\theta_T \tag{8}$$

where the size of the complete instantiation is given by $||\Theta_T|| = \prod_{i=1}^{k} ||\theta_i||$ and with the calculation performed over each partial instantiation, θ_T.

fANOVA induces a random forest [5] to predict the value of the performance measure for a given parameter configuration [8], i.e. $\hat{m}(\theta_\phi)$. Each regression tree defines a partitioning of the configuration space, such that the marginal prediction of the entire forest is the average prediction over each tree.

The importance of each control parameter is interpreted from the fraction of variance of the model associated with that particular subset of control parameters [8]. Control parameters with higher variances are of greater importance, and should have higher priority in the control parameter tuning process.

3   Experimental Procedure

The MGPSO was executed with 30 particles per swarm. The effects of the control parameters was evaluated for the following control parameter values in all combinations: $w \in \{-1.0, -0.933, \cdots, 0.933, 1.0\}$, $c_1 \in \{0.0, 0.0667, \cdots, 1.933, 2.0\}$, $\phi_1 = \lambda c_2$ and $\phi_2 = (1 - \lambda)c_3$ such that $\lambda \in \{0.0, 0.0667, \cdots, 0.933, 1.0\}$ and $c_2 = c_3 = 2$. This produced 31744 control parameter configurations. Each control parameter configuration was evaluated against Eq. (6) to ensure that the control parameter configuration is within the theoretical region of order-2 stability. After performing this check, 19168 control parameter configurations were found to be within this region and evaluated by the MGPSO. The performance of the MGPSO was evaluated on the two-objective Walking Fish Group (WFG) [9] benchmark suite, in a 10 dimensional search space.

Each control parameter configuration was evaluated for 30 independent runs per benchmark problem; each run executed 1000 iterations. The performance measures (or response values), i.e. the inverted generational distance (IGD) [17] and the hypervolume (HV) [7], were calculated using the final archive of each run.

The response values used in the fANOVA procedure were the average objective function value after 1000 iterations. The control parameter importance was determined by analysing the variance in the predicted marginal performance, Eq. (8), with respect to a single objective function at a time. A qualitative analysis of the control parameter was done using response surfaces. These response surfaces were generated by plotting the average objective function value of a particular control parameter and of variations of control parameter configurations, along with the standard deviation over all the variations.

4   Results

4.1   Variance in Predicted Objective Function Values

Each column in Table 1 represents the results of each control parameter independently, respectively with respect to IGD and HV. The results show that the inertia weight, w, had the greatest variance for all of the problems, for both performance measures. On average, the inertia weight contributed to 0.085839 of the variance in the IGD and 0.086773 in the HV. The cognitive acceleration coefficient, c_1, is the second most influential control parameter to tune, accounting for the second highest variance for most of the benchmark functions with respect to both IGD and HV.

Both the social and the archive guide components of Eq. (4) are not solely dependent on a single control parameter, but the product of λc_2 and $(1 - \lambda)c_3$ respectively. Thus these components are analyzed by the products, ϕ_1 and ϕ_2, respectively. An independent analysis of c_2, c_3 and λ is not relevant. The results show that ϕ_2 is on average more important than ϕ_1 with reference to both IGD and HV. This implies that the contribution from the social component is the least important to tune. The archive coefficient accounts for 0.008027 and 0.008381 of the variance with respect to the IGD and HV, respectively. However, for WFG5 and WFG6, the social component is shown to have greater importance than the archive component. Both these problems have a concave POF.

4.2   Response Surface Analysis

Figures 1, 2, 3, 4, 5, 6, 7 and 8 represent the average performance measure associated with particular control parameter configurations, and are referred to as response surfaces. The solid blue line in each plot is the average performance measure and the red area indicates one standard deviation above and below. These figures provide a visual comparison of the control parameter influence the algorithm performance. A larger variability in average performance (blue

Table 1. The proportion of Variance in IGD and HV

	IGD				HV			
	c_1	ϕ_1	ϕ_2	w	c_1	ϕ_1	ϕ_2	w
wfg1	0.010338	0.005814	0.008457	**0.049322**	0.009925	0.005174	0.006862	**0.030629**
wfg2	0.014449	0.009283	0.005665	**0.041766**	0.010825	0.009987	0.007648	**0.023400**
wfg3	0.012384	0.008313	0.010789	**0.031181**	0.009756	0.007365	0.009742	**0.023252**
wfg4	0.013607	0.010366	0.007656	**0.029092**	0.010888	0.003467	0.007822	**0.032223**
wfg5	0.009872	0.00993	0.010337	**0.335960**	0.006032	0.014992	0.011420	**0.379226**
wfg6	0.008541	0.009025	0.009025	**0.042318**	0.012243	0.011366	0.007387	**0.024050**
mean	0.012223	0.008991	0.008027	**0.085839**	0.009376	0.007849	0.008381	**0.086773**
std	0.001800	0.001729	0.002303	0.122819	0.001806	0.0044128	0.001831	0.143330

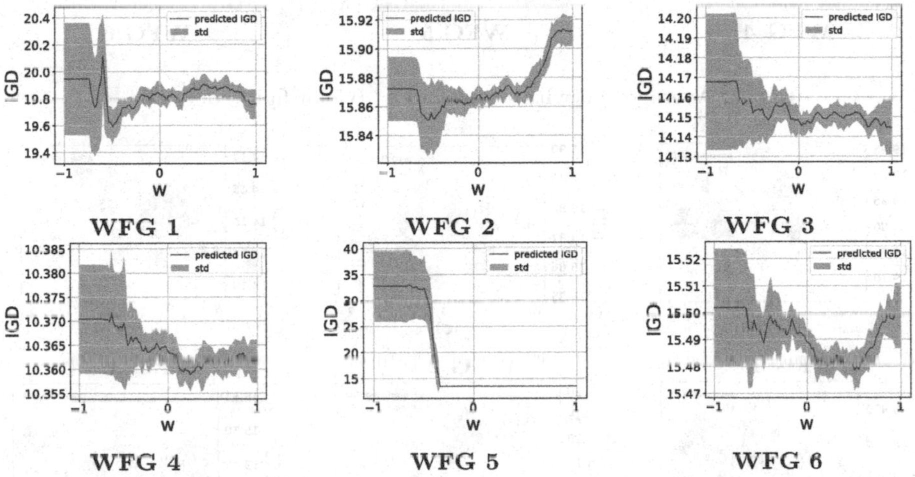

Fig. 1. Average IGD with respect to w. (Color figure online)

line) indicates a greater influence of the particular control parameter on algorithm performance. Similarly, a small standard deviation indicates a low impact in performance when changing the other control parameters. A large standard deviation implies the converse.

Figures 1 and 2 present the response surfaces with respect to w. The inertia weight has the highest variability across all performance measures, which affirms the results determined by fANOVA. The inertia weight also presents with the lowest standard deviation, suggesting that the values of c_1, ϕ_1 and ϕ_2 has less impact on the performance than w. The average IGD and HV associated with c_1, represented in Figs. 3 and 4, varies significantly over the domain with a comparatively low standard deviation. This large variation in performance implies that c_1 has a significant level of importance in MGPSO performance. The low standard deviation suggests that c_1's importance is not affected by changes in the other

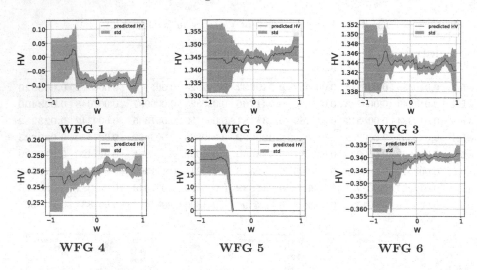

Fig. 2. Average HV with respect to w. (Color figure online)

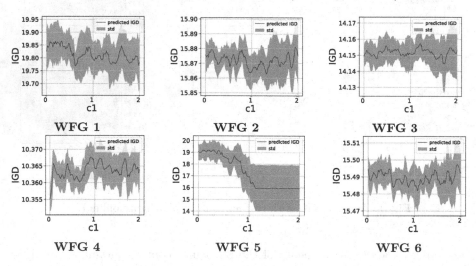

Fig. 3. Average IGD with respect to c_1. (Color figure online)

control parameters. An increasing standard deviation for a performance measure as illustrated in Fig. 3 for c_1 or a large standard deviation as seen for ϕ_1 and ϕ_2 is indicative of the nature of the specific type of problem. The average performance associated with ϕ_1 and ϕ_1 are constant or slow varying, suggesting that the social and archive coefficient is of low importance to MGPSO performance. The multi-modal, concave WFG 4 and WFG 5 show deteriorating performance with an increase in the value of c_1.

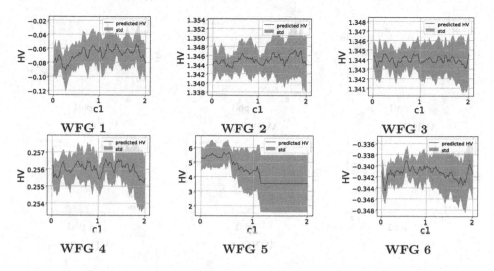

Fig. 4. Average HV with respect to c_1. (Color figure online)

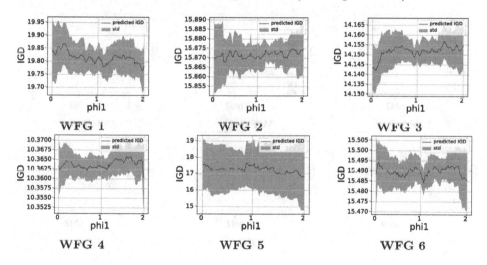

Fig. 5. Average IGD with respect to ϕ_1. (Color figure online)

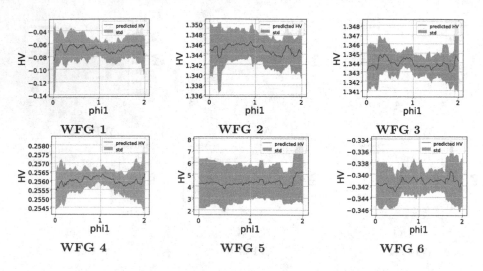

Fig. 6. Average HV with respect to ϕ_1. (Color figure online)

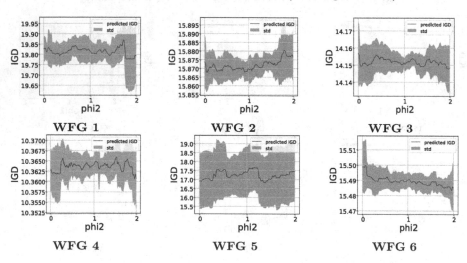

Fig. 7. Average IGD with respect to ϕ_2. (Color figure online)

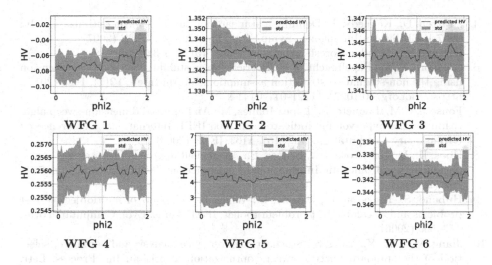

Fig. 8. Average HV with respect to ϕ_2. (Color figure online)

5 Conclusions

This study investigated the relative importance of the multi-guide particle swarm optimization (MGPSO) algorithm control parameters. The MGPSO was initialized under 31744 different control parameter configurations, but evaluated on the Walking Fish Group (WFG) [9] benchmark functions only for configurations within the region of stability as proposed by Scheepers et al. [14]. Each control parameter's importance was determined using function analysis of variance (fANOVA).

The inverted generational distance (IGD) and the hypervolume (HV) were used as performance measures. Results for both measures indicate that the inertia weight accounts for the greatest influence in variance of the performance measures. The social component was the least important control parameter. As such, control parameter tuning efforts for the MGPSO should be focused on adjustment of the inertia weight, followed by the cognitive acceleration coefficient.

Future research will investigate the importance of the MGPSO parameters for higher-dimensional objective functions to determine if the importance ranking will change as the dimensionality of the search landscape increases.

References

1. Beielstein, T., Parsopoulos, K.E., Vrahatis, M.N.: Tuning PSO parameters through sensitivity analysis. Universitätsbibliothek Dortmund (2002)
2. Van den Bergh, F., Engelbrecht, A.P.: A study of particle swarm optimization particle trajectories. Inf. Sci. **176**(8), 937–971 (2006)
3. Bonyadi, M.R., Michalewicz, Z.: Impacts of coefficients on movement patterns in the particle swarm optimization algorithm. IEEE Trans. Evol. Computat. **21**(3), 378–390 (2016)

4. Bratton, D., Kennedy, J.: Defining a standard for particle swarm optimization. In: 2007 IEEE Swarm Intelligence Symposium, pp. 120–127. IEEE (2007)
5. Breiman, L.: Random forests. Machine Learn. **45**(1), 5–32 (2001)
6. Cleghorn, C.W., Engelbrecht, A.P.: Particle swarm stability: a theoretical extension using the non-stagnate distribution assumption. Swarm Intell. **12**(1), 1–22 (2017). https://doi.org/10.1007/s11721-017-0141-x
7. Fonseca, C.M., Paquete, L., López-Ibánez, M.: An improved dimension-sweep algorithm for the hypervolume indicator. In: 2006 IEEE International Conference on Evolutionary Computation, pp. 1157–1163. IEEE (2006)
8. Hoos, H., Leyton-Brown, K., Hutter, F.: An efficient approach for assessing hyperparameter importance. In: International Conference on Machine Learning, pp. 754–762 (2014)
9. Huband, S., Hingston, P., Barone, L., While, L.: A review of multiobjective test problems and a scalable test problem toolkit. IEEE Trans. Evol. Computat. **10**(5), 477–506 (2006)
10. Jiang, M., Luo, Y., Yang, S.: Stochastic convergence analysis and parameter selection of the standard particle swarm optimization algorithm. Inf. Process. Lett. **102**(1), 8–16 (2007)
11. Kennedy, J., Eberhart, R.: Particle swarm optimization. In: Proceedings of ICNN 1995-International Conference on Neural Networks, vol. 4, pp. 1942–1948. IEEE (1995)
12. Raquel, C.R., Naval Jr, P.C.: An effective use of crowding distance in multiobjective particle swarm optimization. In: Proceedings of the 7th Annual Conference on Genetic and Evolutionary Computation, pp. 257–264 (2005)
13. Scheepers, C.: Multi-guided particle swarm optimization: A multi-objective particle swarm optimizer, unpublished thesis (2017)
14. Scheepers, C., Engelbrecht, A.P., Cleghorn, C.W.: Multi-guide particle swarm optimization for multi-objective optimization: empirical and stability analysis. Swarm Intell. **13**(3), 245–276 (2019). https://doi.org/10.1007/s11721-019-00171-0
15. Shi, Y., Eberhart, R.: The 1998 IEEE International Conference On Evolutionary Computation Proceedings (1998)
16. Sobol, I.M.: Sensitivity estimates for nonlinear mathematical models. Math. Model. Comput. Exp. **1**(4), 407–414 (1993)
17. Sun, Y., Yen, G.G., Yi, Z.: IGD indicator-based evolutionary algorithm for many-objective optimization problems. IEEE Trans. Evol. Comput. **23**(2), 173–187 (2018)
18. Tapia, M., Coello, C.: Applications of multi-objective evolutionary algorithms in economics and finance: a survey. In: IEEE Congress on Evolutionary Computation, pp. 532–539 (2007)

Dynamic Response Thresholds: Heterogeneous Ranges Allow Specialization While Mitigating Convergence to Sink States

Annie S. Wu[1][(✉)] and H. David Mathias[2]

[1] University of Central Florida, Orlando, FL, USA
`aswu@cs.ucf.edu`
[2] University of Wisconsin – La Crosse, La Crosse, WI, USA
`dmathias@uwlax.edu`

Abstract. We argue that heterogeneous threshold ranges allow agents in a decentralized swarm to effectively adapt thresholds in response to dynamic task demands while avoiding the pitfalls of positive feedback sinks. Dynamic response thresholds allow agents to dynamically evolve specializations which can improve the responsiveness and stability of a swarm. Dynamic thresholds that adapt in response to previous experience, however, are vulnerable to getting stuck in sink states due to the positive feedback nature of such systems. We show that heterogeneous threshold ranges result in comparable task allocation and improved stability as compared to homogeneous threshold ranges, and that simple static random thresholds should be considered in situations where agent resources are plentiful.

1 Introduction

In this paper, we show that heterogeneous threshold ranges allow agents in a decentralized swarm to effectively adapt thresholds in response to dynamic task demands while avoiding the pitfalls of positive feedback sinks. Response threshold based systems are a biologically inspired approach for generating division of labor in decentralized swarms [1,2,30]. While static thresholds are able to achieve effective task allocation [16,19,34], allowing agents to dynamically adapt their task thresholds over time allows for dynamic specialization which is thought to improve the responsiveness and stability of a swarm. Dynamic thresholds that adapt in response to previous experience, however, are vulnerable to getting stuck in sink states due to the positive feedback nature of such systems [17,30]. We show that varying the threshold ranges of each agent can effectively mitigate the negative effects of sinks while retaining the benefits of dynamic thresholds.

The response threshold approach is an effective method for generating task allocation in decentralized robotic swarms. Each agent possesses a threshold for each task that the agent can potentially take on. An agent's decision as to which

© Springer Nature Switzerland AG 2020
M. Dorigo et al. (Eds.): ANTS 2020, LNCS 12421, pp. 107–120, 2020.
https://doi.org/10.1007/978-3-030-60376-2_9

task, if any, to take on is a function of the agent's threshold for each task and the observed task stimuli. This approach is effective in decentralized systems and is not dependent on inter-agent communication which makes it scalable and useful for problems where stealth is necessary or where agents carry limited power. Response threshold approaches include both static [11,16,19,27,28,33,34] and dynamic thresholds [5–10,12,13,17,18,28,30]. Dynamic thresholds are particularly interesting because they allow a swarm to adjust the distribution of its agent propensities over time. For problems where the distribution of work is not known in advance or may change over time, this adaptability can potentially make the swarm more effective. In addition, dynamic thresholds allow agents to specialize on tasks which improves efficiency by reducing task switching [3,4].

In systems that use dynamic thresholds, agents may adjust their thresholds in response to external[1] [6,14,22–24] or internal factors [20,25,26]. The former tends to be problem dependent and out of the scope of this paper. We study the latter approach, specifically, systems modelled on the concept that previous experience on a task makes an agent more likely to act on that task in the future [12,21,29,30]. This concept is commonly implemented in the form of a learning factor that, in each timestep, lowers the threshold of a task on which an agent is working and a forgetting factor that increases that agent's threshold for all other tasks [5,8,28,30]. While agents in such systems can effectively converge their thresholds into a distribution that meets a given set of task demands, once converged, these systems often have difficulty undoing an expired distribution and re-adjusting to new demands if task demands change [17,18]. The positive feedback structure of this concept results in a tendency for thresholds to evolve to extreme values which are sink states that are difficult to subsequently evolve out of.

We hypothesize that heterogeneous threshold ranges can improve the performance of dynamic response threshold systems by reducing the effects of sink states while still allowing agents to adapt their thresholds and specialize on tasks. Current dynamic response threshold systems assign the same threshold range to all agents. This homogeneity means that once convergence occurs, all agents that have converged will be equally unwilling to revise their thresholds when task demands change. Heterogeneous threshold ranges would result in convergence to different values, allowing some agents to be more willing to revise their thresholds than others. In addition, should agents still get stuck in sink states, the variability in sink states may allow the swarm greater ability to respond to new task demands than if all agents are stuck in the same sink state.

2   Collective Tracking Problem

We test our hypothesis on a collective tracking problem [33,34] which attempts to model a collective task allocation problem similar to that of honeybee thermoregulation [15,31,32]. Where thermoregulation works in a single dimension

---

[1] External factors include but are not limited to task stimuli and observed actions of other agents.

with agents selecting from among two tasks, the tracking problem works in two dimensions with agents selecting from among four tasks.

The collective tracking problem consists of a target that moves in a two dimensional space and a tracker that is collectively controlled by the swarm. The goal of the swarm is to push the tracker such that its movement tracks the target as closely as possible. In each timestep, the individual agents in the swarm select from one of four tasks – *PUSH_NORTH, PUSH_EAST, PUSH_SOUTH, PUSH_WEST* – or remain idle. A positive difference between the target and tracker locations in any direction signifies a task demand in that direction. Each agent can select to push in, at most, one direction in each timestep. The tracker movement in each timestep is calculated by aggregating the decisions of all active agents in that timestep.

The path on which a target moves determines the task demands and how they change over time. For example, constant movement in the northeast direction results in constant equal task demands to the north and east in each timestep. A zigzag path represents task demands that remain stable for a period of time, but occasionally change significantly and abruptly. Serpentine or circular paths, on the other hand, represent constant gradual changes in task demands.

The authors acknowledge that there are more effective and efficient methods to accomplish tracking. We use this collective tracking problem as a testbed because it is a useful example of a decentralized task allocation problem. As the target moves through space, positive difference between the target and tracker in any direction represents a task demand in that direction. The relative number of agents that select to push in each direction determines the aggregate tracker movement; hence, accurate self-allocation of agents to tasks is required to meet task demands. The specification of a target path allows us a systematic way to define dynamic task demands with specific characteristics. The problem is designed such that we are able to quantitatively measure the satisfaction of each task demand individually as well as visually assess the overall performance of the system by comparing the actual target and tracker paths.

3   System Details

We compare the performance of a dynamic response threshold swarm using heterogeneous threshold ranges, termed Dynamic-Heterogeneous, against the performance of two baseline systems. The first baseline system, Dynamic-Homogeneous, is a dynamic response threshold swarm using homogeneous threshold ranges. Dynamic-Homogeneous is representative of how most current dynamic threshold systems work. The second baseline system, Static, is a swarm with static thresholds.

All three systems consist of a population of n decentralized agents, $a_i, i = 0, ..., n$. Each agent has a separate threshold for each task or direction, $\{\theta_{i,N}, \theta_{i,E}, \theta_{i,S}, \theta_{i,W}\}$. These thresholds represent the tolerance of that agent for the corresponding differences, $\{\Delta_N, \Delta_E, \Delta_S, \Delta_W\}$, between target and tracker position. In a given timestep, if the difference in a direction exceeds the agent's threshold for that direction (if $\Delta_j > \theta_{i,j}$), the agent will consider pushing in that direction for that timestep. If more than one task is triggered for an agent, the agent randomly selects one of the triggered tasks on which to act. Note that for this problem, because $\Delta_N = -\Delta_S$ and $\Delta_E = -\Delta_W$ are always true, not more than two tasks will ever be triggered at the same time.

Agent thresholds work as follows in the three systems tested. All thresholds are floating point values. In the two dynamic systems, each threshold, $\theta_{i,j}$, has a range within which it can vary. This range is defined by a minimum, $\theta_{i,j}min$, and maximum, $\theta_{i,j}max$, value. In the Static system, thresholds are static and are initialized uniformly randomly to a value between 0 and R, where R is a user specified parameter indicating the maximum allowed threshold value. In the Dynamic-Homogeneous system, thresholds are dynamic and all thresholds can vary within the range specified by $\theta_{i,j}min = 0$ and $\theta_{i,j}max = R$. The initial value of each threshold is a random value drawn from a uniform distribution between 0 and R. In the proposed Dynamic-Heterogeneous system, thresholds are dynamic and all thresholds vary within a unique range. The lower bound of the range, $\theta_{i,j}min$, is a random value drawn from a uniform distribution between 0 and $\frac{R}{2}$. The upper bound of the range, $\theta_{i,j}max$, is a random value drawn from a uniform distribution between $\frac{R}{2}$ and R. The initial value of each threshold is a random value drawn from a uniform distribution between $\theta_{i,j}min$ and $\theta_{i,j}max$.

For the two dynamic threshold systems, threshold variation occurs the same way as seen in previous work [30]. In each timestep, if an agent is working on a task j, its threshold for that task is decreased by a learning factor ε such that $\theta_{i,j} = \theta_{i,j} - \varepsilon$, and its thresholds for all other tasks are increased by a forgetting factor ψ such that $\theta_{i,j} = \theta_{i,j} + \psi$, where ε and ψ are user specified parameters.

The tracker movement in each timestep is determined by the number of agents pushing in each direction in that timestep. Let $n_j, j \in \{N, E, S, W\}$ be the number of agents pushing in direction j in a given timestep. The distance, d_j, that the tracker moves in direction j is given by $d_j = \frac{n_j}{n} \times \rho$, where ρ is the step ratio. The step ratio specifies the maximum distance that the tracker can move relative to the target in one timestep. Thus, if $\rho = 2.0$, the tracker can move twice as far as the target in one timestep. If $\rho = 0.75$, the tracker can move 75% of the distance that the target can move in one time step.

4   Experimental Details

We compare the performance of the three swarm configurations on four problem scenarios. Each problem scenario is represented as a target path.

– zigzag: Target alternates between moving approximately northeast and moving approximately southeast.
– scurve: Target moves from west to east in a serpentine pattern.
– sharp: Target direction is randomly initialized. In each timestep, target has a 5–10% chance of changing to a random new direction; otherwise, target continues in current direction.

– random: Target direction is randomly initialized. In each timestep, target direction is changed by an angle drawn from a Gaussian distribution.

Table 1. Fixed parameter settings.

Parameter	Value
Population size, n	50
Number of timesteps	500
Maximum threshold range, R	10
Threshold decrease, ε (learning factor)	0.1
Threshold increase, ψ (forgetting factor)	0.033

The zigzag and sharp paths produce significant periods of constant task demands punctuated by occasional abrupt changes. The scurve and random paths produce gradually changing task demands. Because of the randomness in the test problems and system behavior, each experiment is composed of 100 runs. Unless otherwise specified, the results for each experiment are averaged over all 100 runs.

Table 1 gives the parameter settings that remain fixed throughout all experiments reported here. The threshold decrease, ε, and increase, ψ, values are set such that total adjusted threshold is conserved; given four tasks, when an agent's threshold decreases for one task, it increases by one third of that amount for the three other tasks. We examine multiple values of step ratio, from $\rho = 0.75$ to $\rho = 3.0$ in increments of 0.25, to examine the impact of agent availability on system performance.

We evaluate system performance based on three evaluation metrics.

1. Tracker path length: The tracker path length provides a measure of how well the tracker followed the target path. The target path length in all experiments reported here is 500. The optimum value for this measure is 500.
2. Average difference: The average difference is the average of the difference between the target and tracker positions in each timestep of a run. This measures the average deviation of the target and tracker paths over a run. The optimum value for this measure is zero.
3. Number of task switches: The number of task switches is the average number of times that agents change tasks during a run, averaged over all agents in the swarm. A task switch is defined as switching from one task to another as well as switching between idle and acting on a task. One of the expected advantages of dynamic thresholds is that they allow agents to dynamically specialize to one or fewer tasks. Thus, specialization should result in agents focusing on a single or fewer tasks, and reduction in the frequency of task switching. The optimum value for this measure is zero.

4.1   Results

Figure 1 compares the performance of the three swarm systems with respect to tracker path length. The top row of plots give the results for the two regular paths, `zigzag` and `scurve`. The bottom row of plots give the results for the two random or irregular paths, `sharp` and `random`. The x-axis of each plot indicates the step ratio, ρ. The y-axis of each plot indicates tracker path length. The optimum path length is 500, as indicated by the dashed line.

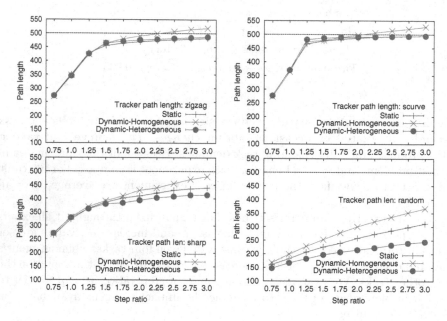

Fig. 1. Average and 95% confidence interval of the tracker path length, averaged over 100 runs. The optimal path length is 500, as indicated by the dashed line. The confidence intervals are extremely tight but they are plotted. (Color figure online)

Comparing the two dynamic systems, the red line and the aqua line, we see that the Dynamic-Homogeneous tracker tends to travel longer paths than the Dynamic-Heterogeneous tracker. As the step ratio increases (as we have more extra agents) this difference increases. On the two regular paths, `zigzag` and `scurve`, Dynamic-Homogeneous overshoots more and more as the step ratio increases. This indicates that more agents than necessary are specializing on tasks and the swarm is likely repeatedly over-shooting and over-correcting the tracker path. Once there are sufficient agents to meet task demands, Dynamic-Heterogeneous and Static both converge gradually toward the optimum path length without over-shooting as extra agent resources increase. On the irregular paths, `sharp` and `random`, Dynamic-Homogeneous generates path lengths closer to the optimum path length than Dynamic-Heterogeneous. Examination of actual paths, however, reveals that both systems generate similar quality solutions. Figure 2 shows example `sharp` runs for step ratio 3.0, the value at which Dynamic-Homogeneous shows the greatest improvement over Dynamic-Heterogeneous. Both systems track the target similarly well and the extra length of the Dynamic-Homogeneous path is actually due to over-correction choppiness.

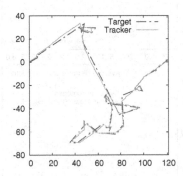 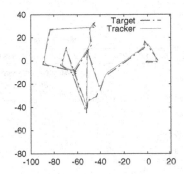

Fig. 2. Target and tracker paths. The left plot is an example Dynamic-Homogeneous run. The right plot is an example Dynamic-Heterogeneous run. Both runs are on the `sharp` path and have a step ratio of 3.0.

Figure 3 compares the performance of the three swarm systems with respect to the average distance between the target and tracker throughout a run. The x-axis of each plot indicates the step ratio, ρ. The y-axis of each plot indicates distance. Comparing the two dynamic systems, we see that when the step ratio is low (there are little or no extra agents), Dynamic-Heterogeneous performs better than Dynamic-Homogeneous, keeping the tracker closer to the target during the run. As step ratio increases (the number of extra agents increase), Dynamic-Homogeneous becomes the better performer. Static continues to perform well relative to the dynamic systems, achieving the best or close to best performance of the three. All three systems performed similarly overall; on a path of length 500 units, all three systems maintained average distances within two units or less of each other for each step ratio value.

Figure 4 compares the performance of the three swarm systems with respect to the average number of task switches per agent per run. The x-axis of each plot indicates the step ratio, ρ. The y-axis of each plot indicates number of switches. In all paths except for `zigzag`, Dynamic-Homogeneous performs significantly worse than either Dynamic-Heterogeneous or Static. In the `zigzag` path, the performance of the two dynamic systems is similar when the step ratio is low, and Dynamic-Heterogeneous becomes significantly better as step ratio increases. Static performs significantly better than either dynamic system on the regular paths. Static's advantage is less consistent on the irregular paths where Dynamic-Heterogeneous outperforms it (undergoes significantly fewer task switches) on the `random` path.

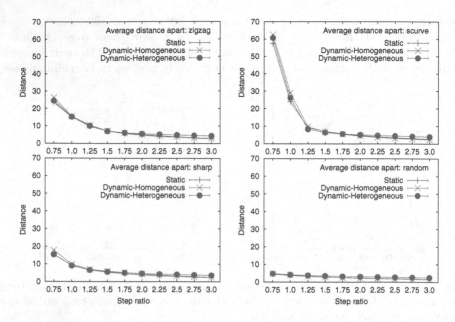

Fig. 3. Average and 95% confidence interval of the average distance between target and tracker during a run, averaged over 100 runs. The confidence intervals are extremely tight but they are plotted.

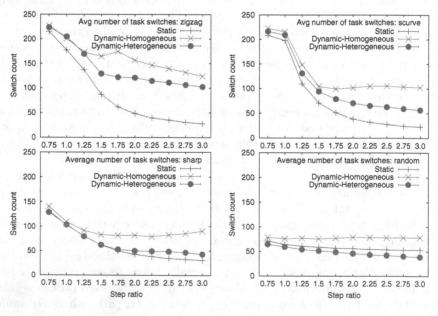

Fig. 4. Average and 95% confidence interval of the average number of task switches per agent during a run, averaged over 100 runs. The confidence intervals are extremely tight but they are plotted.

4.2 Agent Thresholds and Actions

The previous results suggest that Dynamic-Homogeneous and Dynamic-Heterogeneous are able to track the target with similar skill, with Dynamic-Heterogeneous forming more stable specializations. To verify this conclusion, we need to examine how agents act and adapt their thresholds over the course of a run.

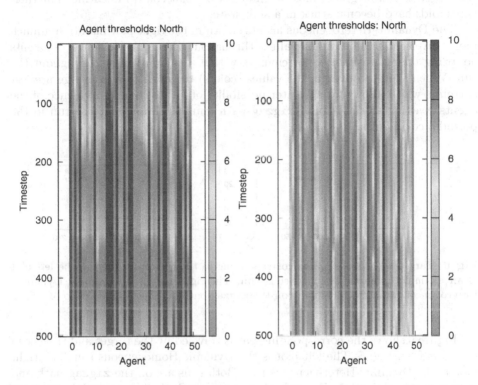

Fig. 5. Threshold values for $\theta_{i,N}$ for all agents over the course of a run. The left plot is an example Dynamic-Homogeneous run. The right plot is an example Dynamic-Heterogeneous run. Both runs are on the `zigzag` path and have a step ratio of 1.5. (Color figure online)

Figure 5 shows how the $\theta_{i,N}$ threshold of all agents in the swarm change over time in two example runs. The left plot is an example Dynamic-Homogeneous run. The right plot is an example Dynamic-Heterogeneous run. Both runs are on the `zigzag` path and have a step ratio of 1.5. The x-axis of both plots indicates agent number, i. The y-axis of both plots indicates timestep. Each column shows the values for one agent's $\theta_{i,N}$ (threshold for pushing north) and how they change over time. Green indicates low threshold (quick to act) and red indicates high threshold (unlikely to act).

Recall that initial thresholds are randomly generated in both systems. Accordingly, there is a mix of colors in the top rows of both plots. As the runs proceed, the Dynamic-Homogeneous agents shown in the left plot clearly converge to extreme threshold values as indicated by the bright red and green values in the second half of the plot. Although early on (top half of the plot), color changes within a column indicates that there are agents that are adapting their thresholds, instances of color changes diminish as the run proceeds and the bottom half of the plot shows much less evidence of threshold adaptation. Once converged to red or green, most agents stay on that color, indicating that their thresholds have become stuck in a sink state.

The Dynamic-Heterogeneous agents shown in the right plot maintain a much more diverse distribution of values throughout the run. Evidence of agents adapting their threshold (color changes within a column) exist throughout the run. When agents converge, the values (colors) to which they converge are less extreme, which allows for a greater possibility of future change. Evidence of the agents reacting to the regular `zigzag` path remains throughout the run in the periodic color shifts.

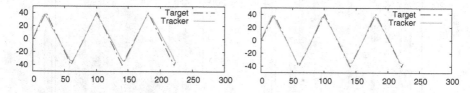

Fig. 6. Target and tracker paths corresponding to the runs from Fig. 5. The left plot is an example Dynamic-Homogeneous run. The right plot is an example Dynamic-Heterogeneous run. Both runs are on the `zigzag` path and have a step ratio of 1.5.

Figure 6 shows the corresponding paths traveled by the target and tracker in the runs from Fig. 5. The left plot is the Dynamic-Homogeneous run. The right plot is the Dynamic-Heterogeneous run. Both runs are on the `zigzag` path and have a step ratio of 1.5. While both systems track the target well, we can see in the left plot that, as Dynamic-Homogeneous agent thresholds converge, the system's tracking ability declines. Notably, Dynamic-Homogeneous continues to track well when travelling northeast, the direction for which its thresholds first begin to adapt. Its ability to track in the southeast direction declines over time, likely due to agent threshold having converged to a distribution optimized for the first set of tasks it encountered. Dynamic-Heterogeneous agents, on the other hand, track the target well throughout the run in both directions while also generating fewer task switches (as indicated in Fig. 4).

5   Conclusions

In this paper, we test the hypothesis that using heterogeneous threshold ranges instead of homogeneous threshold ranges will allow dynamic response threshold swarms to adapt agent thresholds in response to changing task demands while mitigating the problem of convergence to and inability to leave sink states that occurs with homogeneous threshold ranges. We compare the performance of the proposed Dynamic-Heterogeneous approach with two baseline approaches: the existing Dynamic-Homogeneous approach where all agents have the same threshold ranges and the basic Static approach where all agents are assigned uniformly random static thresholds that do not change.

We test these three systems on a collective tracking problem that is modelled after a honeybee thermoregulation task allocation problem. We test four instances of this problem. Two instances generate regular repeated task demands over time. Two instances generate irregular, somewhat random, task demands over time. In each pair of instances, one illustrates periods of stable task demand punctuated by occasional abrupt change, the other illustrates constant gradual change in task demands.

Our results indicate that, in most situations, Dynamic-Heterogeneous performs as well or better than Dynamic-Homogeneous in terms of allocating appropriate numbers of agents to each task demand over time. The Dynamic-Heterogeneous approach results in a significantly more stable swarm in that it significantly reduces the number of times agents switch tasks. This stability is due in part to the fact that the Dynamic-Heterogeneous approach reduces the likelihood of agent thresholds converging and becoming stuck in extreme values or sink states. Avoidance of those sink states allows agents greater ability to re-adapt their thresholds if task demands change. Examination of how agent thresholds adapt over the course of an example run finds that Dynamic-Heterogeneous maintains a more diverse and more adaptable distribution of thresholds than Dynamic-Homogeneous. As seen in previous work, Dynamic-Homogeneous thresholds tend to converge in response to the first set of task demands encountered and have difficulty re-adapting to new task demands. Dynamic-Heterogeneous threshold remain responsive to changes in task demand while converging enough to lower task switching and increase stability.

An interesting and unexpected result that we have not yet discussed is the fact that swarms in which agents are assigned static uniformly distributed thresholds matches or outperforms both dynamic threshold approaches in a large number of the scenarios that we tested. It is this result that prompted us to examine a range of step ratio values. In trying to understand when dynamic thresholds are necessary, we hypothesize that dynamic thresholds are more crucial in systems without extra agent resources. In such systems, an appropriate distribution of thresholds is necessary in order for the swarm to address all task demands in a timely manner. In systems that do have excess agents, inappropriate distributions of thresholds (and agents that stubbornly refuse to leave tasks that do not need attending) have less of an effect because there are plenty of extra agents to take on unaddressed task demands. This hypothesis is borne out in the data from Figs. 1, 3, and 4 that show that Static's performance advantage over Dynamic-Heterogeneous and Dynamic-Homogeneous is always significantly reduced at lower step ratio values where the systems have few to no extra agents.

In summary, our results suggest two general conclusions with respect to swarms that use dynamic response thresholds. First, heterogeneous threshold ranges effectively mitigate the problem of convergence to sink states that occurs with homogeneous threshold ranges, while still retaining the benefits of threshold adaptation. Second, if a swarm is expected to have excess agents, static uniformly distributed thresholds are a simple and effective approach that deserve serious consideration.

Acknowledgements. This work was supported by the National Science Foundation under Grant No. IIS1816777.

References

1. Bonabeau, E., Theraulaz, G., Deneubourg, J.L.: Quantitative study of the fixed threshold model for the regulation of division of labor in insect societies. Proc. Royal Soc. London Biol. Sci. **263**(1376), 1565–1569 (1996)
2. Bonabeau, E., Theraulaz, G., Deneubourg, J.L.: Fixed response thresholds and the regulation of division of labor in insect societies. Bull. Math. Biol. **60**, 753–807 (1998). https://doi.org/10.1006/bulm.1998.0041
3. Brutschy, A., Pini, G., Pinciroli, C., Birattari, M., Dorigo, M.: Self-organized task allocation to sequentially interdependent tasks in swarm robots. Auton. Agent. Multi-Agent Syst. **25**, 101–125 (2014)
4. Brutschy, A., et al.: Costs and benefits of behavioral specialization. Robot. Auton. Syst. **60**, 1408–1420 (2012)
5. Campos, M., Bonabeau, E., Theraulaz, G., Deneubourg, J.: Dynamic scheduling and division of labor in social insects. Adapt. Behav. **8**, 83–96 (2000)
6. Castello, E., et al.: Adaptive foraging for simulated and real robotic swarms: the dynamical response threshold approach. Swarm Intell. **10**, 1–31 (2018). https://doi.org/10.1007/s11721-015-0117-7
7. Castello, E., Yamamoto, T., Nakamura, Y., Ishiguro, H.: Task allocation for a robotic swarm based on an adaptive response threshold model. In: Proceedings of the 13th IEEE International Conference on Control, Automation, and Systems, pp. 259–266 (2013)
8. Cicirello, V.A., Smith, S.F.: Distributed coordination of resources via wasp-like agents. In: Truszkowski, W., Hinchey, M., Rouff, C. (eds.) WRAC 2002. LNCS (LNAI), vol. 2564, pp. 71–80. Springer, Heidelberg (2003). https://doi.org/10.1007/978-3-540-45173-0_5
9. de Lope, J., Maravall, D., Quinonez, Y.: Response threshold models and stochastic learning automata for self-coordination of heterogeneous multi-task distribution in multi-robot systems. Robot. Auton. Syst. **61**, 714–720 (2013)
10. de Lope, J., Maravall, D., Quinonez, Y.: Self-organizing techniques to improve the decentralized multi-task distribution in multi-robot systems. Neurocomputing **163**, 47–55 (2015)
11. dos Santos, F., Bazzan, A.L.C.: An ant based algorithm for task allocation in large-scale and dynamic multiagent scenarios. In: Proceedings of the Genetic and Evolutionary Computation Conference, pp. 73–80 (2009)
12. Gautrais, J., Theraulaz, G., Deneubourg, J., Anderson, C.: Emergent polyethism as a consequence of increase colony size in insect societies. J. Theoret. Biol. **215**, 363–373 (2002)

13. Goldingay, H., van Mourik, J.: The effect of load on agent-based algorithms for distributed task allocation. Inf. Sci. **222**, 66–80 (2013)
14. Jones, C., Mataric, M.J.: Adaptive division of labor in large-scale minimalist multi-robot systems. In: Proceedings of the IEEE/RSJ International Conference on Intelligent Robots and Systems, pp. 1969–1974 (2003)
15. Jones, J.C., Myerscough, M.R., Graham, S., Oldroyd, B.P.: Honey bee nest thermoregulation: diversity promotes stability. Science **305**(5682), 402–404 (2004)
16. Kanakia, A., Touri, B., Correll, N.: Modeling multi-robot task allocation with limited information as global game. Swarm Intell. **10**(2), 147–160 (2016). https://doi.org/10.1007/s11721-016-0123-4
17. Kazakova, V.A., Wu, A.S.: Specialization vs. re-specialization: effects of Hebbian learning in a dynamic environment. In: Proceedings of the 31st International Florida Artificial Intelligence Research Society Conference, pp. 354–359 (2018)
18. Kazakova, V.A., Wu, A.S., Sukthankar, G.R.: Respecializing swarms by forgetting reinforced thresholds. Swarm Intell. **14**(3), 171–204 (2020). https://doi.org/10.1007/s11721-020-00181-3
19. Krieger, M.J.B., Billeter, J.B.: The call of duty: self-organised task allocation in a population of up to twelve mobile robots. Robot. Auton. Syst. **30**, 65–84 (2000)
20. Labella, T.H., Dorigo, M., Deneubourg, J.: Division of labor in a group of robots inspired by ants' foraging behavior. ACM Trans. Auton. Adapt. Syst. **1**(1), 4–25 (2006)
21. Langridge, E.A., Franks, N.R., Sendova-Franks, A.B.: Improvement in collective performance with experience in ants. Behav. Ecol. Sociobiol. **56**, 523–529 (2004). https://doi.org/10.1007/s00265-004-0824-3
22. Lee, W., Kim, D.E.: Local interaction of agents for division of labor in multi-agent systems. In: Tuci, E., Giagkos, A., Wilson, M., Hallam, J. (eds.) SAB 2016. LNCS (LNAI), vol. 9825, pp. 46–54. Springer, Cham (2016). https://doi.org/10.1007/978-3-319-43488-9_5
23. Lee, W., Kim, D.: History-based response threshold model for division of labor in multi-agent systems. Sensors **17**, 1232 (2017)
24. Lerman, K., Jones, C., Galstyan, A., Mataric, M.J.: Analysis of dynamic task allocation in multi-robot systems. Int. J. Robot. Res. **25**, 225–241 (2006)
25. Liu, W., Winfield, A., Sa, J., Chen, J., Dou, L.: Strategies for energy optimisation in a swarm of foraging robots. In: Şahin, E., Spears, W.M., Winfield, A.F.T. (eds.) SR 2006. LNCS, vol. 4433, pp. 14–26. Springer, Heidelberg (2007). https://doi.org/10.1007/978-3-540-71541-2_2
26. Liu, W., Winfield, A., Sa, J., Chen, J., Dou, L.: Towards energy optimisation: Emergent task allocation in a swarm of foraging robots. Adapt. Behav. **15**, 289–305 (2007)
27. Meyer, B., Weidenmuller, A., Chen, R., Garcia, J.: Collective homeostasis and time-resolved models of self-organised task allocation. In: Proceedings of the 9th EIA International Conference on Bio-inspired Information and Communication Technologies, pp. 469–478 (2015)
28. Price, R., Tiňo, P.: Evaluation of adaptive nature inspired task allocation against alternate decentralised multiagent strategies. In: Yao, X., et al. (eds.) PPSN 2004. LNCS, vol. 3242, pp. 982–990. Springer, Heidelberg (2004). https://doi.org/10.1007/978-3-540-30217-9_99
29. Ravary, F., Lecoutey, E., Kaminski, G., Chaline, N., Jaisson, P.: Individual experience alone can generate lasting division of labor in ants. Curr. Biol. **17**, 1308–1312 (2007)

30. Theraulaz, G., Bonabeau, E., Deneubourg, J.: Response threshold reinforcement and division of labour in insect societies. Proc. Royal Soc. B **265**, 327–332 (1998)
31. Weidenmüller, A.: The control of nest climate in bumblebee (*Bombus terrestris*) colonies: interindividual variability and self reinforcement in fanning response. Behav. Ecol. **15**, 120–128 (2004)
32. Weidenmüller, A., Chen, R., Meyer, B.: Reconsidering response threshold models—short-term response patterns in thermoregulating bumblebees. Behav. Ecol. Sociobiol. **73**(8), 1–13 (2019). https://doi.org/10.1007/s00265-019-2709-5
33. Wu, A.S., Mathias, H.D., Giordano, J.P., Hevia, A.: Effects of response threshold distribution on dynamic division of labor in decentralized swarms. In: Proceedings of the 33rd International Florida Artificial Intelligence Research Society Conference (2020)
34. Wu, A.S., Riggs, C.: Inter-agent variation improves dynamic decentralized task allocation. In: Proceedings 31st International Florida Artificial Intelligence Research Society Conference, pp. 366–369 (2018)

Grey Wolf, Firefly and Bat Algorithms: Three Widespread Algorithms that Do Not Contain Any Novelty

Christian Leonardo Camacho Villalón[(✉)][iD], Thomas Stützle[iD], and Marco Dorigo[iD]

IRIDIA, Université Libre de Bruxelles, Brussels, Belgium
{ccamacho,stuetzle,mdorigo}@ulb.ac.be

Abstract. In this paper, we carry out a review of the grey wolf, the firefly and the bat algorithms. We identify the concepts involved in these three metaphor-based algorithms and compare them to those proposed in the context of particle swarm optimization. We provide compelling evidence that the grey wolf, the firefly, and the bat algorithms are not novel, but a reiteration of ideas introduced first for particle swarm optimization and reintroduced years later using new natural metaphors. These three algorithms can therefore be added to the growing list of metaphor-based algorithms—to which already belong algorithms such as harmony search and intelligent water drops—that are nothing else than repetitions of old ideas hidden by the usage of new terminology.

1 Introduction

Algorithms inspired by natural or artificial metaphors have become a common place in the stochastic optimization literature [4]. Despite being invariably presented as novel methods, many of these algorithms do not seem to be proposing novel ideas; rather, they reintroduce well-known concepts already proposed in previously published algorithms [23]. That is, the same ideas developed in the context of local search (LS) heuristics, evolutionary algorithms (EAs), and ant colony optimization (ACO), to mention a few, appear in these "novel" algorithms, although presented using new terminology. In addition to this, it is often the case that rather than clearly expressing new ideas in plain algorithmic terms and highlighting differences with what has already been proposed in the literature, authors of these algorithms focus on aspects such as the novelty and beauty of the new inspiring source.

Several are the undesirable consequences of this practice. Perhaps the most detrimental one is that it has generated a lot of confusion in the literature, since using different terminologies for referring to concepts already defined makes it difficult to compare algorithms—both conceptually and experimentally—hindering our understanding. Additionally, presenting ideas using unconventional terminology instead of the normal one used in optimization, adds an unnecessary extra effort to distinguish between what is novel and what is not.

© Springer Nature Switzerland AG 2020
M. Dorigo et al. (Eds.): ANTS 2020, LNCS 12421, pp. 121–133, 2020.
https://doi.org/10.1007/978-3-030-60376-2_10

A number of studies [2,3,12,17,24,26,27] have shown that the use of metaphors has served the sole purpose of hiding the similarities among different methods, thus allowing copies of well-known approaches to be misrepresented as new. One of the first examples of this was provided in 2010 by Weyland [26,27] in a rigorous analysis of the harmony search (HS) algorithm. In this analysis, Weyland found that this seemingly new method was the same as a particular evolutionary algorithm, called "evolution strategy $(\mu + 1)$" [20], which was proposed about 30 years before the HS algorithm. Similarly, other studies in the same direction have shown that the black holes optimization algorithm is a simplification of particle swarm optimization (PSO) [17], and, more recently, that the intelligent water drops algorithm is a special case of ACO [2,3]. Even though it has been shown that a number of these algorithms are a reiteration of well-known ideas and it has been suggested that the whole trend is unhealthy and damaging for the entire research field [23,24], new algorithms and new variants of metaphor-based algorithms continue to be published with alarming high frequency.

In this paper, we review the concepts utilized in three highly-cited algorithms proposed for continuous optimization problems: the grey wolf, the firefly and the bat algorithms. We provide evidence that (i) *using new metaphors*, (ii) *changing the terminology*, and (iii) *presenting the algorithm in a confusing way* allowed to overlook the fact that they were all PSO variants.

The rest of the paper is organized as follows. In Sect. 2, we briefly review the basic concepts of PSO and some of its most popular variants. In Sect. 3, we describe the three metaphor-based algorithms that we analyze using standard PSO terminology. In this way, it becomes immediately apparent the fact that these algorithms are indeed PSO variants. In Sect. 4, we conclude the paper by highlighting some aspects that make the analysis of these 'novel' metaphor-inspired algorithms a challenging endeavor.

2   Particle Swarm Optimization

Particle swarm optimization is arguably the most popular swarm intelligence algorithm to tackle continuous optimization problems. It was introduced by Kennedy and Eberhart [7,9] in 1995, and is inspired by the observation of social dynamics in bird flocks. The optimization capabilities of PSO come from the repeated application of a set of rules that allow particles, which represent problem solutions, in the swarm to identify and move in promising directions in the search space that they estimate from locations that they previously visited and that correspond to good solutions [11]. Particles employ simple memory structures to keep track of important information collected during the search process, such as their own position and velocity and the position of the best among neighboring particles.

In the standard PSO (SPSO) [21,22] algorithm, each particle i knows at every iteration t its current position $\boldsymbol{x}_t^i$, its velocity $\boldsymbol{v}_t^i$, its personal best position $\boldsymbol{p}_t^i$, and the position $\boldsymbol{l}_t^i$ of its local best particle, where the local best particle is the particle j in the neighborhood of particle i that has the best personal best position p_t^j. Many different types of neighborhood are possible among which star, ring, and lattices are typical options; when the neighborhood consists of all the particles in the swarm, then the local best particle is called global best and its position, which is the same for all particles i, is indicated by g_t.

The rules to update a particle's position and velocity in SPSO are:

$$x_{t+1}^i = x_t^i + v_{t+1}^i \tag{1}$$

$$v_{t+1}^i = \omega v_t^i + \varphi_1 a_t^i \odot (p_t^i - x_t^i) + \varphi_2 b_t^i \odot (l_t^i - x_t^i) \tag{2}$$

where ω is the inertia weight that controls the effect of the velocity at time t, φ_1 and φ_2 are the acceleration coefficients that weigh the relative influence of the personal best and local best position, a_t^i and b_t^i are two random vectors used to provide diversity to particle's movement, and $\odot$ indicates the Hadamard (pointwise) product between two vectors.

Over the years, many variants of PSO have been proposed to improve the optimization capabilities of the algorithm. One of them is SPSO-2011 [5,30], that was developed with the goal of preventing the issue of rotation variance that affects SPSO. In this variant, the velocity update rule—Eq. 2 above—was modified as follows:

$$v_{t+1}^i = \omega v_t^i + {x'}_t^i - x_t^i \tag{3}$$

where ${x'}_t^i$ is a randomly generated point in the hypersphere $\mathcal{H}_i(c_t^i, |c_t^i - x_t^i|)$ with center c_t^i and radius $|c_t^i - x_t^i|$, and $|\cdot|$ is the Euclidean norm (L2).

The center c_t^i is computed as

$$c_t^i = (L_t^i + P_t^i + x_t^i)/3 \tag{4}$$

where

$$\begin{aligned} P_t^i &= x_t^i + \varphi_1 a_t^i \odot (p_t^i - x_t^i) \\ L_t^i &= x_t^i + \varphi_2 b_t^i \odot (l_t^i - x_t^i) \end{aligned} \tag{5}$$

The two key concepts proposed in SPSO-2011 consist of (i) the definition of the new point c_t^i in the search space defined as a function of the position of three particles in the swarm: x_t^i, p_t^i and l_t^i (Eq. 4), and (ii) the use of c_t^i to generate a point ${x'}_t^i$ that is then used to update the particles' velocities; the point ${x'}_t^i$ is randomly selected from a hyper-spherical distribution with center c_t^i and radius $|c_t^i - x_t^i|$, which is invariant to rotation (Eq. 3).

Another interesting variant is the fully informed PSO (FiPSO) [13], in which the authors propose to use the whole swarm to influence particles' new velocities:

$$v_{t+1}^i = \chi \left(v_t^i + \sum_{k \in T_t^i} \varphi a_{kt}^i \odot (p_t^k - x_t^i) \right) \tag{6}$$

where $\chi = 0.7298$ is a constant value called constriction coefficient and defined in [6], T_t^i is the set of particles in the neighborhood of i, and φ is a parameter.

The main innovation in this variant is a generalization of some of the algorithmic ideas used in PSO. In particular, while in most previous PSO variants only one particle (the local best) contributed to update the particles' velocities, here the number of particles that influence the velocity update becomes a design choice typically referred as model of influence.

One last example of a PSO variant relevant for our analysis in the next section is the "bare-bones" PSO [8]. This variant belongs to a class of PSO algorithms typically referred to as velocity-free variants in which the rule to update the position of particles does not include a velocity vector:

$$x_{t+1}^i = \mathcal{N}\left(\frac{p_t^i + l_t^i}{2}, |p_t^i - l_t^i|\right) \tag{7}$$

Although in this variant the particles' new position is obtained by sampling values from a normal distribution whose center $(p_t^i + l_t^i)/2$ and dispersion $|p_t^i - l_t^i|$ are computed as a function of p_t^i and l_t^i, many other ways have been proposed to compute x_{t+1}^i in velocity-free PSOs. For example, in [15,16], it was proposed a way of instantiating various velocity-free variants from a generalized position update rule given by

$$x_{t+1}^i = x_t^i + \epsilon(y - x_t^i) \tag{8}$$

where ϵ is an acceleration coefficient and y is a vector which is obtained combining information from other particles, for example:

$$y = \frac{u_1 p_t^1 + u_2 p_t^2}{u_1 + u_2} \tag{9}$$

where u_1 and u_2 are random values drawn from $\mathcal{U}[0,1]$, and p_t^1 and p_t^2 are the personal best positions of two neighbor particles of i chosen according to some criterion. Velocity-free PSO variants are relatively less common in the literature of PSO, and they vary in all kind of aspects, including the way in which the current position of a particle is taken into account in the computation of x_{t+1}^i.

3   The Grey Wolf, Firefly, and Bat Algorithms—Explained

The three algorithms that we analyze in this paper—grey wolf, firefly, and bat algorithms—are taken from the evolutionary computation bestiary [4]. We chose them not only because they were amenable to the analysis that we present in this paper, but also because they were highly cited[1], which gives a reasonable indication of the impact these algorithms have had on the research community.

In the reminder of this section, we present a detailed description of the grey wolf, firefly and bat algorithms using terminology and concepts belonging to PSO. The idea is to recreate together with the reader these three algorithms from concepts that he/she is already familiar with, and to give the details of the

---

[1] Grey Wolf Optimizer [14]: 3656 citations; Firefly Algorithm [28]: 3018 citations; and Bat Algorithm [29]: 3549 citations. Source: Google Scholar. Retrieved: July 10, 2020.

metaphor employed by the authors only after the algorithm has been described in plain algorithmic terms. By doing so, it will be easier to understand if the metaphor was necessary (or useful) to understand the proposed algorithm and if the introduced concepts were indeed new or just hidden by the unconventional terminology used.

3.1 Grey Wolf Optimizer (GWO)

GWO as a PSO Algorithm. The grey wolf optimizer (GWO) [14] is an algorithm in which, in PSO terms, the three best particles in the swarm are used to bias the movement of the remaining swarm particles. This idea is implemented in GWO by defining three vectors s_t^k as follows:

$$
\begin{aligned}
s_t^1 &= g_t^1 - (2\,\varphi_t - 1)r_t^1 \left|2\,q_t^1 \odot g_t^1 - x_t^i\right| \\
s_t^2 &= g_t^2 - (2\,\varphi_t - 1)r_t^2 \left|2\,q_t^2 \odot g_t^2 - x_t^i\right| \\
s_t^3 &= g_t^3 - (2\,\varphi_t - 1)r_t^3 \left|2\,q_t^3 \odot g_t^3 - x_t^i\right|
\end{aligned}
\tag{10}
$$

where g_t^1, g_t^2 and g_t^3 indicate the position of the three best particles in the swarm at iteration t, r_t^k, q_t^k ($k = 1, 2, 3$) are random vectors with values drawn from $\mathcal{U}[0, 1]$ that will induce perturbation to the components of s_t^k, and φ_t is a decreasing acceleration coefficient that goes from 2 to 0.

The position update rule combining the information of the three best particles s_t^k is defined as follows:

$$
x_{t+1}^i = (s_t^1 + s_t^2 + s_t^3)/3
\tag{11}
$$

How Does GWO Compare to PSO? The values s_t^k in Eq. 10 are defined in a very similar way to P_t^i and L_t^i of Eq. 5. The main difference is that instead of defining the vectors in terms of x_t^i (as in SPSO-2011), in GWO they are defined in terms of the three values g_t^k (see Eq. 10). The kind of perturbation induced by q_t^k in Eq. 10 is equivalent to the one induced by vectors a_t^i or b_t^i in PSO (see Eq. 2); however, the one induced by r_t^k is different because the entries of r_t^k are multiplied by $(2\,\varphi_t - 1)$ producing both positive and negative values. Computing the Euclidean norm of $(2\,q_t^k \odot g_t^k - x_t^i)$ to generate new random points in a radius $|2\,q_t^k \odot g_t^k - x_t^i|$ is the same idea as proposed in SPSO-2011 to generate a random point around the hypersphere center c_t^i (see Eq. 3). To compute a φ_t that linearly decreases from 2 to 0, GWO uses the same mechanism proposed in the "self-organizing hierarchical PSO with time-varying acceleration coefficients" [19] for computing φ_1, with the only difference that the lower bound in GWO is set to 0 instead of 0.5 as done in [19].

The position update rule introduced in Eq. 11 is an extension of the recombination rule in velocity-free PSOs (Eqs. 8 and 9) which uses the three best particles in the swarm. Similarly to how it was done in bare-bones PSO (Eq. 7), where the authors employed the recombination operator shown in Eq. 9 assuming $u_1 = u_2$ for computing the center of the normal distribution, the authors of

GWO employed the same recombination operator (also assuming $u_1 = u_2 = u_3$) for computing the particles' position.

The Metaphor of Grey Wolves Hunting. The authors of GWO say in their original paper published in 2014 [14] that they were inspired by the way in which grey wolves organize their hunting following a strict social hierarchy in which they divide—from top to bottom—their pack: α, β, δ and ω wolves. The authors of GWO mention that there are three phases during hunting, each one composed of a number of steps: (i) *tracking, chasing*, and *approaching* the prey; (ii) *pursuing, encircling*, and *harassing* the prey until it stops moving; and (iii) *attacking* towards the prey. However, GWO does not consider 5 of the 7 steps mentioned, and seems to take inspiration only from two steps respectively in phase (i) and (iii): *encircling* and *attacking*.

In GWO, a solution to the problem being tackled is called a "wolf", the optimum of the problem is referred to as the "prey" that the wolves are hunting, and the three best solutions and the remaining particles are named as α_t, β_t, δ_t and ω_t respectively, in analogy to the levels in the wolves social hierarchy. The GWO algorithm then consists in the "wolves encircling and attacking the prey".

To model encircling the authors used Eq. 11 while attacking the prey was modeled by linearly decreasing the value of φ from 2 to 0 in Eq. 10. In fact, in the imaginary of the metaphor, when φ_t is lower than 1, wolves can concentrate around the prey (therefore attacking it); and when it is greater than 1, they search for other preys.[2] The authors mentioned that the use of q_t^k, as done in Eq. 10, emphasizes the *search* behavior of wolves in a similar way in which $\varphi_t > 1$ does it, although, in their view, q_t^k represents "the effect of obstacles when wolves approach a prey in nature."

Unfortunately, as it should be clear to the reader by now, the wolf hunting metaphor is neither necessary nor useful to the definition and understanding of the way GWO works. In fact, it is not at all clear what is the optimization process in the wolf hunting that is translated in effective choices in the design of the optimization algorithm. While there is not a PSO variant that exactly matches GWO, as we have shown above, all the concepts introduced in GWO are related to existing concepts already proposed in the PSO literature and the only contribution given by the use of novel terms such as "wolf", "prey", "attacking", and so on is to create confusion and to hinder understanding.

3.2 Firefly Algorithm (FA)

FA as a PSO Algorithm. The firefly algorithm (FA) [28] is, in PSO terminology, an algorithm in which the swarm of particles is fully-connected and the particles movement is influenced only by those other particles in the swarm that have a higher quality. This means that the movement of the best particle is

---

[2] Although search is not an activity in the hunting phases of wolves, the authors explain it as "the divergence among wolves during hunting in order to find a *fitter prey*" [14, p. 50].

not influenced by any other particle. In FA, at each iteration particles are sorted according to their quality; the particle position update is then applied iteratively starting with worst quality particle and ending with the best quality particle. When particle i updates its position, it has to determine the set $W_t^i \subseteq T_t^i$ (where T_t^i is the set of particles in the neighborhood of i) that contains the $|W_t^i|$ particles with quality higher than its own. Updating a particle's position for the next iteration $t + 1$ requires $|W_t^i|$ movements of the particle (one for each particle in W_t^i), where the position of the particle obtained in movement $s - 1$ (indicated by $m_{t,s-1}^i$) is the starting position for the next one ($m_{t,s}^i$). The initial position of the particle is set to $m_{t,s=0}^i = x_t^i, \forall i \, \forall t$.

The position update rule of FA is given by the following two equations:

$$x_{t+1}^i = m_{t,s=|W_t^i|}^i \tag{12}$$

$$m_{t,s}^i = m_{t,s-1}^i + \varphi_t^{w_{t,s}^i, m_{t,s-1}^i} \left(w_{t,s}^i - m_{t,s-1}^i \right) + \xi r_{t,s}^i \tag{13}$$

where $w_{t,s}^i$ is an element of the ordered set W_t^i, $\varphi_t^{w_{t,s}^i, m_{t,s-1}^i}$ is an acceleration coefficient[3] whose value depends on the Euclidean distance between the two intermediate points $w_{t,s}^i$ and $m_{t,s-1}^i$, and $r_{t,s}^i$ is a vector whose components are random numbers drawn from the uniform distribution $\mathcal{U}[0,1]$ multiplied by a real scalar ξ.

The acceleration coefficient $\varphi^{w,m}$ is computed as follows:

$$\varphi^{w,m} = \alpha \cdot e^{-\gamma|w-m|^2} \tag{14}$$

where $|w - m|$ is the Euclidean distance between the position of two particles w and m, γ is a parameter that allows to control the weight given to $|w - m|^2$, and α a parameter that controls the weight of the exponential function. Because of the way $\varphi^{w,m}$ is computed, solutions have larger displacements when they are located close to each other and smaller ones when they are far away.

How Does FA Compare to PSO? To better understand how FA is a combination of known PSO concepts, we consider the case in which $|W_t^i| = 1$. In this case, particle i updates its position performing only one movement. This allows us to rewrite Eqs. 12 and 13 as follows:

$$x_{t+1}^i = x_t^i + \varphi_t^{i,w_t^i} \left(w_t^i - x_t^i \right) + \xi r_t^i \tag{15}$$

Equation 15 can be obtained from Eq. 8 by setting $\epsilon = 1$, $y = w_t^i$ and by adding ξr_t^i at the end of the equation. While setting the value of ϵ and adding ξr_t^i are typical design choices for velocity-free PSO variants, using the current position w_t^i of a neighbor instead of the neighbor's personal best position is not a common design choice for an implementation using Eqs. 8 and 9. In practice,

---

[3] Note that in the following we will use the shorter notation $\varphi_t^{w,m}$ when the meaning is clear from the context.

using the neighbor's current position may increase the diversity of the solutions in the algorithm since a particle's position changes more often than its personal best position, which is updated only when a new better quality position is found. The last term ξr_t^i in Eq. 15—a random perturbation—is used to increase the exploration of the algorithm and also allows the global best solution to move from its initial position in the search space.

The Metaphor of Fireflies Flashing. The author of the FA algorithm, first published in 2009 [28], says he was inspired by the flashing behavior of fireflies. Because of the metaphor used, he introduced the following terms: "fireflies" to indicate solutions of the considered problem, and "brightness" to indicate a function that computes the value of the acceleration coefficient $\varphi^{w,m}$. The acceleration coefficient $\varphi^{w,m}$ weighs the distance between two solutions according to their positions in the search space—in the context of the fireflies flashing metaphor, this is meant to model the fact that fireflies are attracted to other "brighter" fireflies.

Most of the metaphor of fireflies flashing is explained in terms of the different behaviors that can be obtained varying the value of parameter γ, for which the author considered two limit cases: $\gamma \to 0$ and $\gamma \to \infty$. When $\gamma \to 0$, the value of $\varphi^{w,m} \to 1$ and the attraction among fireflies becomes constant regardless of their distance in the search space. In the metaphor of fireflies flashing, this is the case when "the light intensity does not decay in an idealized sky" and "fireflies can be seen anywhere in the domain" [28, p. 174].

For the other limiting case, when $\gamma \to \infty$, the value of $\varphi^{w,m} \to 0$ (making the attractiveness among fireflies negligible) and new solutions can only be created by means of the random vector $\xi r_{t,s}^i$ (see Eq. 13). According to the metaphor, this is the case when fireflies are either "short-sighted because they are randomly moving in a very foggy region", or (for reasons not explained in the paper) "they feel almost zero attraction to other fireflies."

As can be seen from the explanations given for the use of the metaphor, its usefulness in describing and understanding the proposed algorithm is very doubtful. The only contribution of the metaphor of fireflies flashing seems to be the idea of using an exponential function based on the distance between two particle to compute the value of φ. However, this ideas was also explored before in the context of PSO in a variant called extrapolation PSO (ePSO) [1], published around 2 years before FA, in late 2007.

In ePSO, a particle i experiences a stronger attraction toward g_t when $f(g_t) \ll f(x_t^i)$, where $f(\cdot)$ is the objective function of a minimization problem, and a weak attraction when $f(g_t) \approx f(x_t^i)$. Note that, although it is the same idea, it is applied with opposite goals in the two algorithms, that is, in ePSO particles are more attracted towards particles that are far away while in FA they are attracted more to particles that are closer. Also, the distance is defined differently, since ePSO uses the distance with regard to the function evaluation and FA uses the Euclidean distance.

3.3 Bat Algorithm (BA)

BA as a Hybrid PSO and Simulated Annealing Algorithm. The bat algorithm (BA) [29] is an algorithm in which (i) particles in the swarm move by identifying good search directions exploiting the location of the global best particle, and (ii) there is the occasional introduction of new random solutions around the global best solution that are accepted using a simulated annealing like criterion. Using PSO terminology, the BA algorithm can be explained as follows.

Each particle employs two parameters: the probability ρ_t^i—increasing over time—of randomly generating a solution around g_t, and the probability ζ_t^i—decreasing over time—to accept the new solution generated. At each iteration t and with probability ρ_t^i, a particle generates a random point around g_t and keeps it in a variable z_t^i, which will be accepted as the new position of the particle if two conditions are verified: (i) the quality of z_t^i must be higher than that of g_t, that is, $f(z_t^i) < f(g_t)$, where $f(\cdot)$ is the objective function;[4] (ii) z_t^i is accepted with probability ζ_t^i. Therefore, for z_t^i to be accepted the following variable Accept must be true: Accept $= ((f(z_t^i) < f(g_t)) \wedge (\mathcal{U}[0,1] < \zeta_t^i))$.

If the random particle around g_t is not generated (this happens with probability $(1 - \rho_t^i)$) or when Accept is false[5] (i.e., z_t^i was rejected), particles update their position by adding a velocity vector to their current position.

The process described above is mathematically modeled as follows:

$$x_{t+1}^i = \begin{cases} g_t + \hat{\zeta}_t\, r_t^i, & \text{if Generate} \wedge \text{Accept} \\ x_t^i + v_{t+1}^i & \text{if (Generate} \wedge (\neg\text{Accept})) \vee (\neg\text{Generate}) \end{cases} \tag{16}$$

where $\hat{\zeta}_t$ is the average of the parameters ζ_t^i of all the particles in the swarm, r_t^i is a vector with values randomly distributed in $\mathcal{U}[-1,1]$, and Generate is a logical variable which is true when the algorithm decides, with probability ρ_t^i, to create a random solution around g_t.

The equations to update the probabilities ρ_t^i and ζ_t^i are:

$$\rho_{t+1}^i = \rho_{t=0}(1 - e^{-\beta_1 t'})$$

$$\zeta_{t+1}^i = \begin{cases} \beta_2\, \zeta_t^i & \text{if Generate} \wedge \text{Accept} \\ \zeta_t^i & \text{otherwise} \end{cases} \tag{17}$$

where $\beta_1 > 0$ and $0 < \beta_2 < 1$ are parameters, t' is an iteration counter that is updated every time Generate $\wedge$ Accept $=$ TRUE in Eq. 16, and $\rho_{t=0}$ is the initial value of parameter ρ. As it is clear from Eq. 17, the value of ρ_t^i tends to $\rho_{t=0}$ and the value of ζ_t^i tends to 0. Also, note that since the value ζ_t^i decreases with

---

[4] In this paper, we consider minimization problems; the obvious adaptation should be made in case of maximization problems.

[5] Due to the constraint that both conditions have to be met, it may be the case that z_t^i is rejected even when its quality is higher than that of g_t.

the number of iterations, so does $\hat{\zeta}_t$; this means that for increasing t values the solutions generated in the first case of Eq. 16 will be closer and closer to g_t.

As mentioned above and indicated in Eq. 16, when $(\texttt{Generate} \wedge (\neg\texttt{Accept})) \vee (\neg\texttt{Generate})$ the particles update their positions by adding a velocity vector to their current position, which is defined as follows:

$$v_{t+1}^i = v_t^i + d_t^i \odot (g_t - x_t^i) \tag{18}$$

where, except for d_t^i, the components of Eq. 18 are the same as those of SPSO (see Eq. 2). The vector d_t^i is computed as follows:

$$d_t^i = \varphi_1 + a_t^i(\varphi_2 - \varphi_1) \tag{19}$$

where φ_1 and φ_2 are parameters and a_t^i is the same random vector as in Eq. 2.

How Does BA Compare to PSO and Simulated Annealing?[6] The velocity update rule of BA—Eq. 18—is a special case of SPSO—Eq. 2. It can be easily seen that, if we set $\omega = 1$ and $\varphi_1 = 0$ in the velocity update rule of SPSO, it simplifies to the BA velocity update rule in Eq. 18. The only difference is that, in BA, the magnitude of the random vector a_t^i depends on the value of parameters φ_1 and φ_2.

The parameter ζ_t^i is very similar to the temperature acceptance criterion T first introduced in simulated annealing (SA) [10]. Two minor differences are that (i) in BA, the value of ζ_t^i is updated only when a solution is accepted, while in SA the value of T is typically updated at the end of each iteration; and (ii) that BA only accepts solutions with better quality than that of the global best solution, while SA can accept both improving and worsening solutions.

The Metaphor of Bats Echolocation. BA introduces a rather technical terminology, in which "bats" are the solutions to the considered problem, the range of "frequencies" in which bats emit their sound are φ_1 and φ_2 (defined in Eq. 19), the "loudness" of bats' sound is the acceptance criterion (ζ_t^i), and the "pulse emission rate" of their sound is the probability ρ_t^i of starting the process in which new solutions are generated around g_t.

The author of BA says he was inspired by the echolocation that some bat species use to find their way in the dark by producing sound waves that echo when they are reflected off an object. In order to develop the bat algorithm, the author strongly simplified several aspects of this process. In the words of the author, it was assumed that: (i) "bats are able to differentiate in some magical way between food/prey and background barriers"; (ii) "bats can automatically adjust the frequency and rate in which they are emitting sound"; and (iii) "the loudness of their sound can only decrease from a large value to almost 0."

---

[6] Note that, although in this paper we compared BA with PSO and SA, BA could also be interpreted as a variant of differential evolution (DE) [25]. This is because the probability ρ_t^i and the Accept criterion in BA are used in the same way as the mutation probability and the acceptance between donor and trial vectors in DE [18].

The author imagined that bats have two different flying modes, which correspond to the two cases in Eq. 16. In the first flying mode, bats fly randomly adjusting their "pulse emission rate" ρ_t^i and "loudness" ζ_t^i. According to the metaphor, bats decrease the pulse emission rate and produce louder sounds when they are randomly searching for a prey; and vice-versa when they have found one. Bats adjusting their pulse emission rate and loudness was modeled using Eq. 17. In the second flying mode, modeled by Eqs. 18 and 19, bats control their step size and range of movement by adjusting their sound frequency (vector d_t^i in Eq. 19) and by moving towards the best bat in the swarm.

As it should be clear to the reader at this point, the metaphor of bats echolocation seems to be an odd and confusing way of explaining the algorithm. This is because there are so many simplifications and unrealistic assumptions in the way in which the metaphor was translated into algorithmic terms that metaphor and algorithm seem to be two completely different things. In fact, except for generalities, the metaphor described by the author in his article does not provide a convincing basis for the choices made in the design of the resulting algorithm.

4   Conclusions

In this article, we have rewritten three highly-cited, metaphor-based algorithms in terms of PSO. We have shown that, perhaps contrary to the goal of the authors, the metaphors of *grey wolves hunting*, *fireflies flashing*, and *bats echolocation* do not facilitate understanding the corresponding GWO, FA and BA algorithms; rather, they create confusion because they hide their strong similarities with existing PSO algorithms. Even though with the help of imagination it is possible to vaguely understand how some of the ideas coming from the metaphor were used to match the corresponding algorithms, it is hard to see how such metaphors are useful at all.

After reviewing the GWO, FA and BA algorithms, we found that none of them propose truly novel ideas. In fact, they can all be seen as variants of existing PSO algorithms. Therefore, we conclude that these three algorithms are unnecessary since they do not add anything new to the tools that can be used to tackle optimization problems. In future work, we intend to experimentally compare GWO, FA and BA with other PSO variants and to analyze the impact that the specific design choices used in these algorithms have on their performance.

The problem of well-known concepts being reintroduced using new terminology has been spreading in the literature for over 15 years and is currently one of the main criticisms of metaphor-based algorithms. Rigorous analyses [3,12,17,24,26] have shown that a number of these algorithms are equivalent, or differ minimally, from well-known methods. Yet, instead of being proposed as variants of existing algorithms, they are often introduced as completely novel approaches—just as it was the case for the three algorithms studied in this paper.

Acknowledgments. Christian Leonardo Camacho Villalón, Thomas Stützle and Marco Dorigo acknowledge support from the Belgian F.R.S.-FNRS, of which they are, respectively, research fellow and research directors.

References

1. Arumugam, M.S., Murthy, G.R., Rao, M., Loo, C.X.: A novel effective particle swarm optimization like algorithm via extrapolation technique. In: International Conference on Intelligent and Advanced Systems, pp. 516–521. IEEE (2007)
2. Camacho-Villalón, C.L., Dorigo, M., Stützle, T.: Why the Intelligent Water Drops Cannot Be Considered as a Novel Algorithm. In: Dorigo, M., Birattari, M., Blum, C., Christensen, A.L., Reina, A., Trianni, V. (eds.) ANTS 2018. LNCS, vol. 11172, pp. 302–314. Springer, Cham (2018). https://doi.org/10.1007/978-3-030-00533-7_24
3. Camacho-Villalón, C.L., Dorigo, M., Stützle, T.: The intelligent water drops algorithm: why it cannot be considered a novel algorithm. Swarm Intell. **13**, 173–192 (2019). https://doi.org/10.1007/s11721-019-00165-y
4. Campelo, F.: Evolutionary computation bestiary. https://github.com/fcampelo/EC-Bestiary (2017). Accessed 22 Jan 2018
5. Clerc, M.: Standard particle swarm optimisation from 2006 to 2011. Open archive HAL hal-00764996, HAL (2011)
6. Clerc, M., Kennedy, J.: The particle swarm-explosion, stability, and convergence in a multidimensional complex space. IEEE Trans. Evol. Comput. **6**(1), 58–73 (2002)
7. Eberhart, R., Kennedy, J.: A new optimizer using particle swarm theory. In: Proceedings of the Sixth International Symposium on Micro Machine and Human Science, pp. 39–43 (1995)
8. Kennedy, J.: Bare bones particle swarms. In: Proceedings of the 2003 IEEE Swarm Intelligence Symposium, SIS 2003 (Cat. No. 03EX706), pp. 80–87. IEEE (2003)
9. Kennedy, J., Eberhart, R.: Particle swarm optimization. In: Proceedings of ICNN 1995 International Conference on Neural Networks, vol. 4, pp. 1942–1948. IEEE (1995)
10. Kirkpatrick, S.: Optimization by simulated annealing: quantitative studies. J. Stat. Phys. **34**(5–6), 975–986 (1984). https://doi.org/10.1007/BF01009452
11. Lones, M.A.: Metaheuristics in nature-inspired algorithms. In: Igel, C., Arnold, D.V. (eds.) Proceedings of the Genetic and Evolutionary Computation Conference, GECCO 2014. pp. 1419–1422. ACM Press, New York (2014)
12. Melvin, G., Dodd, T.J., Groß, R.: Why 'GSA: a gravitational search algorithm' is not genuinely based on the law of gravity. Natural Comput. **11**(4), 719–720 (2012). https://doi.org/10.1007/s11047-012-9322-0
13. Mendes, R., Kennedy, J., Neves, J.: The fully informed particle swarm: simpler, maybe better. IEEE Trans. Evol. Comput. **8**(3), 204–210 (2004)
14. Mirjalili, S., Mirjalili, S.M., Lewis, A.: Grey wolf optimizer. Adv. Eng. Softw. **69**, 46–61 (2014)
15. Peña, J.: Simple dynamic particle swarms without velocity. In: Dorigo, M., Birattari, M., Blum, C., Clerc, M., Stützle, T., Winfield, A.F.T. (eds.) ANTS 2008. LNCS, vol. 5217, pp. 144–154. Springer, Heidelberg (2008). https://doi.org/10.1007/978-3-540-87527-7_13
16. Peña, J.: Theoretical and empirical study of particle swarms with additive stochasticity and different recombination operators. In: Ryan, C. (ed.) Proceedings of the Genetic and Evolutionary Computation Conference, GECCO 2008, pp. 95–102. ACM Press, New York (2008)
17. Piotrowski, A.P., Napiorkowski, J.J., Rowinski, P.M.: How novel is the "novel" black hole optimization approach? Inf. Sci. **267**, 191–200 (2014)
18. Price, K.V., Storn, R.M., Lampinen, J.A.: Differential Evolution. NCS. Springer, Heidelberg (2005). https://doi.org/10.1007/3-540-31306-0

19. Ratnaweera, A., Halgamuge, S.K., Watson, H.C.: Self-organizing hierarchical particle swarm optimizer with time-varying acceleration coefficients. IEEE Trans. Evol. Comput. **8**(3), 240–255 (2004)
20. Rechenberg, I.: Evolutionsstrategie: Optimierung technischer Systeme nach Prinzipien der biologischen Evolution. Frommann-Holzboog, Stuttgart, Germany (1973)
21. Shi, Y., Eberhart, R.: A modified particle swarm optimizer. In: Simpson, P.K., Haines, K., Zurada, J., Fogel, D. (eds.) Proceedings of the 1998 IEEE International Conference on Evolutionary Computation, ICEC 1998, pp. 69–73. IEEE Press, Piscataway (1998)
22. Shi, Y., Eberhart, R.: Empirical study of particle swarm optimization. In: Proceedings of the 2009 Congress on Evolutionary Computation (CEC 2009), pp. 1945–1950. IEEE Press, Piscataway (2009)
23. Sörensen, K.: Metaheuristics–the metaphor exposed. Int. Trans. Oper. Res. **22**(1), 3–18 (2015). https://doi.org/10.1111/itor.12001
24. Sörensen, K., Arnold, F., Palhazi Cuervo, D.: A critical analysis of the "improved Clarke and wright savings algorithm". Int. Trans. Oper. Res. **26**(1), 54–63 (2019)
25. Storn, R., Price, K.: Differential evolution - a simple and efficient heuristic for global optimization over continuous spaces. J. Global Optim. **11**(4), 341–359 (1997). https://doi.org/10.1023/A:1008202821328
26. Weyland, D.: A rigorous analysis of the harmony search algorithm: how the research community can be misled by a "novel" methodology. Int. J. Appl. Metaheuristic Comput. **12**(2), 50–60 (2010)
27. Weyland, D.: A critical analysis of the harmony search algorithm: how not to solve Sudoku. Oper. Res. Pers. **2**, 97–105 (2015)
28. Yang, X.-S.: Firefly algorithms for multimodal optimization. In: Watanabe, O., Zeugmann, T. (eds.) SAGA 2009. LNCS, vol. 5792, pp. 169–178. Springer, Heidelberg (2009). https://doi.org/10.1007/978-3-642-04944-6_14
29. Yang, X.S.: A new metaheuristic bat-inspired algorithm. Nature inspired cooperative strategies for optimization (NICSO 2010). In: González, J.R., Pelta, D.A., Cruz, C., Terrazas, G., Krasnogor, N. (eds.) Nature Inspired Cooperative Strategies for Optimization (NICSO 2010). Studies in Computational Intelligence, vol. 284, pp. 65–74. Springer, Heidelberg (2010). https://doi.org/10.1007/978-3-642-12538-6_6
30. Zambrano-Bigiarin, M., Clerc, M., Rojas, R.: Standard particle swarm optimisation 2011 at cec-2013: a baseline for future pso improvements. In: Proceedings of the 2013 Congress on Evolutionary Computation (CEC 2013), pp. 2337–2344. IEEE Press, Piscataway (2013)

Guerrilla Performance Analysis for Robot Swarms: Degrees of Collaboration and Chains of Interference Events

Heiko Hamann[1]([✉])[iD], Till Aust[1], and Andreagiovanni Reina[2][iD]

[1] Institute of Computer Engineering, University of Lübeck, Lübeck, Germany
hamann@iti.uni-luebeck.de
[2] Department of Computer Science, University of Sheffield, Sheffield, UK
a.reina@sheffield.ac.uk

Abstract. Scalability is a key feature of swarm robotics. Hence, measuring performance depending on swarm size is important to check the validity of the design. Performance diagrams have generic qualities across many different application scenarios. We summarize these findings and condense them in a practical performance analysis guide for swarm robotics. We introduce three general classes of performance: linear increase, saturation, and increase/decrease. As the performance diagrams may contain rich information about underlying processes, such as the degree of collaboration and chains of interference events in crowded situations, we discuss options for quickly devising hypotheses about the underlying robot behaviors. The validity of our performance analysis guide is then made plausible in a number of simple examples based on models and simulations.

1 Introduction

In a world of growing businesses and growing populations the question of how to collaborate effectively and how to form efficient groups is important. Groups that are too large can become inefficient as the cost needed by the group members to coordinate their actions is greater than the benefits the collaboration would bring. For example, rumor has it that Jeff Bezos limits group sizes by the amount its members can eat (so-called 'Two Pizza Rule' [33]) and Brooks's law says "adding manpower to a late software project makes it later" [5]. A scientific result is the Ringelmann effect describing the decreasing productivity of individuals with increasing group size [38]. However, certain systems can exploit collaboration at their advantage to obtain a superlinear increase in group performance, that is, the work completed by the group is more than the sum of work each individual could perform alone. Superlinear increase in group performance, commonly found in swarm robotics [18,30], can also be found in collaborating humans [41] and in distributed computing [9].

In engineered systems, collaboration between the units composing the system can be constrained by limited shared resources, for example, memory access in

© Springer Nature Switzerland AG 2020
M. Dorigo et al. (Eds.): ANTS 2020, LNCS 12421, pp. 134–147, 2020.
https://doi.org/10.1007/978-3-030-60376-2_11

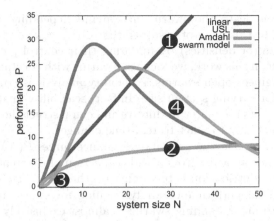

Fig. 1. Schematic plot of qualitatively different swarm performance $P(N)$ curves as a function of the system size N. For large swarm sizes, the performance can (i) increase linearly (purple line, zone ❶), (ii) saturate to a constant value (blue line, zone ❷), or (iii) initially increase (zone ❸ of green and orange lines) and subsequently decrease (zone ❹ of green and orange lines). In scalability analysis, constant increase corresponds to Gustafson's law[10], saturation corresponds to Amdahl's law [1], and increase/decrease to Gunther's universal scalability law (USL) [7]. We show two curves of increase/decrease: The initial phase (zone ❸) can show a slow (orange) or quick (green) increase of performance. The final phase, zone ❹, can show high order (green) or almost linear (orange) decrease. These visual cues can give insights about swarm behavior and efficient design. (Colour figure online)

computing [17] or physical space in swarm robotics [12]. The system performance varies when increasing system size or reducing resources. What is measured by swarm performance P depends on the particular application and scenario. Performance P is a quantification of a task-specific feature that is commonly agreed as a valid measure of success. For example, in foraging that can be the number of collected items [42], in emergent taxis the traveled distance of the swarm's barycenter [2], and in collective decision-making a combination of speed and accuracy [44]. A scalable system is supposed to work efficiently for different load and/or system size [29]. In swarm robotics, scalability of system size is supposed to be a common feature of a properly engineered swarm system [13]. However, robots have a physical body and their movement can interfere with others when the swarm density $\rho = N/A$ (number of robots N per area A) is high [11]. Increasing area A (shared resource) with swarm size N would keep the density ρ constant. This experiment design would provide no information gain about the system's scalability. Instead, we are interested in measuring swarm performance P over swarm density ρ. In most published experiments this, in turn, means measuring swarm performance P over system size N because usually the provided area A is constant.

A promising feature of robot swarms is that they can form an open system ('open swarms' [35]) that have potential for scalability in real time. That is, robots can join and leave the swarm on demand depending on the needs of the moment [27]. In this type of systems, the robots can collectively adapt to varying swarm size (or densities) by updating their control parameters in real time [34,45]. While in this work, we focus on swarms with constant size within an experiment, to engineer open swarms, scalability analysis is crucial to quantify the performance for varying system size. In fact, scalability analysis may reveal that adding more units is counterproductive and can instruct the swarm engineer on the most efficient way to react to real-time changes.

Our main motivation is that swarm performance curves $P(N)$ seem to possess generic qualities that appear across a wide collection of different swarm scenarios [11,12,14]. Our contribution is to summarize these findings here and to turn them into a practical performance analysis guide for swarm robotics. In this work, our access to understanding swarms is almost exclusively phenomonemological and macroscopic. Still, such findings can help to understand essential qualitative features of the swarm and to develop approaches to resolve performance issues. Although deriving microscopic properties (e.g., required behaviors of individual robots) from macroscopic properties is difficult [15,36], we are able to indicate some micro-macro links that may even be generic. For example, we show how the macroscopic performance curve can indicate whether small or big groups of robots interact in beneficial or detrimental ways.

The term 'guerrilla' in the title is a tribute to Gunther [7] who wrote the renowned book 'Guerrilla Capacity Planning' [8] to provide industry managers with a simple framework for scalabilty planning. Here we let the term represent the rather practical and phenomenological top-down approach to performance in swarm robotics. We provide a practical guide to quickly understand fundamental scalability features of a studied swarm based on elementary insights about superficial characteristics of their $P(N)$ plot. We present three classes of performance: linear increase, saturation, and increase/decrease. Then we focus on the increase/decrease class and discuss how the performance curve can explain the relationship between collaboration and interference among the robots.

2   Three General Classes of Performance System Behavior

Analyzing the system performance $P(N)$ reveals three qualitatively different types of scalability classes: linear increase, saturation, and increase/decrease.

2.1   Linear Increase

If we observe a sustained trend of performance $P(N) \propto N$ up to large values of N (see purple line ❶ in Fig. 1), then we observe the scalability class of *linear increase*. This situation is advantageous as the swarm performance improves by increasing the number of robots. However, we should note that it cannot be considered the ideal case as also superlinear performance scaling can be observed in swarm robotics and computing systems [9,12,28], as represented by the rapid initial increase of the green curve in Fig. 1.

2.2 Saturation

We observe the *saturation* class when performance $P(N)$ approaches a maximum $P(N \to \infty) = s^*$ (see blue curve ❷ in Fig. 1). Therefore, such a regime has no performance peak and is equivalent to Amdahl's law [1] that was originally formulated to (pessimistically) describe the scalability of parallel computers. While Amdahl's law has demonstrated its applicability [8], we argue that this saturation scenario is rare in swarm robotics or ignores costs (see Sect. 2.4). Physical interference due to high robot densities usually has a significant impact on swarm performance causing an increase/decrease situation.

2.3 Increase/Decrease

In swarm robotics, the representative scalability class is *increase/decrease*, characterized by increasing performance for small N, a performance peak at a critical swarm size N_c, and decreasing performance for $N > N_c$. Performance $P(N < N_c)$ increases because robots efficiently collaborate or work in parallel to perform the task, and performance $P(N > N_c)$ decreases because robots interfere with each other.

Gunther [7,9] proposed the universal scalability law (USL) to describe this increase/decrease class as observed in computing. The USL is based on performance improvements S (speedup) for size N compared to the minimal system $N = 1$. The USL is

$$S(N) = \frac{P(N)}{P(1)} = \frac{N}{1 + \sigma(N - 1) + \kappa N(N - 1)} , \tag{1}$$

with parameter σ describing the influence of contention (e.g., queues for shared resources) and parameter κ describing a coherency delay (e.g., distributing and synchronizing information). The USL properly parameterized by σ and κ covers all scalability classes (*linear increase, saturation, increase/decrease*). In Fig. 1, the green line (labeled USL) gives an example for *increase/decrease*. Another model developed especially for the scalability analysis of robot swarms [11,12] is

$$P(N) = aN^b \exp(cN) , \tag{2}$$

for constants $a > 0$, $b > 0$, and $c < 0$. The function can be understood as a dichotomous pair of a term for potential of collaboration N^b and a term for interference $\exp(cN)$. In Fig. 1 the orange curve is an example of Eq. 2 in the *increase/decrease* regime (labeled 'swarm model').

2.4 Ambiguous Definition of Swarm Performance

In Sect. 1 we argued that the definition of swarm performance $P(N)$ for a particular application scenario should be an agreed measure of success. This introduces degrees of subjectivity in our scalability analysis and ultimately ambiguity in the observed results. While it seems unlikely that this can be resolved in a fully generic way, we propose four simple guidelines of how to improve the scalability analysis and avoid common mistakes: constant task, full range, added cost, and marginal performance.

Constant Task. In any performance analysis, but specifically for large system size, when the performance curve keeps growing as $P(N) \propto N$ (see ❶ in Fig. 1) the swarm performance analysis practitioner should question if the performance has been measured on the same task T for any swarm size N. By adapting the task to large system sizes, the performance may not provide useful indications on the system's scalability as two parameters (size N and task T) have been changed at the same time. We recommend to consider as part of the task a constant working area $A \in T$. Increasing size N of a swarm on constant area has the effect of increasing swarm density $\rho = N/A$ that can increase physical interference among robots. Physical interference is expected to have a negative impact on the swarm performance $P(N)$. While we acknowledge that certain tasks—e.g., area coverage or movement-free tasks based on communication only—could exploit increased density to improve the performance [6], we argue that linear increase is a pathological case that should be carefully interpreted. As the performance should measure the completion of a fixed task, it could be expected that it would, at least, saturate for large sizes N.

Full Range. Another typical shortcoming of performance analysis that could explain the observation of a linear increase of performance for large sizes N, is a short range of N. Considering only relatively small sizes of N would only show a partial picture of the system behavior. An incomplete scalability analysis could be harmful as the system behavior would not be fully understood. For example, in cleaning or object collection tasks, it is reasonable that performance saturates once dirty areas get scarce or most objects have been collected respectively.

Added Cost. A performance curve that does not decrease for large system sizes (e.g., see ❶ and ❷ in Fig. 1 for linear increase and saturation respectively) suggests minimal interference among robots. For example, in an area coverage task, the more robots are added to the swarm, the better the area gets covered until performance saturation is observed [31]. Scalability analysis should support the system designer in making decisions about the optimal swarm size in terms of its internal function and real-world factors, such as deployment cost. Hence, system performance $P(N)$ should be complemented with the cost of added units to select the 'best' swarm size N. In the above coverage task, the saturation of the performance puts the scalability analyst in a situation where performances of large swarms cannot be distinguished anymore (e.g., $P(N) \approx P(10^3 N)$). We would ignore effects of diminishing returns. In addition, one may be tempted to add more robots to increase redundancy and robustness (redundancy-induced robustness). The lack of any cost suggests that 'bigger is better' as there is no immediate negative impact of interference and performance P increases monotonically with N or interference may even be a feature [6].

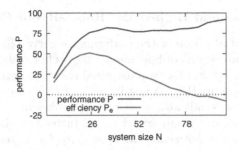

Fig. 2. Saturating performance P based on data from simulations of a coverage task by Özdemir et al. [31] and the efficiency measure $P_e = P(N) - C(N)$ for $C(N) = N$.

We recommend to always complement the study of swarm performance $P(N)$ with the study of cost $C(N)$, to analyze the efficiency $P_e(N) = P(N) - C(N)$, that can be more informative than $P(N)$ alone. Cost $C(N)$ should account for relevant aspects, such as economical (purchase of additional robots [40]) or logistic costs (covering the environment with robots would reduce the space for other type of activities). For example, this would allow to usefully balance the cost and benefits of redundancy-induced robustness. In Fig. 2b, we show the effect of adding a constant cost per unit $C(N) = c\,N$ to the performance data of an area coverage study by Özdemir et al. [31]. P_e shows a peak and can hence indicate an optimal swarm size N. A designer seeking robustness can quantify the decrease in efficiency and choose an appropriate swarm size N.

Marginal Performance. Another measure that can improve scalability analysis is marginal performance $P_m(N) = P(N) - P(N-1) = dP(N)/dN$. Considering added swarm performance per unit can help deciding the swarm size. The measure $P_m(N)$ can be particularly useful when compared with the marginal cost $C_m(N) = C(N) - C(N-1) = dC(N)/dN$. For $P_m(N) < C_m(N)$, adding robots to the system would decrease swarm performance. Similarly, one could consider the mean individual performance $I(N) = P(N)/N$. In a more holistic way, here the entire swarm shares the benefits of an added robot. Also in this case, the measure $I(N)$, that indicates the performance contribution of each robot, can be compared with the individual cost $I_c(N) = C(N)/N$ in order to appropriately scale the system.

3   From Eye-Catchers to a Practical Performance Analysis

The performance class that is most frequently observed in swarm robotics is *increase/decrease*. For this class we provide a guide how to quickly interpret $P(N)$ diagrams in terms of two features: shape of the curve for small system sizes (see ❸ in Fig. 1) and shape of the curve for large systems (see ❹ in Fig. 1).

3.1   Increase: Low- and High-Order Robot-Robot Collaboration

By looking at the initial phase of the performance curve (❸ in Fig. 1, for $N <$ 15), we can obtain indications of how much robot-robot collaboration is done to complete the task (cf. other, more sophisticated efforts to derive group sizes from macroscopic measurements [15]). A fast increase of $P(N)$ for smallest values $N \in$ $\{1,2,3\}$, shows that a small swarm is already sufficient to complete at least parts of the task (e.g., green curve of Fig. 1). Instead, if the curve has a slow start and $P(N)$ shows a noticeable increase only for larger sizes N, it could indicate the necessity of robot-robot collaboration in larger groups. In most published swarm performance measurements, the initial increase of performance is approximately linear (fast increase). However, there are rare cases of published datasets showing a nonlinear (curved) and slow increase [26]. Note that we do not focus on distinguishing between super- and sub-linear performance increases, instead we try to understand when to expect linear and when nonlinear increases.

Both scalability functions described in Sect. 2.3 can represent both linear (fast) and nonlinear (slow) increase despite their simplicity. Interestingly, similar nonlinear system behaviors can be observed in models from not directly related fields, such as PT2 lag elements in control theory, or residence times in cascades of stirred-tank reactors (tanks in series) [23]. In both of these examples, sequences of events or higher order time-delays introduce the observed nonlinearity. Comparable effects emerge in robot swarms when several robots need to collaborate in order to perform the given task.

To support our above claims, we show two minimal examples in which observing the system performance curve for small system sizes (❸, $N < 10$, in Fig. 1) allows us to estimate the necessary amount of robot-robot interactions to complete the task. If robots can perform the task without any/much help from other robots, then the initial increase is steep and close to linear. We say that robots require low-order interactions. If robots require considerable help from other robots to perform the task, then the initial performance remains low for small sizes N and shows a curved (nonlinear) increase. We say that robots require high-order interactions. We give evidence for this conjecture through two simple analyses: a simple combinatorial argumentation and empirical observations in simulations of an abstract system inspired by the stick pulling experiment [18].

Our combinatorial consideration is based on the precondition for robot-robot collaboration: robots need to be in close proximity to each other. In swarm robotics, robot movement is often based on random motion [3]. We consider the probability that collaboration among k robots takes place as a stochastic event proportional to k and swarm density ρ. Assuming a simple grid environment where collaboration takes place between neighboring robots, we can derive the probability of having at least k robots in Moore neighborhoods, 3×3, of $m = 9$ cells. Swarm density ρ indicates the (independent) probability of finding a robot in a given cell. The probability Γ_k of finding at least k robots in a Moore neighborhood of $m = 9$ cells corresponds to $\Gamma_k = \sum_{i=k}^{m} \binom{m}{i} \rho^i (1-\rho)^{(m-i)}$. In Fig. 3a we show Γ_k as a function of ρ for $k \in \{1,2,3,4\}$. As expected, the probability that at least k robots meet (our assumed precondition for collabo-

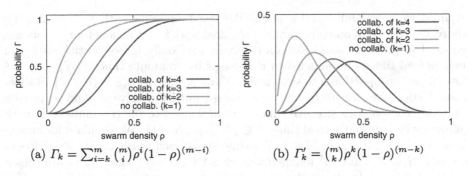

(a) $\Gamma_k = \sum_{i=k}^{m} \binom{m}{i} \rho^i (1 - \rho)^{(m-i)}$ (b) $\Gamma'_k = \binom{m}{k} \rho^k (1 - \rho)^{(m-k)}$

Fig. 3. Combinatorial explanation of chances for collaboration, collaboration probability Γ_k for $k \in \{1, 2, 3, 4\}$ and neighborhood size $m = 9$ (Moore neighborhood), and swarm density ρ; two scenarios: (a) without and (b) with interference.

ration) decreases by increasing k. Looking at the initial part of the curves, for low density values, larger groups have a slow (nonlinear) increase. Instead, small groups (e.g., $k = 1$ or $k = 2$) have fast and almost linear increases. In Fig. 3a, we assume no interference between robots, thus values larger than k still allow for collaboration without overhead. In Fig. 3b, we assume that values larger than k would prohibit collaboration. The shown probability is $\Gamma'_k = \binom{m}{k} \rho^k (1 - \rho)^{(m-k)}$. Despite the different shapes for high densities (see Sect. 3.2), the initial part shows the same type of shapes for varying k.

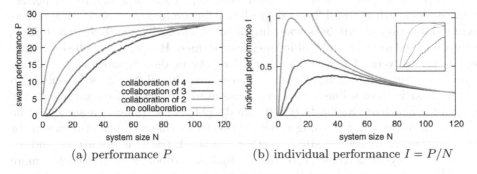

(a) performance P (b) individual performance $I = P/N$

Fig. 4. Stick pulling scenario: (a) Swarm performance P and (b) individual performance I measured in an abstract stochastic model of the robot swarm stick pulling scenario where one (no collaboration), two, three, or four robots are required to pull a stick. Performance normalized to equalize for $P(N = 120)$. Inset shows $N \in \{1, 2, \ldots, 15\}$ for all performance maxima normalized to one.

Our second argumentation is based on a minimal simulation of a simplistic abstract model inspired by the stick pulling scenario [18] that was published before [12]. We have a swarm of N robots and $M = 20$ stick sites containing one stick each. In the original experiment, collaboration of $k = 2$ robots is required to successfully pull a stick. We test four cases with $k \in \{1, 2, 3, 4\}$, where $k = 1$ means no collaboration required and $k = 4$ means four robots are required to pull one stick. Robots commute randomly between stick sites and their arrival times are modeled in time steps by commute times $\tau(N) = N + \xi$ (i.e., linearly proportional to swarm size N) for a noise term $\xi \in \{0, 1, 2\}$. Robots wait at stick sites for up to seven time steps until they give up and leave to a randomly chosen stick site. All robots are initialized to the commute state with uniformly distributed arrival times $\tau \in \{0, 1, \ldots, N-1\}$. We simulate for swarm sizes $N \in \{1, 2, \ldots, 120\}$, for 1,000 time steps each, with constant stick site number $M = 20$, and 10^4 independent runs for each N. The results are shown in Fig. 4. Figure 4a shows the normalized swarm performance P for required collaboration $k \in \{1, 2, 3, 4\}$. Swarm performance P saturates because commute times τ scale linearly with swarm size N. The nonlinear effect of higher-order collaboration ($k = 3$ and $k = 4$) is obvious. Figure 4b shows the normalized individual performance $I = P/N$. Again showing the nonlinear effect of higher-order collaboration. In addition, we also see qualitative differences in the curves for high swarm sizes $N > 30$: curved for $k = 1$ and almost linear for $k = 4$.

3.2   Decrease: Low- and High-Order Robot-Robot Interaction

Now we study the decrease in swarm performance for sizes N bigger than the swarm size N_c for peak performance (see ❹ in Fig. 1). In published works reporting swarm performance diagrams, the plot of $P(N)$ is sometimes almost linear [16, 25], sometimes slightly curved [22, 43, 47], and sometimes curved [18, 39, 42] for sizes $N > N_c$. For example, Llenas et al. [25] report performance plots with graceful linear degradation for a foraging scenario. The underlying simulation of Kilobots was simplified, temporarily small clusters formed that dissolved quickly, and traffic lanes were formed. Hence, most collision avoidance actions were of first order, that is, robots made a transition to collision avoidance but didn't trigger collision avoidance in others. This is similar to traffic models were a linear decrease is assumed classically, for example in the Lighthill–Whitham–Richards (LWR) model [24]. The traffic is assumed to be fully synchronized with strong serial dependencies due to lanes (1-d space) for system size N_c. If system size is further increased, traffic is disturbed, and for too crowded systems traffic jams emerge. In swarm robotics the situation is more complex as space is 2-d and it is unknown which robots in collision avoidance state may trigger collision avoidance in others. Another analogy are transport systems [19]. There viscosity or mechanical impedance increases nonlinearly with concentration (cf. interference in Eq. 2). For robots that translates to number of collision avoidance events.

Performance for big sizes (for $N > 20$ as seen at ❹ in Fig. 1) is our focus now. If robots interfering with each other manage to resolve the interference (e.g., by avoidance movements) and return to productive mode quickly, then the performance decrease is low and close to linear. We say they show low-order interference. If robots by trying to resolve interference, trigger cascades of collision avoidance, then the performance decrease is steep and curved. We say they show high-order interference. To support our claims, we present empirical evidence based on a simulation. The main idea of this experiment is to control the number of collision-avoidance events that a robot triggers. We define as first order interference the collision avoidance that is triggered by two robots moving close to each other. During *collision avoidance* (CA), the robots perform a set of maneuvers to avoid physical collision. If during the execution of this set of avoidance maneuvers, the robot triggers collision avoidance in another robot, we define it as second order interference. Therefore, when these robots performing collision avoidance (in state CA) trigger another j^{th} robot, such event corresponds to the jth order interference, for j robots involved. This is related to the basic reproduction number R_0 in the SIR epidemic model [20], where R_0 defines the average number of infections that each infected individuals causes. Considering R_0 the average number of collision avoidance events that each robot in state CA triggers, we have that with $R_0 = 1$ each robot in state CA 'infects' one other robot with the 'collision-avoidance disease.' With $R_0 > 1$ each robot in state CA triggers more than one collision avoidance, its growth is exponential, and the resulting decrease of performance $P(N)$ is nonlinear.

We use the Webots simulation environment [46] for our experiments on interference. The simulated robot is the Thymio II [37] operating as a swarm of size N in a 2 m × 2 m arena. We simulate a simple multi-robot navigation task. The arena has four bases (north, south, east, west). The robots' goal is to reach the respective opposite base (e.g., from north to south and vice versa). At the beginning of each run, we randomly distribute $N = 1$ to $N = 55$ robots (at least 10 cm away from any wall) depending on the density we want to test. Then the robots do a random walk until they touch a base. They use it then as their reference base (from where they started) to set the vis-a-vis base as their next target (north and south, east and west). When a robot detects an obstacle (wall or robot) or its target base, it turns in a random direction for a random time, and moves straight again. When they touch the target, the performance counter is increased by one (and the new target is the opposite base). One run takes 6 simulated minutes. We do two types of simulation runs. In full avoidance runs, all robots follow the standard procedure and remain in the arena at all times. In first order avoidance runs, we limit the effect of interference by limiting collision avoidance triggering cascades. A robot in state CA that triggers a transition to CA in another robot is allowed to trigger only this one CA event but is then temporarily removed with probability P_{remove} for the time it stays in state CA. Once its CA behavior has been completed, it is put back into the arena if the spot is empty; otherwise it is put back later once the spot is empty. We vary probability $P_{remove} \in \{0.4, 0.7, 1\}$ where $P_{remove} = 1$ means the robot is always removed once it has triggered CA in another robot.

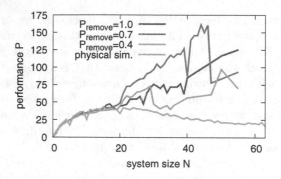

Fig. 5. Interference in swarms of N robots: simple navigation task, with probability $P_{\text{remove}} \in \{0.4, 0.7, 1\}$ we remove robots that triggered collision avoidance of others ($n = 100$ independent simulation runs per data point).

The results are shown in Fig. 5.[1] The data is noisy despite the invested computational effort of more than 200 CPU days. However, an overall trend can be noticed. With P_{remove} the swarm density is regulated by removing robots that cannot be put back into the arena immediately because their spot is taken. Due to a nontrivial and not further discussed interplay of effects, the setup $P_{\text{remove}} = 1$ is better for $N \geq 47$ but outperformed by setup $P_{\text{remove}} = 0.7$ for $20 < N < 47$. Either robots are taken out too quickly slowing down their travels ($P_{\text{remove}} = 1$, $N < 47$) or robots are left in the system increasing collision avoidance events ($P_{\text{remove}} = 0.7$, $N \geq 47$). Hence, independent of the task a robot swarm can artificially be pushed from increase/decrease to the saturation scenario. This is similar to other approaches where simulated physics (embodied systems not allowing to pass through other bodies) was turned on/off [39, 42]. Also behaviors in ants mitigate overcrowded situations to avoid the increase/decrease situation in favor of a saturation scenario [4, 21, 32]. With our experiment we investigated the impact of interference on performance by modulating probability P_{remove}.

4   Conclusion

We have given a practical guide to analyze swarm performance and scalability. Swarm performance plots contain rich information about underlying processes. The left part of the swarm performance plot can give hints on the level of collaboration necessary to solve the task. The right part of the plot is a reflection of the ratio between marginal cost and performance. Performance scales in qualitatively different ways depending on the task. Tasks that are not limited by physical interference (e.g., area coverage) show no collapse of performance for increased swarm sizes. However, usually physical interference has a negative effect in a variety of tasks. These qualitative differences vanish once we a apply a benefit-cost analysis (BCA) that reveals the relation between the marginal performance

---

[1] See http://doi.org/10.5281/zenodo.3947822 for videos, screenshot, and data.

(added swarm performance of an added robot) and the relative marginal cost. An important design choice is about the redundancy-induced robustness. Swarm robotics is commonly assumed to be robust to failures because of its high degree of redundancy. In a homogeneous swarm, robots are exchangeable and serve as mutual replacements. Through BCA and marginal cost/performance analysis the designer can make a more informed choice to balance the efficiency-robustness tradeoff. Following our practical ('guerrilla') performance analysis guide allows swarm scalability analysts to quickly formulate hypotheses about the underlying system behaviors and consequently to speedup the design and studies in swarm robotics.

References

1. Amdahl, G.M.: Validity of the single processor approach to achieving large scale computing capabilities. In: AFIPS Conference Proceedings, pp. 483–485. ACM (1967)
2. Bjerknes, J.D., Winfield, A., Melhuish, C.: An analysis of emergent taxis in a wireless connected swarm of mobile robots. In: Shi, Y., Dorigo, M. (eds.) IEEE Swarm Intelligence Symposium, pp. 45–52. IEEE Press, Los Alamitos (2007)
3. Dimidov, C., Oriolo, G., Trianni, V.: Random walks in swarm robotics: an experiment with kilobots. In: Dorigo, M., et al. (eds.) ANTS 2016. LNCS, vol. 9882, pp. 185–196. Springer, Cham (2016). https://doi.org/10.1007/978-3-319-44427-7_16
4. Dussutour, A., Fourcassié, V., Helbing, D., Deneubourg, J.L.: Optimal traffic organization in ants under crowded conditions. Nature 428, 70–73 (2004)
5. Frederick, P., Brooks, J.: The Mythical Man-Month. Addison-Wesley, Boston (1995)
6. Goldberg, D., Matarić, M.J.: Interference as a tool for designing and evaluating multi-robot controllers. In: Kuipers, B.J., Webber, B. (eds.) Proceedings of the Fourteenth National Conference on Artificial Intelligence (AAAI 1997), pp. 637–642. MIT Press, Cambridge (1997)
7. Gunther, N.J.: A simple capacity model of massively parallel transaction systems. In: CMG National Conference, pp. 1035–1044 (1993)
8. Gunther, N.J.: Guerrilla Capacity Planning. Springer, Heidelberg (2007). https://doi.org/10.1007/978-3-540-31010-5
9. Gunther, N.J., Puglia, P., Tomasette, K.: Hadoop super-linear scalability: the perpetual motion of parallel performance. ACM Queue 13(5), 46–55 (2015)
10. Gustafson, J.L.: Reevaluating Amdahl's law. Commun. ACM 31(5), 532–533 (1988). https://doi.org/10.1145/42411.42415
11. Hamann, H.: Towards swarm calculus: urn models of collective decisions and universal properties of swarm performance. Swarm Intell. 7(2–3), 145–172 (2013). https://doi.org/10.1007/s11721-013-0080-0
12. Hamann, H.: Superlinear scalability in parallel computing and multi-robot systems: shared resources, collaboration, and network topology. In: Berekovic, M., Buchty, R., Hamann, H., Koch, D., Pionteck, T. (eds.) ARCS 2018. LNCS, vol. 10793, pp. 31–42. Springer, Cham (2018). https://doi.org/10.1007/978-3-319-77610-1_3
13. Swarm Robotics: A Formal Approach. Lecture Notes in Computer Science. Springer, Cham (2018). https://doi.org/10.1007/978-3-319-74528-2_7
14. Hamann, H., Reina, A.: Scalability in computing and robotics. arXiv, June 2020. https://arxiv.org/abs/2006.04969

15. Hamann, H., Valentini, G., Khaluf, Y., Dorigo, M.: Derivation of a micro-macro link for collective decision-making systems. In: Bartz-Beielstein, T., Branke, J., Filipič, B., Smith, J. (eds.) PPSN 2014. LNCS, vol. 8672, pp. 181–190. Springer, Cham (2014). https://doi.org/10.1007/978-3-319-10762-2_18

16. Hayes, A.T.: How many robots? Group size and efficiency in collective search tasks. In: Asama, H., Arai, T., Fukuda, T., Hasegawa T. (eds.) Distributed Autonomous Robotic Systems, vol. 5, pp. 289–298. Springer, Tokyo (2002). https://doi.org/10.1007/978-4-431-65941-9_29

17. Hill, M.D.: What is scalability? ACM SIGARCH Comput. Archit. News **18**(4), 18–21 (1990)

18. Ijspeert, A.J., Martinoli, A., Billard, A., Gambardella, L.M.: Collaboration through the exploitation of local interactions in autonomous collective robotics: the stick pulling experiment. Auton. Robots **11**, 149–171 (2001). https://doi.org/10.1023/A:1011227210047

19. Jensen, K.H., Kim, W., Holbrook, N.M., Bush, J.W.M.: Optimal concentrations in transport systems. J. Roy. Soc. Interface **10**(83), 20130138 (2013)

20. Keeling, M.J., Rohani, P.: Modeling Infectious Diseases in Humans and Animals. Princeton University Press, Princeton (2011)

21. Laure-Anne, P., Sebastien, M., Jacques, G., Buhl, J., Audrey, D.: Experimental investigation of ant traffic under crowded conditions. eLife **8**, e48945 (2019)

22. Lerman, K., Galstyan, A.: Mathematical model of foraging in a group of robots: effect of interference. Auton. Robots **13**, 127–141 (2002)

23. Levenspiel, O.: Chemical reaction engineering. Ind. Eng. Chem. Res. **38**(11), 4140–4143 (1999)

24. Lighthill, M.J., Whitham, G.B.: On kinematic waves. II. A theory of traffic flow on long crowded roads. Proc. Roy. Soc. London **A229**(1178), 317–345 (1955)

25. Font Llenas, A., Talamali, M.S., Xu, X., Marshall, J.A.R., Reina, A.: Quality-sensitive Foraging by a robot swarm through virtual pheromone trails. In: Dorigo, M., Birattari, M., Blum, C., Christensen, A.L., Reina, A., Trianni, V. (eds.) ANTS 2018. LNCS, vol. 11172, pp. 135–149. Springer, Cham (2018). https://doi.org/10.1007/978-3-030-00533-7_11

26. Mateo, D., Kuan, Y.K., Bouffanais, R.: Effect of correlations in swarms on collective response. Sci. Rep. **7**, 10388 (2017). https://doi.org/10.1038/s41598-017-09830-w

27. Mayya, S., Pierpaoli, P., Egerstedt, M.: Voluntary retreat for decentralized interference reduction in robot swarms. In: International Conference on Robotics and Automation (ICRA), pp. 9667–9673, May 2019. https://doi.org/10.1109/ICRA.2019.8794124

28. Mondada, F., Gambardella, L.M., Floreano, D., Nolfi, S., Deneubourg, J.L., Dorigo, M.: The cooperation of swarm-bots: physical interactions in collective robotics. IEEE Robot. Autom. Mag. **12**(2), 21–28 (2005)

29. Neuman, B.C.: Scale in distributed systems. In: Readings in Distributed Computing Systems. IEEE Computer Society Press (1994)

30. O'Grady, R., Gross, R., Christensen, A.L., Mondada, F., Bonani, M., Dorigo, M.: Performance benefits of self-assembly in a swarm-bot. In: IEEE/RSJ International Conference on Intelligent Robots and Systems (IROS). pp. 2381–2387, October 2007. https://doi.org/10.1109/IROS.2007.4399424

31. Özdemir, A., Gauci, M., Kolling, A., Hall, M.D., Groß, R.: Spatial coverage without computation. In: International Conference on Robotics and Automation (ICRA), pp. 9674–9680. IEEE (2019)

32. Poissonnier, L.A., Motsch, S., Gautrais, J., Buhl, J., Dussutour, A.: Experimental investigation of ant traffic under crowded conditions. eLife **8**, e48945 (2019). https://doi.org/10.7554/eLife.48945

33. Pratt, E.L.: Virtual teams in very small classes. Virtual Teamwork 91 (2010)

34. Rausch, I., Reina, A., Simoens, P., Khaluf, Y.: Coherent collective behaviour emerging from decentralised balancing of social feedback and noise. Swarm Intell. 321–345 (2019). https://doi.org/10.1007/s11721-019-00173-y

35. Reina, A.: Robot teams stay safe with blockchains. Nat. Mach. Intell. **2**, 240–241 (2020). https://doi.org/10.1038/s42256-020-0178-1

36. Reina, A., Miletitch, R., Dorigo, M., Trianni, V.: A quantitative micro-macro link for collective decisions: the shortest path discovery/selection example. Swarm Intell. **9**(2–3), 75–102 (2015)

37. Riedo, F., Chevalier, M., Magnenat, S., Mondada, F.: Thymio II, a robot that grows wiser with children. In: IEEE Workshop on Advanced Robotics and its Social Impacts (ARSO 2013), pp. 187–193. IEEE (2013)

38. Ringelmann, M.: Recherches sur les moteurs animés: Travail de l'homme. Annales de l'Institut National Agronomique, 2nd series **12**, 1–40 (1913)

39. Rosenfeld, A., Kaminka, G.A., Kraus, S.: A study of scalability properties in robotic teams. In: Scerri, P., Vincent, R., Mailler, R. (eds.) Coordination of Large-Scale Multiagent Systems, pp. 27–51. Springer, Boston (2006). https://doi.org/10.1007/0-387-27972-5_2

40. Salman, M., Ligot, A., Birattari, M.: Concurrent design of control software and configuration of hardware for robot swarms under economic constraints. PeerJ Comput. Sci. **5**, e221 (2019). https://doi.org/10.7717/peerj-cs.221

41. Sornette, D., Maillart, T., Ghezzi, G.: How much is the whole really more than the sum of its parts? 1 ⊞ 1 = 2.5: superlinear productivity in collective group actions. PLOS ONE **9**(8), 1–15 (2014). https://doi.org/10.1371/journal.pone.0103023

42. Talamali, M.S., Bose, T., Haire, M., Xu, X., Marshall, J.A.R., Reina, A.: Sophisticated collective foraging with minimalist agents: a swarm robotics test. Swarm Intell. **14**(1), 25–56 (2019). https://doi.org/10.1007/s11721-019-00176-9

43. Trianni, V., Groß, R., Labella, T.H., Şahin, E., Dorigo, M.: Evolving aggregation behaviors in a swarm of robots. In: Banzhaf, W., Ziegler, J., Christaller, T., Dittrich, P., Kim, J.T. (eds.) ECAL 2003. LNCS (LNAI), vol. 2801, pp. 865–874. Springer, Heidelberg (2003). https://doi.org/10.1007/978-3-540-39432-7_93

44. Valentini, G.: Achieving Consensus in Robot Swarms. SCI, vol. 706. Springer, Cham (2017). https://doi.org/10.1007/978-3-319-53609-5

45. Wahby, M., Petzold, J., Eschke, C., Schmickl, T., Hamann, H.: Collective change detection: adaptivity to dynamic swarm densities and light conditions in robot swarms. Artif. Life Conf. Proc. **31**, 642–649 (2019). https://doi.org/10.1162/isal_a_00233

46. Webots: version r2020a by Cyberbotics Ltd. (2020). https://cyberbotics.com

47. Zahadat, P., Hofstadler, D.N.: Toward a theory of collective resource distribution: a study of a dynamic morphogenesis controller. Swarm Intell. (1), 347–380 (2019). https://doi.org/10.1007/s11721-019-00174-x

Heterogeneous Response Intensity Ranges and Response Probability Improve Goal Achievement in Multi-agent Systems

H. David Mathias[1]($\boxtimes$), Annie S. Wu[2], and Laik Ruetten[1]

[1] University of Wisconsin - La Crosse, La Crosse, WI, USA
dmathias@uwlax.edu
[2] University of Central Florida, Orlando, FL, USA
aswu@cs.ucf.edu

Abstract. Inter-agent variation is well-known in both the biology and computer science communities as a mechanism for improving task selection and swarm performance for multi-agent systems. Response threshold variation, the most commonly used form of inter-agent variation, desynchronizes agent actions allowing for more targeted agent activation. Recent research using a less common form of variation, termed dynamic response intensity, demonstrates that modeling levels of agent experience or varying physical attributes and using these to allow some agents to perform tasks more efficiently or vigorously, significantly improves swarm goal achievement when used in conjunction with response thresholds. Dynamic intensity values vary within a fixed range as agents activate for tasks. We extend previous work by demonstrating that adding another layer of variation to response intensity, in the form of heterogeneous ranges for response intensity values, provides significant performance improvements when response is probabilistic. Heterogeneous intensity ranges break the coupling that occurs between response thresholds and response intensities when the intensity range is homogeneous. The decoupling allows for increased diversity in agent behavior.

1 Introduction

Swarms of artificial agents, which model, among other things, natural colonies of insects, are comprised of some number of software or hardware agents working in concert to achieve a goal. The agents accomplish the goal by completing, usually repeatedly, one or more tasks. The swarms in this work are decentralized. Thus, there is no leader or central control of the swarm and the agents do not communicate. Each agent chooses which tasks to perform and when. Work modeling natural swarms with artificial swarms dates back two decades [14].

Agents determine which tasks to undertake by considering environmental stimuli. When agents respond to these stimuli in the same way, their actions are synchronized. This synchrony often results in poor goal achievement. Swarm performance can be improved via inter-agent variation: differences in how and

© Springer Nature Switzerland AG 2020
M. Dorigo et al. (Eds.): ANTS 2020, LNCS 12421, pp. 148–160, 2020.
https://doi.org/10.1007/978-3-030-60376-2_12

when agents select and perform tasks. Common forms of inter-agent variation include response thresholds [6,15] and response probabilities [15].

Response thresholds desynchronize agents' actions by varying the stimulus required for an agent to act. This models the non-determinism inherent in natural swarms. Systems typically utilize response thresholds in one of two ways: probabilistically or deterministically. Probabilistic response, introduced by Bonabeau, et al. [1,2], uses a formula based on a task stimulus τ and an agent's response threshold θ to determine whether the agent activates for the task. The probability of activation increases with τ, approaching 1.0 when $\tau \gg \theta$. When $\tau = \theta$, the probability is 0.5. Deterministic response [7,13,20] activates an agent if $\tau \geq \theta$. Agent actions are desynchronized only if threshold values are heterogeneous. In the biology literature, Weidenmüller [15] suggests that use of heterogeneous response thresholds together with probabilistic response can further improve diversity in agent behaviors. Wu, et al. studied this effect in artificial swarms, confirming the benefit of non-determinism with heterogeneity [19].

Isolation of probabilistic response into a separate form of inter-agent variation, one that can be set and tuned independent of response thresholds, leads to a form of inter-agent variation known as *response probability* [13,17–19]. A first-order effect of decreased response probability is a decrease in the number of agents that activate for a task. Perhaps more importantly, a second-order effect is that inaction by frequent actors, agents with low response thresholds, may allow other agents to gain experience with a task [15,19]. This results from increased need due to the reduction in agents performing the task. Increased need exceeds the response threshold for additional agents, allowing them to participate. The experience gained by these agents may be important to the swarm if, for example, an extinction eliminates frequent actors.

The need for redundancy in artificial swarms has been acknowledged for many years as a way to mitigate the effects of agent failure or loss [4]. Similar effects are common in natural swarms in which agents are lost due to age, predators, or competitors. If these agents are frequent actors for a task, their loss may create significant short-term difficulty for the swarm as less experienced agents must fill the void. If, however, frequent actors sometimes remain idle due to decreased response probability, agents with higher response thresholds would gain experience with the task, making the swarm more tolerant of agent loss.

Response intensity is a less known form of inter-agent variation, particularly for artificial swarms. Response intensity models differences in quantities such as a natural agent's physical size, strength or stamina, attributes that may allow the agent to work more vigorously or more efficiently. Biologists have documented variation in response intensity [3,11]. For example, in the natural world some insects are known to change their response intensity as necessary to meet the needs of their colony [5]. Response intensity may also model an agent's experience on a task. We are not aware of previous work, prior to this year, that attempts to model this natural phenomenon in artificial swarms [10].

Experience is known to impact not only individual task efficiency but also individual task selection as well as collective colony performance [9,12]. In

Cerapachys biroi ants, individuals that find early success in foraging activities choose to forage again, whereas those individuals that were unsuccessful are more likely to choose to care for young in the nest [12]. In *Leptothorax albipennis* ants, task repetition improved colony performance for emigration, the task of moving the colony to a new nesting location [9]. Because the entire colony is exposed during emigration, and therefore in danger, efficient emigration is highly desirable.

In artificial agents, response intensity may model a decrease in output due to wear and tear or the increase in the output of a new and improved device. Importantly, heterogenous response intensities, when paired with heterogeneous response thresholds, play a role in determining which agents undertake a task and, therefore, gain experience and proficiency in that task.

Mathias *et al.* [10] demonstrated that dynamic, heterogeneous response intensities significantly improve swarm task achievement when combined with heterogeneous response thresholds and result in increased agent specialization. Dynamic response intensities vary within a specified range over the course of a run, increasing when an agent activates for a task and decreasing when it does not, modeling an agent's experience with the task. The range within which response intensities vary is homogeneous.

One consequence of combining heterogeneous response thresholds with dynamic, heterogeneous response intensities is that the values couple. This occurs because agents with low thresholds for a task activate more frequently for that task. Each activation increases the response intensity for the task, within the specified range. Thus, over time, an agent's response threshold for a task correlates with its response intensity for that task. Further, if the work performed by frequent actors is sufficient to meet task demand, agents with higher thresholds are denied the opportunity to gain experience for that task. This is potentially harmful to the swarm.

In this work, we demonstrate that using dynamic, heterogeneous response intensities that vary within *heterogeneous ranges*, rather than homogeneous ranges, improves swarm performance as response probability decreases. This occurs because heterogeneous intensity ranges and decreased response probability serve to *decouple* response thresholds and response intensities. We show that this makes the swarm more resistant to the effects of extinctions of experienced agents.

2   Model and Testbed Problem

We extend previous work on response intensities in two significant ways. First, we augment the dynamic, heterogeneous response intensities with heterogeneous intensity ranges. Thus, rather than all agents sharing a common range within which their intensities vary with experience, each agent has a unique intensity range. Second, we incorporate response probability. Response probability allows an agent to fail to undertake a task when the agent's response threshold for that task is met. The response probability values used here are homogeneous.

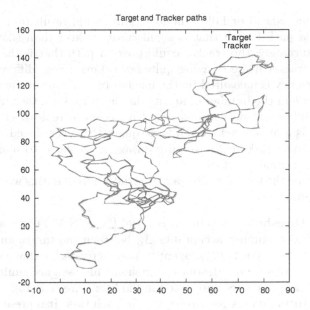

Fig. 1. An example random target path (purple) and corresponding tracker movement (blue) over 500 time steps. (Colour figure online)

Our testbed is a 2D tracking problem. This problem consists of a *target* object that moves in the plane and a *tracker* object. A swarm controls the tracker, pushing it to stay as close as possible to the target, which moves at random or according to one of several predefined paths. The paths are unknown to the agents. Each agent is capable of performing all tasks required of the swarm. The tasks are: push_N, push_E, push_S, or push_W. Agents may also remain idle if none of their response thresholds are met or due to the response probability. An example random target path is illustrated in Fig. 1.

A simulation consists of a predefined number of time steps. The target moves a fixed distance in each time step. The target's direction of travel can change as often as each time step, allowing frequently changing task demands. Agents are aware of the distance from the tracker to the target, defined by: $\Delta x =$ target.$x -$ tracker.x and $\Delta y =$ target.$y -$ tracker.y. In each time step, each agent chooses a task to perform from among those tasks for which the agent's response thresholds are met.

Swarm goal achievement is measured according to two criteria:

Goal 1. Minimize the average positional difference, per time step, between the target location and the tracker location.

Goal 2. Minimize the difference between total distance traveled by target and the total distance traveled by the tracker.

We note that neither criterion alone is sufficient to gauge the swarm's success. Consider using only Goal 1. The tracker could remain close to the target while alternately racing ahead or falling behind. This would result in a good average difference but a path length that is significantly greater than that traveled by the target. Alternately, the tracker could travel a path that is the same length as that of the tracker while straying quite far, taking a very different path.

Swarm efficiency is measured by the number of times agents switch from one task to another and the number of agents that activate for a task in a time step. Both task switches and activations may have costs in real-world applications, thus, a swarm is more efficient when these quantities are reduced. For example, undertaking a new task might require an agent to move to a new location, incurring costs in time and fuel.

Here we define the forms of inter-agent variation used in this work. Let a_i, $i \in \{1, \ldots, n\}$ be an agent.

- **Response threshold**: A value $\theta_{i,D}$ ($D \in \{N, E, S, W\}$) for each task that represents the maximum acceptable Δ_D between the target and tracker for that task. If Δ_D exceeds $\theta_{i,D}$, agent a_i may activate for that push_D. These values are heterogeneous. Response thresholds are assigned uniformly at random in $[0.0..1.0]$, a choice supported in the literature [7,8,16].
- **Response intensity**: A multiplier $\gamma_{i,D}$ for each task. It represents the factor by which the experience of a_i for task push_D differs from the default value of 1.0. This manifests as increased/decreased pushing power, equal to $\gamma_{i,D}$. These values are dynamic and heterogeneous. They are initialized uniformly at random within the agent's response intensity range for that task.
- **Response intensity range**: Intensity multipliers increase or decrease with an agent's experience for a task. The values for task push_D for agent a_i are bounded within a range $[\gamma_{i,D}min, \gamma_{i,D}max]$. Ranges may be homogeneous or heterogeneous. See Sect. 3 for a more detailed discussion.
- **Response probability**: A value p that represents the probability that an agent activates for a task. This value is homogeneous across all agents and tasks. It is a parameter to our system and is varied between runs.

3   Experimental Design

As response intensity and response probability are the focus of this work, we run experiments with two different types of intensity ranges – homogeneous and heterogeneous – and 7 response probability values, $[0.4..1.0]$ in increments of 0.1.

To model the loss of agents and our system's ability to recover from such events, we implement three different forms of agent extinction. kill-0, in which no agents are killed; kill-20-100-0, in which 20 agents are killed at time step 100; and kill-20-100-100, in which 20 agents are killed every 100 time steps beginning at time step 100. In each case, the agents chosen for extinction are those that were idle in the fewest time steps. This means that we kill those agents with the most experience and examine how well the swarm is able to recover.

Homogeneous intensity ranges are fixed at $[0.5, 2.0]$. Heterogeneous intensity range for agent a_i and task push_D is assigned by first choosing a size d uniformly

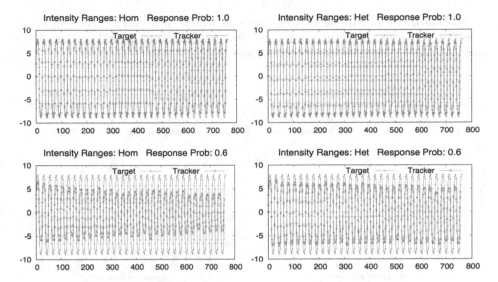

Fig. 2. A tracker performance comparison for four representative runs on path s-curve. The rows differ in response probability with 1.0 above 0.6. The columns are different response intensity ranges with homogeneous followed by heterogeneous. With response probability 0.6, the tracker performs substantially better with heterogeneous ranges. (Color figure online)

at random in $[0.6, 1.6]$. Offset $\gamma_{i,D}min$ is then chosen uniformly at random in $[0.3, (2.4 - d)]$. $\gamma_{i,D}max = \gamma_{i,D}min + d$ yielding range $[\gamma_{i,D}min, \gamma_{i,D}max] \subset [0.3, 2.4]$. These values are also determined empirically under the same conditions listed above. We note that the upper and lower endpoint values for both homogeneous and heterogeneous ranges are empirically determined to optimize behavior for the respective intensity range type for runs in which the response probability is 1.0 and no agent extinctions occur.

We test our system on three target paths: random, s-curve, and sharp. Random paths are generated by calculating an angle change, in radians, at every time step. The change is Gaussian $\mathcal{N}(0.0, 1.0)$. S-curve is a periodic curve seen in Fig. 2. Sharp is a randomized path in which a new heading and probability q of changing direction are chosen in every time step. The heading is chosen uniformly in $[0, 2\pi]$ and q is uniform in $[0.2, 0.6]$. Thus, turns are sharper than in the random path. All three paths create changing task demands though, random and sharp change more dramatically.

The variations discussed in this section produce 126 experiments for testing, 42 for each of the target paths. For each experiment we perform 100 runs. Each run lasts 500 time steps. In each time step, the target moves 3 distance units for a total path length of 1500. The swarm consists of 200 agents each of which is capable of performing all four tasks.

At each time step, we record the tracker's distance from the target. In addition, we record the total distance traveled by the target, total distance traveled by the tracker, the number of time steps in which each agent pushes in each direction, the number of times an agent does not perform a task (remains idle), and the number of times an agent switches from one task to another.

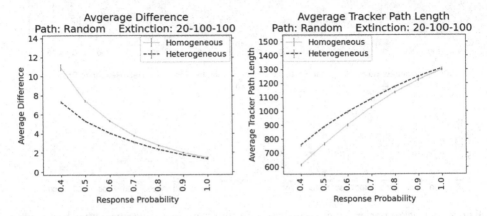

Fig. 3. Average positional difference and tracker path length for homogeneous and heterogeneous intensity ranges for target path random for 100 runs. Error bars are shown in red. Both quantities are improved for heterogeneous ranges. (Color figure online)

4   Experimental Results

In this section, we report the results of the experiments described in the previous section. These results support our central argument: Heterogeneous response intensity ranges improve swarm performance, when agents respond probabilistically, due to increased inter-agent variability and the decoupling of response threshold values and response intensity values. In addition, our results support those of previous work in demonstrating that response probabilities $p < 1.0$ allow swarms to recover more quickly from agent extinctions.

The data support the following performance improvements for heterogeneous intensity ranges, relative to homogeneous intensity ranges, for lower response probabilities and paths with frequently changing task demands:

- reduced average positional difference between the target and the tracker
- reduced variability, within a run, in average positional difference between the target and the tracker
- reduced difference between the target and tracker path lengths
- more accurate target path tracking
- reduced task switching

Figure 2 illustrates the effect of heterogeneous intensity ranges on target tracking when response probability is reduced. The top row shows that when response probability $p = 1.0$, homogeneous and heterogeneous intensity ranges

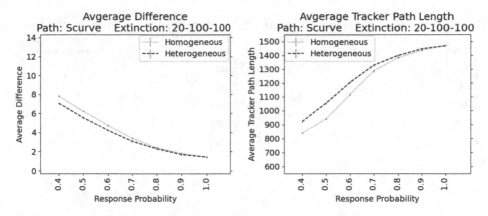

Fig. 4. Average positional difference and tracker path length for homogeneous and heterogeneous intensity ranges for target path s-curve for 100 runs. Error bars are shown in red. Both quantities are improved for heterogeneous ranges. (Color figure online)

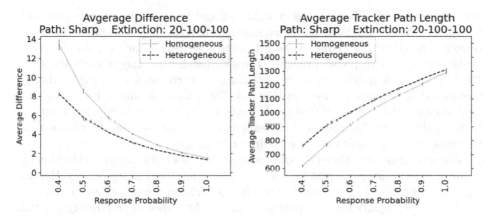

Fig. 5. Average positional difference and tracker path length for homogeneous and heterogeneous intensity ranges for target path sharp. Error bars are shown in red. Both quantities are improved for heterogeneous ranges. (Colour figure online)

produce similar results, with the tracker (red) staying close to the target (blue) throughout the run. Note that there is minimal degradation of performance as agents are killed at 100 time step intervals. In the bottom row, the response probability $p = 0.6$. Thus, there is probability 0.4 that an agent fails to act when its response threshold is met. As a consequence, system performance suffers – recall that parameter values are optimized for $p = 1.0$. We note that tracking is significantly better for heterogeneous intensity ranges than for homogeneous ranges when $p = 0.6$.

Figures 3, 4, and 5 provide data for the tracking effects observed in Fig. 2 for paths random, s-curve, and sharp, respectively. In each figure, the left plot shows

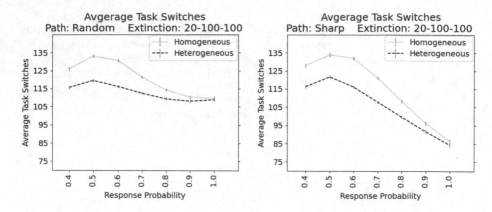

Fig. 6. Average task switches, per agent, for homogeneous and heterogeneous intensity ranges, for target paths random and sharp. These paths have frequently changing task demands. In both cases, heterogeneous ranges reduce the number of switches.

average positional difference between target and tracker for response probabilities $p \in [0.4, 1.0]$ for both homogeneous and heterogeneous response intensity ranges. The right plot shows average tracker path length for the same response probabilities and response intensity ranges. Recall that target path length is 1500. Each data point represents 100 runs. 95% confidence intervals, though quite small in most cases, are shown in red. The data show that at lower response probabilities, the average difference is lower and tracker path length is closer to target path length for runs with heterogeneous response intensity ranges than for runs with homogeneous response intensity ranges.

Figure 6 illustrates the effect of heterogeneous intensity ranges on the average number of switches per agent for paths random (left) and sharp (right). Heterogeneous intensity ranges allow the swarm to perform fewer task switches, particularly when agents respond probabilistically. At response probability $p = 0.6$, the difference is approximately 15 fewer task switches per agent or 3000 fewer switches during a run for a population of 200 agents. Because task switches can incur a cost in real-world applications, this is a significant improvement.

The observed trends are explained as follows. With homogeneous intensity ranges, all frequent actors for a task have similar response intensity values due to the common maximum value. As frequent activation results from low response thresholds, this results in a coupling of the two values: small $\theta \rightarrow$ large γ. In contrast, heterogeneous intensity ranges have different sizes and different minimum and maximum values. The smallest range size $d = 0.6$ and the smallest possible $\gamma_{i,D}min = 0.3$ resulting in a minimum intensity range of $[0.3, 0.9]$. The largest possible $\gamma_{i,D}max = 2.4$. As with homogeneous intensity ranges, frequent actors may reach the maximum intensity value in their range, however, these maximum values vary considerably making the swarm in general, and the group of frequent actors in particular, more diverse. In this way, $\gamma_{i,D}$ is far less dependent on $\theta_{i,D}$. Thus, heterogeneous intensity ranges decouple response intensity values from response threshold values.

This decoupling has multiple effects. First, it allows the swarm to better adapt to frequently changing task demands. This occurs because when task demands change frequently, agents are unlikely to maximize their intensity values through activation. This may result in insufficient intensities, for those agents that activate due to low thresholds, to maintain a small positional difference with the target. The greater diversity of intensity ranges can mitigate this. Second, it helps regulate swarm behavior, in the short-term, after an agent extinction because survivors – agents with higher response thresholds – may have higher response intensities than is possible with homogeneous intensity ranges. Thus, the swarm is better able to meet task demand. Of course, the random nature of intensity range creation could result in a swarm with too few agents with high intensity ranges but due to the population size used, this is unlikely.

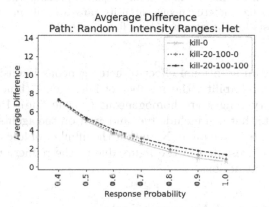

Fig. 7. Average positional difference heterogeneous intensity ranges with each agent extinction implementation for target path random. This demonstrates that extinction type does not significantly affect the swarm's ability to track the target.

The results presented above are for runs using extinction `kill-20-100-100` in which 20 agents are killed every 100 time steps. Extinction types `kill-0` and `kill-20-100-0` are also used in our experiments. Figure 7 illustrates why we focus the discussion on a single extinction type. The figure shows that average positional difference does not vary significantly with changes in extinction. The same trend is observed for average path length. Therefore, we choose to concentrate the analysis on `kill-20-100-100` to simplify the presentation. The y-axis range in Fig. 7 is the same as in Figs. 3, 4, and 5 to facilitate comparison.

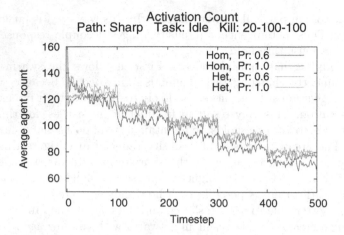

Fig. 8. Idle agent counts for target path sharp. Homeogeneous and heterogeneous response intensity ranges are compared for response probabilities 0.6 and 1.0. For response probability 0.6, heterogeneous intensity ranges result in an increase in idle agents, reducing costs for the swarm. (Color figure online)

Figure 8 shows an additional effect of heterogeneous intensity ranges. With reduced response probability, the number of idle agents (blue line) is greater than when intensity ranges are homogeneous (purple line). Recall that agent activation has costs that can include fuel and wear on the agents. Thus, a higher number of idle agents is desirable. Note that the number of idle agents decreased through the runs represented in the figure due to the reduction in the number of agents through extinctions.

5   Conclusions and Future Work

In this work we explore the effects of a little-studied and promising form of inter-agent variation: response intensity. Expanding on previous work that shows the benefit of heterogeneous response intensity values that vary within a homogeneous range, we implement response intensity values that vary within heterogeneous response intensity ranges. Our system also uses homogeneous response probability and heterogeneous response thresholds.

We find that heterogeneous response intensity ranges provide significant improvement over homogeneous response intensity ranges for decreased response probabilities and problems with frequently changing task demands for a 2-D tracking problem. The improvement is seen in all measures of swarm performance: average positional difference, average tracker path length, and average number of task switches. The observed improvements are due to the decoupling effect that heterogeneous intensity ranges have on response intensity values and response probability values. This results in far more diversity among frequent actors and the backup agents that replace them when agent extinctions occur.

In future work, we will test our model on a more complex task allocation problem and explore additional forms of inter-agent variation. In addition, we plan to investigate heuristic methods for initializing the values of response thresholds and response intensities.

Acknowledgement. This work is supported by the National Science Foundation under grant IIS1816777.

References

1. Bonabeau, E., Theraulaz, G., Deneubourg, J.L.: Quantitative study of the fixed threshold model for the regulation of division of labor in insect societies. In: Proceedings: Biological Sciences, pp. 1565–1569 (1996)
2. Bonabeau, E., Theraulaz, G., Deneubourg, J.L.: Fixed response thresholds and the regulation of division of labor in insect societies. Bull. Math. Biol. **60**, 753–807 (1998). https://doi.org/10.1006/bulm.1998.0041
3. Dornhaus, A., Holley, J., G.Pook, V., Worswick, G., Franks, N.R.: Why do not all workers work? Colony size and workload during emigrations in the ant temnothorax albipennis. Behav. Ecol. Sociobiol. **63**, 43–51 (2008). https://doi.org/10.1007/s00265-008-0634-0
4. Hackwood, S., Beni, G.: Self-organization of sensors for swarm intelligence. In: Proceedings of the IEEE International Conference on Robotics and Automation, pp. 819–829 (1992)
5. Jeanne, R.L.: Regulation of nest construction behavior in Polybia occidentalis. Animal Behav. **52**, 473–488 (1996)
6. Jones, J.C., Myerscough, M.R., Graham, S., Oldroyd, B.P.: Honey bee nest thermoregulation: diversity promotes stability. Science **305**(5682), 402–404 (2004)
7. Krieger, M.J.B., Billeter, J.B.: The call of duty: self-organised task allocation in a population of up to twelve mobile robots. Robot. Auton. Syst. **30**, 65–84 (2000)
8. Krieger, M.J.B., Billeter, J.B., Keller, L.: Ant-like task allocation and recruitment in cooperative robots. Nature **406**, 992–995 (2000)
9. Langridge, E.A., Franks, N.R., Sendova-Franks, A.B.: Improvement in collective performance with experience in ants. Behav. Ecol. Sociobiol. **56**, 523–529 (2004). https://doi.org/10.1007/s00265-004-0824-3
10. Mathias, H.D., Wu, A.S., Ruetten, L., Coursin, E.: Improving multi-agent system coordination via intensity variation. In: Proceedings of the 33rd International Florida Artificial Intelligence Research Society Conference (2020)
11. Oster, G.F., Wilson, E.O.: Caste and Ecology in the Social Insects. Princeton University Press, Princeton (1978)
12. Ravary, F., Lecoutey, E., Kaminski, G., Chaline, N., Jaisson, P.: Individual experience alone can generate lasting division of labor in ants. Curr. Biol. **17**, 1308–1312 (2007)
13. Riggs, C., Wu, A.S.: Variation as an element in multi-agent control for target tracking. In: Proceedings of the IEEE/RSJ International Conference on Intelligent Robots and Systems, pp. 834–841 (2012)
14. Theraulaz, G., Goss, S., Gervet, J., Deneubourg, J.L.: Task differentiation in Polistes wasp colonies: a model for self-organizing groups of robots. In: Proceedings of the 1st International Conference on Simulation of Adaptive Behavior: From Animals to Animats, pp. 346–355 (1991)

15. Weidenmüller, A.: The control of nest climate in bumblebee (*Bombus terrestris*) colonies: Interindividual variability and self reinforcement in fanning response. Behav. Ecol. **15**, 120–128 (2004)
16. Wu, A.S., Mathias, H.D., Giordano, J., Hevia, A.: Effects of response threshold distribution on dynamic division of labor in decentralized swarms. In: Proceedings of the 33rd International Florida Artificial Intelligence Research Society Conference (2020)
17. Wu, A.S., Wiegand, R.P., Pradhan, R.: Using response probability to build system redundancy in multi-agent systems. In: Proceedings of the 12th International Conference on Autonomous Agents and Multiagent Systems, pp. 1343–1344 (2013)
18. Wu, A.S., Wiegand, R.P., Pradhan, R.: Building redundancy in multi-agent systems using probabilistic action. In: Proceedings of the 29th International Florida Artificial Intelligence Research Society Conference. pp. 404–409 (2016)
19. Wu, A.S., Wiegand, R.P., Pradhan, R.: Response probability enhances robustness in decentralized threshold-based robotic swarms. Swarm Intell. (2020). https://doi.org/10.1007/s11721-020-00182-2
20. Wu, A.S., Wiegand, R.P., Pradhan, R., Anil, G.: The effects of inter-agent variation on developing stable and robust teams. In: Proceedings of the AAAI 2012 Spring Symposium: AI, The Fundamental Social Aggregation Challenge, and the Autonomy of Hybrid Agent Groups (2012)

HuGoS: A Multi-user Virtual Environment for Studying Human–Human Swarm Intelligence

Nicolas Coucke[1,2]([⊠]) [iD], Mary Katherine Heinrich[1] [iD], Axel Cleeremans[2] [iD],
and Marco Dorigo[1] [iD]

[1] IRIDIA, Université Libre de Bruxelles, Brussels, Belgium
{nicolas.coucke,mary.katherine.heinrich,axcleer,mdorigo}@ulb.ac.be
[2] CO3, Center for Research in Cognition and Neurosciences,
Université Libre de Bruxelles, Brussels, Belgium

Abstract. The research topic of human–human swam intelligence includes many mechanisms that need to be studied in controlled experiment conditions with multiple human subjects. Virtual environments are a useful tool to isolate specific human interactions for study, but current platforms support only a small scope of possible research areas. In this paper, we present HuGoS—'Humans Go Swarming'—a multi-user virtual environment in Unity, as a comprehensive tool for experimentation in human–human swarm intelligence. We identify possible experiment classes for studying human collective behavior, and equip our virtual environment with sufficient features to support each of these experiment classes. We then demonstrate the functionality of the virtual environment in simple examples for three of the experiment classes: human collective decision making, human social learning strategies, and agent-level human interaction with artificial swarms, including robot swarms.

1 Introduction

Human–human swarm intelligence is a broad field of study [20], including topics such as crowd dynamics [31], online social networks [22], and collective problem solving [39]. While some studies of human group behavior use data collection from real-world systems, such as social networks [33], many study types require controlled experiment conditions. As self-organization in human groups normally occurs within the context of other mechanisms and influences, a comprehensive tool for studying human–human swarm intelligence must enable the experimenter to artificially limit human capabilities of perception and communication, according to the given experiment. Virtual environments have been proposed as tools to isolate and study specific aspects of human interaction [1].

In this paper, we develop a virtual environment for experiments with multiple human subjects. To be comprehensive, the environment needs to support studies in three main topics of human–human swarm intelligence. First, humans often use simple mechanisms and strict self-organization to coordinate (e.g., in

© Springer Nature Switzerland AG 2020
M. Dorigo et al. (Eds.): ANTS 2020, LNCS 12421, pp. 161–175, 2020.
https://doi.org/10.1007/978-3-030-60376-2_13

human crowds [31]), displaying behaviors similar to those observed in artificial swarms and animal groups. Second, humans also use more complex mechanisms (e.g., advanced negotiation, or hierarchical social structures) that are often formed via self-organization. For instance, hierarchy can be self-organized according to response speeds of individuals [21], or strengths of preexisting interpersonal ties [7]. Third, comparative studies between human groups and artificial swarms are relevant even for complex mechanisms, as self-organized leadership and hierarchy have recently been studied not only in humans, but also in groups of non-human animals [13] and groups of robots [24]. In this paper, we propose HuGoS—'Humans Go Swarming'—a multi-user virtual environment that supports research and experimentation in each of these three topics. In HuGoS, human participants interact via avatars in a controlled experiment setup, capable of supporting both simple and complex interactions. HuGoS also supports avatars controlled by artificial agents, enabling comparative studies between human and artificial behaviors.

1.1 Related Work

A number of multi-user virtual environments have been developed for studies with human groups, in two main categories. The first category of environments use collective human gameplay or other interactions as tools for solving computationally intensive problems [2,6,9,18,23,44], rather than studying underlying cognitive or behavioral mechanisms. The second category of virtual environments are those developed primarily to study the mechanisms of collective human behavior. For collective decision making, the *UNUM* platform [40,41], also referred to as Swarm AI®, supports a group of participants that collaboratively explore a decision space [26]. For physical coordination, a Unity implementation supports human-like avatars with first-person view for the study of crowd behaviors [32,45,53]. Finally, a third category supports the study of leadership—specifically, the impact of better informed individuals on implicit leadership (e.g., the *HoneyComb* game for human crowd movement [4]). These existing environments are mostly developed for a specific task. For instance, in the *UNUM* platform, each player controls a 'magnet' that exerts influence on a 'puck.' This platform could not be easily re-purposed to investigate, for instance, a task involving environment exploration. To our knowledge, there are no existing platforms that are versatile enough to be used for a wide variety of topics in human–human swarm intelligence. In this paper, we target the contribution of a platform that can comprehensively cover this field of study. For instance, in addition to the implicit leadership studied in [4], a comprehensive platform should also be able to study explicit leadership, such as 'follower' functionalities developed in online trading networks [19]. Furthermore, the existing platforms log experiment data, such as positions of avatars in the virtual environment, but conduct analysis externally. We target a platform in which recorded data can also be analyzed internally, allowing real-time feedback to be incorporated in the experiment. This capability enables the study of, for instance, the relationship between group behavior and different types of performance feedback.

Existing tools used for robot simulation support 3D environments in which artificial agents can interact both with each other and with the environment. ARGoS [37] is a tool built specifically for robot swarms, while other tools such as ROS [38] or Webots [27] are built for robots generally. These versatile tools can be adapted to a wide variety of experiment scenarios, and many types of data collection and analysis. However, they have limited applicability to the study of human behavior. A few studies have looked at human–swarm interaction using general tools for robots (e.g., using ROS [49] or Webots [48]). In these setups, humans are able to give high-level directions to individual robots. However, the humans cannot act independently of the robots, as they cannot control their own avatars with first-person view. Also, the approaches do not demonstrate multiple human users in one environment.

2 Design of HuGoS: 'Humans Go Swarming'

We target the design of a multi-user virtual environment—HuGoS: 'Humans Go Swarming'—that can be used as a tool to study human collective behavior generally, including collective decision-making, collaborative task performance, and the emergence of leadership. HuGoS should also support the study of differences and similarities between human swarm intelligence and artificial swarm intelligence, and the interactions between human and artificial agents.

2.1 Experimentation Scope for Human–Human Swarm Intelligence

There are several classes of experiments that HuGoS must support, to facilitate comprehensive study of human–human swarm intelligence. We base this experimentation scope on existing studies of robot and artificial swarms, such that these behaviors could also be studied in humans. The first class of experiments is physical coordination between individuals, as in flocking and self-assembly (e.g., [42]). In HuGoS, this would require a minimal environment in which participants control avatars, whose positions are continuously recorded for analysis. The second class involves observation of environmental features. In cooperative navigation, for example, agents might extract and share information to find the shortest path in an environment (e.g., [8]). In best-of-n decision-making, a swarm might choose the best of several options based on observations of the environment (e.g., [46]). In HuGoS, this requires that the environment be populated with game objects that act as obstacles, or represent environment features with observable discrete or continuous properties. In order to study decision making based on external information, of the type studied in human groups in platforms such as *UNUM* [40], landmarks in the environment could be labelled with each option, and participants could use their avatar positions in relation to the landmarks to indicate their opinions. The third class involves the agents making changes to the environment, such as in *stigmergic* communication (e.g., [12]) or in the performance of a task such as collective construction (e.g., [51]). In HuGoS, this requires game objects that can be manipulated or modified by avatars—for instance, immovable environment features with modifiable properties such as color, or movable objects such as construction blocks. All classes require HuGoS to enable various types of direct and indirect communication between players, in ways that can be expressly limited. For example, avatars' view capabilities may be limited such that players can see only their immediate neighbors, not all avatars. Or, in a task such as collective construction, players may be able to see only the construction blocks, not the other avatars. In all classes, studies might include a comparison between human and artificial behaviors, or collaboration between human and artificial agents. This requires, in addition to human-controlled avatars, that HuGoS supports artificial agents whose avatars may be indistinguishable for human players. As it might be fruitful to replicate these experiments in real setups, we also need to integrate robot models into HuGoS (such as those developed for the ARGoS multi-robot simulator [37]).

2.2   Features of the HuGoS Virtual Environment

Unity 3D Game Engine. HuGoS is built in Unity, a 3D game development platform that can support intelligent agents in a physically realistic game environment [14]. In Unity, basic building blocks of virtual environments are termed *game objects*. Each game object represents a physical 3D object within the game environment that is subject to physics engines and is linked to specific C# back-end scripts. Through these scripts, each game object in HuGoS can be: i) passive, ii) controlled by simple rule-based behaviors but immobile, iii) mobile and equipped with a controller to act as an artificial agent, or iv) mobile and controlled by a human player. We refer to game objects in HuGoS as *avatars* if they act as artificial agents or are controlled by human players. We refer to game objects as *landmarks* if they are immobile, whether passive or controlled by simple rule-based behaviors (e.g., change display color to match that of neighboring landmark). Using Unity's networking capabilities, we organize the multi-user architecture of HuGoS as follows.

HuGoS initiates on a server, and each new player joins as client on that server. Throughout the experiments, the server logs all data recorded from HuGoS. The activities taking place on the server are divided into three modules: the player module, the environment module, and the task module (Fig. 1(a)). The environment module tracks landmarks, including changes made to them by either players or controllers, and passes those updates to the clients. The player module tracks player actions, and mediates any communication between clients. The task module tracks and analyzes the progress of the specific experiment, which can optionally be shared with player clients.

Avatar Capabilities. Each player controls an avatar that is situated in the virtual environment. The capabilities of the avatars in a given experiment setup are defined in the player module. In HuGoS, players have a third-person view of their avatar through a virtual camera that follows the avatar position and rotation. Players move their avatar by pressing four user-customizable keys (e.g.,

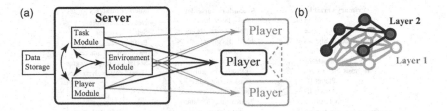

Fig. 1. (a) Program architecture. The server contains the task, player, and environment modules. Communication with player clients is managed by modules (dark arrows represent communication between modules and player; lighter arrows illustrate that same communication to multiple players; dashed lines represent communication between players, as mediated by the player module). (b) Example of layered player networks. Layer 1 is fully connected (visibility of other avatars), while layer 2 is partly connected (visibility of the other avatars' color opinions)

WSAD), and rotate via the left/right arrow keys or cursor movement. Depending on the experiment scenario, specific additional actions can be activated for the avatars. For example, the player can be permitted to manipulate the environment by clicking on landmarks to grab them, then moving and releasing the cursor to move them. Indirect communication between players can occur via changes to the environment, for instance by moving landmarks, or by changes to display features of the player's avatar, such as color. Direct communication can also be permitted—and limited as desired—by sending symbols, or written or spoken messages. In the player module of HuGoS, the players' environment perception is controlled firstly by changing the field of view (FOV) of the player. A player that has a limited top-down avatar FOV (Fig. 3(c)) can only perceive the environment in a small perimeter, while a player that has an oblique avatar FOV (Fig. 3(a)) can see a much greater proportion of the environment, in the viewed direction. Unlimited top-down FOV is also possible, giving a player global view (Fig. 3(b)). Additionally, avatars and landmarks can be programmed to be invisible to players, or to be visible only for a subset of players.

Player interactions can be modulated by changing the structure of the *player networks* in the player module, which are directed graphs. If a player network is fully connected, for instance, then every player can interact with all other players in the way associated to that network. Player networks manages different types of interaction, and have independently defined structures. For example, a fully connected network might be defined for viewing avatar positions, while a sparsely connected network might be defined for viewing avatar colors (Fig. 1(b)). Player networks also govern explicit message passing between players. As connections are directed (e.g., player 1 might be able to see player 2, while player 2 cannot see player 1), the information privileges of players can be made hierarchical. Certain players can have higher node indegrees or outdegrees. The structure of player networks can be changed during experiment runtime, and can optionally be triggered by the players. For instance, players might be permitted to 'follow' another player by clicking on its avatar, causing their own decisions to

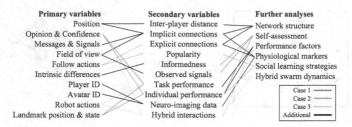

Fig. 2. Primary variables from the environment can be used for analysis specific to the experimental conditions, to calculate secondary variables and conduct further analysis.

automatically copy those of the followed player, until that player is unfollowed (see Sect. 3.2). The ability to control communication links between players also allows for comparison between limited communication networks and fully connected communication networks. This can facilitate the study of information cascades, bias in the group, or dysfunctional dynamics that may lead to low performance.

Data Types (Primary Variables). Data about the players, environment, and task are logged for analysis. Each player has a unique anonymized player ID, and each avatar has an avatar ID. These two IDs are important in cases where players switch avatar identities between trials, so that the behaviors of specific players can be analyzed separately from the features accumulated by a shared avatar. Additionally, the player IDs are important in experiment setups that involve players' physical environments (e.g., players occupy the same physical environment and can communicate, or players' physiological data is monitored, such as EEG). Avatar capabilities, positions, FOVs, and actions are all logged, according to avatar ID. These logs enable calculation of other simple data about avatars, such as which other avatars are in one avatar's FOV. Messages passed by avatars are also logged, including the content, time, sender avatar ID, and receiver avatar ID. All other player interactions are also tracked and logged as events—for instance, a player choosing to follow another player—again including content, time, and sender and receiver IDs. Changes in the environment are also logged, including positions and states of landmarks. When artificial agents such as robots are included in a setup, any data specific to that agent and experiment is logged. For instance, in a setup with models of e-puck robots [29], proximity sensing and motor control might be logged.

Analysis Types (Secondary Variables). The data logged as primary variables (i.e., recorded directly) allow many secondary variables to be calculated and analyzed during runtime or during post-processing (Fig. 2). Here, we use task performance as an illustrative example. Task performance can be continuously calculated by the task module, according to the specific scenario. In a flocking scenario, the task performance would depend on player positions; in decision-making, on player opinions. Once task performance is calculated, additional analysis might assess, for instance, how this performance relates to the in-game behavior of players. Player behavior in this case might be represented by distances between avatars, the network of implicit connections between individuals that occur when avatars enter each other's FOVs, or the network of direct messages between players with connection weights representing message frequency. The primary variables also allow analysis of individual behavior, that can be used to give feedback to players during the experiment. For instance, in a collective decision-making scenario, comparing individual opinion to overall task performance yields the relative player performance. If this is provided as feedback to players, players can use it to determine and display their opinion confidence. If the calculated player performance is not provided to the player when the player determines opinion confidence, then a comparison of these two variables will yield the player's *self-assessment* (i.e., the ability to evaluate their own performance). Using player IDs, out-of-game data can also be used in post-analysis. For example, each player might be asked to fill in a questionnaire about personality traits or subjective experience during the game. In an extended out-of-game setup, gameplay could even be linked to real-time physiological recordings, such as eye-tracking, ECG or EDA tracking of stress, or neural recordings via EEG or fMRI. Such extensions could be used to analyze the connection between individual cognitive mechanisms and collective performance during gameplay.

3   Demonstration of HuGoS Features

We demonstrate the suitability of HuGoS for the defined experimentation scope in three case studies. First, we demonstrate human players in a basic scenario previously studied in robot swarms [47]. Second, we demonstrate an aspect of human behavior that is outside typical swarm intelligence studies—specifically, the establishment of leader–follower relationships, similar to behavior mimicking in human trading networks [19]. Third, we demonstrate artificial agent avatars in a basic swarm intelligence scenario, and demonstrate interaction between human-controlled avatars and artificial agent avatars (specifically, robot avatars).

3.1   Case 1: Collective Decision Making

Collective decision making is widely studied in artificial swarms (e.g., [41,47]). We implement a setup based on that of [47] in HuGoS, but with consensus to be reached by human players rather than kilobots (Fig. 3). The task is to reach a consensus about the predominant color in the environment, which is populated with cylindrical landmarks that are randomly distributed and colored red or blue (Fig. 3(b)). The difficulty of the task can be adjusted by changing the color ratio and the density of landmarks. In this example there are 1 000 landmarks, in a color ratio of 55%-45%. At initiation of the run, the task module randomly determines whether red or blue will be the most prevalent. Each player controls an avatar that initiates at a random position. Half of the avatars initiate with

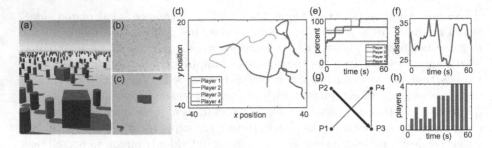

Fig. 3. (a) Player view on display monitor. The player controls an avatar (blue cube in the center of the screen) and can see the avatar of another player (red cube), in addition to cylindrical landmarks. (b) Top view of the environment with random distribution of blue and red cylinders. (c) Player view with limited information—views avatar from the top and can see only local information. (d–h) Trial of collective decision making with four human-controlled avatars. (d) Trajectories traveled by the avatars. (e) Percentage of the environment seen by each player. (f) Average Euclidean distance between avatars. (g) Network of player–player view (time 0–60 s); connection weights indicate total view time. (h) Number of avatars displaying the correct color opinion. (Color figure online)

the opinion red, and the other half blue, assigned randomly. The players can switch their color opinions by pressing a key, and can move and rotate without restriction. The player has a third-person avatar view through a virtual camera that follows the avatar (Fig. 3(a)). By moving and rotating the avatar, and thus the FOV, a player can explore the environment from different perspectives and make a subjective estimate of the majority color in the playing field. A player can see the movements and color changes of all avatars in the FOV. On the server, the player module passes avatar colors to the task module, where task performance is calculated, according to the homogeneity of avatar opinions, and whether the majority opinion matches the dominant color in the environment. Gameplay ends when all players have the same color, or when a time limit is reached. Variables are logged and calculated during runtime, including the task performance (Fig. 3(h)), avatar FOV, and avatar positions (Fig. 3(d)). The average euclidean distance between every two avatars is calculated (Fig. 3(f)), as is the percentage of cylinders cumulatively observed in the FOV (Fig. 3(e)), and the directed graph of all avatars' appearances in others' FOVs (Fig. 3(g)).

3.2 Case 2: Social Learning Strategies

In swarm intelligence in artificial agents, interactions are typically not chosen by individuals, but rather happen indiscriminately through random encounters. In humans and certain other animals, selective interaction can have a substantial impact on collective behavior, and has been widely studied in the context of *social learning* (cf. [17,52]). Humans in social learning scenarios use this selectivity to choose when, from whom, and what to learn [17]. In a collective decision making scenario, the dynamics of this selectivity would have a significant impact on the

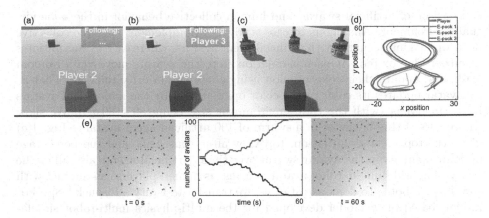

Fig. 4. (a–b) In Case 2: Explicit leader–follower relationships, in the view of player 2. (a) Player 2 acting individually. (b) Player 2 following player 3. (c–d) In Case 3: Human–artificial interaction. (c) Human player-controlled avatar (blue cube) followed by artificial agent avatars (e-puck robots). (d) Trajectories of a human player (in blue) being followed by three e-puck robots. (e) In Case 3: 100 artificial agent avatars performing random walk and using majority rule for color opinions.

outcome of collective behavior. Player strategies for selective interaction could be studied implicitly after calculating the information seen in the avatar FOV, as in Figs. 3(e,g). Strategies for selective interaction could be studied more explicitly, via a function for the self organization of explicit leader–follower relationships between players. Individuals might choose, for example, to follow the most prestigious individual (i.e., who already has most followers) [5, 17], or the most successful individual (i.e., highest individual performance, for the task) [11, 17]. We therefore add events for leader–follower relationships to HuGoS. During gameplay, players choose either to act for themselves (Fig. 4(a)), or to 'follow' another avatar and copy its actions (Fig. 4(b)). A player can choose to follow another by clicking on that player's avatar. Once the relationship is established, the following status is displayed to the follower player in a dialogue box, and a tiara appears above the leader avatar in the follower's display (Fig. 4(b)). While the relationship exists, the follower player no longer is in control of the avatar, which automatically copies the behavior of the leader avatar. In a scenario similar to Case 1, this would mean that the follower's motion copies that of the leader, and the follower automatically adopts the color of the leader. A follower can at any time decide to follow a different leader by clicking on the corresponding avatar, or decide to act independently again by clicking on the ground plane.

3.3   Case 3: Interaction with Artificial Agents

HuGoS also includes mobile avatars that are artificial agents, rather than being controlled by human players, as well as immobile landmarks equipped with rule-based behaviors for their display features. This enables straightforward comparison of artificial swarms and human collective behavior in the same virtual environment. Analogous to case study 1, we implement a simple collective decision making scenario with artificial agent avatars, where each avatar initiates as either blue or red. At each simulation step, the avatars move via a random walk, and update their color opinion using a simple majority rule [30]. That is, an avatar updates its color to the color opinion held by the majority of avatars in a 5 m radius (with the total environment size being $80 \times 80\,\text{m}^2$). We show the results of this behavior in a swarm of 100 artificial agent avatars—Fig. 4(e) gives the top view at initiation, top view after consensus, and the percentage of color opinions in the avatar swarm over time. Artificial agents also allow the study of hybrid human–robot avatar swarms, as human players can interact with simulated robots. To demonstrate this, we transfer to Unity a Pi-puck [28] robot model, based on a model developed for the multi-physics multi-robot simulator ARGoS [37]. As Unity includes built-in physics engines, existing ARGoS kinematics and dynamics robot models could also be transferred into HuGoS, including those that have already been calibrated to real hardware (e.g., [36]). To demonstrate interaction between players and robots, we programmed the simulated robots to detect a nearby human player avatar, and to move in the direction of that avatar when the distance becomes too large (Fig. 4(c)). The robot avatars also sense their distance to other robots and move away from each other if they get too close. We show the trajectories of all avatars in a setup with one human player avatar and three robot avatars (Fig. 4(d)).

4   Discussion

We have introduced a novel multi-player virtual environment that is suited for studying human behavior in swarm intelligence scenarios. HuGoS logs primary variables related to actions of individual players and multiple players, and variables that depend on the environment. These primary variables, recorded directly, are then used to calculate secondary variables (e.g., the collective performance of avatars for a given task).

Many research topics proposed here are typically studied with in-person experimental setups in which participants interact directly. Such setups allow for many types of interaction that are not possible in HuGoS, such as eye contact, speech characteristics, and body language. We do not propose that HuGoS is a replacement for in-person studies. Rather, our objective is to use the simplifications of a virtual environment for complementary studies that provide new insights by isolating certain aspects of previously studied dynamics. Isolating specific aspects can be useful in studying human behavior [1], for instance by helping to disentangle different forms of interaction that would be too closely related in an in-person setup.

As HuGoS targets human players, some open questions remain. As pointed out in [3], it is possible that human players will pay less attention to other individuals when there are less communication channels available or when too many other individuals are present. Individual behavior is also highly dependent on whether or not players are convinced that other avatars are actually human-controlled [3]. Another challenge with human players is to keep players engaged, and motivated to perform well in the task. Performance trackers such as leaderboards might motivate players [50]. However, in many cases, these external rewards do not increase players' intrinsic motivation [25], and in some cases have been shown to not improve performance [35]. For players to be optimally engaged—that is, making the game intrinsically motivating—there should be a clear goal, players should feel in control of the outcome and receive regular feedback on their actions, and the task should be neither too easy nor too difficult [15,34]. Another problem might be reproducibility. Humans have a wide range of individual behaviours, and, the more varieties of behavior possible in a task, the less likely it might be that players would follow similar behavior over several trials. Players might also start with widely different prior skills on a task; some players might be more familiar with video games than others. Also, personality differences might play a role in the game. For example, players that are more socially dominant might be less inclined to follow other players. These variations can be assessed with questionnaires prior to the experiment [16]. Human participants that are recruited in one experiment setting are also likely to have a mostly homogeneous cultural background, which has been shown to have a substantial impact on behavior [10]. Interesting changes in behaviors might be explored when players from multiple backgrounds collaborate in a game. Finally, recruiting participants might be a challenge. If experiments are conducted in one shared computer client room, there is the advantage of control and overview of participant behavior (cf. [53]). If we alternatively run HuGoS as an online browser game, it might make it easier to recruit large numbers of participants, but can make participant behavior less controllable [6,49].

5 Conclusions

We have designed and presented HuGoS, a multi-user virtual environment that supports the study of human–human interactions and group behaviors relevant to the topic of swarm intelligence. HuGoS supports collection of the primary data types required to analyze relevant aspects of human behavior. The environment's flexibility allows for implementation of a wide variety of swarm intelligence scenarios. We have demonstrated three simple cases of such scenarios, demonstrating support for: 1) studying human behavior in tasks typically studied in artificial swarms, such as best-of-n collective decision making; 2) studying new behaviors that may be especially relevant to human collective behavior, such as the self-organization of hierarchical social structures; and 3) studying direct comparisons between human swarms and artificial swarms, as well as interaction between human swarm agents and artificial swarm agents, such as robots.

Acknowledgements. This work was partially supported by the program of Concerted Research Actions (ARC) of the Université libre de Bruxelles. M.K. Heinrich, A. Cleeremans and M. Dorigo acknowledge support from the F.R.S.-FNRS, of which they are, respectively, postdoctoral researcher and research directors.

References

1. Bailenson, J.N., Beall, A.C., Loomis, J., Blascovich, J., Turk, M.: Transformed social interaction: decoupling representation from behavior and form in collaborative virtual environments. Presence Teleoperators Virtual Environ. **13**(4), 428–441 (2004). https://doi.org/10.1162/1054746041944803
2. Barrington, L., et al.: Crowdsourcing earthquake damage assessment using remote sensing imagery. Ann. Geophys. **54**(6) (2011). https://doi.org/10.4401/ag-5324
3. Blascovich, J., Loomis, J., Beall, A., Swinth, K., Hoyt, C., Bailenson, J.: Immersive virtual environment technology as a methodological tool for social psychology. Psychol. Inq. **13**, 103–124 (2002)
4. Boos, M., Pritz, J., Lange, S., Belz, M.: Leadership in moving human groups. PLoS Comput. Biol. **10**(4), e1003541 (2014). https://doi.org/10.1371/journal.pcbi.1003541
5. Cheng, J.T., Tracy, J.L., Foulsham, T., Kingstone, A., Henrich, J.: Two ways to the top: evidence that dominance and prestige are distinct yet viable avenues to social rank and influence. J. Pers. Soc. Psychol. **104**(1), 103–125 (2013). https://doi.org/10.1037/a0030398
6. Cooper, S., et al.: Predicting protein structures with a multiplayer online game. Nature **466**(7307), 756–760 (2010). https://doi.org/10.1038/nature09304
7. De Montjoye, Y.A., Stopczynski, A., Shmueli, E., Pentland, A., Lehmann, S.: The strength of the strongest ties in collaborative problem solving. Sci. Rep. **4**, 5277 (2014)
8. Ducatelle, F., et al.: Cooperative navigation in robotic swarms. Swarm Intell. **8**(1), 1–33 (2013). https://doi.org/10.1007/s11721-013-0089-4
9. Eberhart, R., Palmer, D., Kirschenbaum, M.: Beyond computational intelligence: blended intelligence. In: 2015 Swarm/Human Blended Intelligence Workshop (SHBI). IEEE (2015). https://doi.org/10.1109/shbi.2015.7321679
10. Henrich, J., Heine, S.J., Norenzayan, A.: The weirdest people in the world? Behav. Brain Sci. **33**(2–3), 61–83 (2010). https://doi.org/10.1017/s0140525x0999152x
11. Heyes, C.: Who knows? Metacognitive social learning strategies. Trends Cogn. Sci. **20**(3), 204–213 (2016). https://doi.org/10.1016/j.tics.2015.12.007
12. Hunt, E.R., Jones, S., Hauert, S.: Testing the limits of pheromone stigmergy in high-density robot swarms. Roy. Soc. Open Sci. **6**(11), 190225 (2019). https://doi.org/10.1098/rsos.190225
13. Ioannou, C.C.: Swarm intelligence in fish? The difficulty in demonstrating distributed and self-organised collective intelligence in (some) animal groups. Behav. Process. **141**(2), 141–151 (2017)
14. Juliani, A., et al.: Unity: a general platform for intelligent agents. arXiv preprint arXiv:1809.02627 (2018). https://arxiv.org/pdf/1809.02627.pdf
15. Jung, J.H., Schneider, C., Valacich, J.: Enhancing the motivational affordance of information systems: the effects of real-time performance feedback and goal setting in group collaboration environments. Manage. Sci. **56**(4), 724–742 (2010). https://doi.org/10.1287/mnsc.1090.1129
16. Kalma, A.P., Visser, L., Peeters, A.: Sociable and aggressive dominance: personality differences in leadership style? Leadersh. Quart. **4**(1), 45–64 (1993). https://doi.org/10.1016/1048-9843(93)90003-c
17. Kendal, R.L., Boogert, N.J., Rendell, L., Laland, K.N., Webster, M., Jones, P.L.: Social learning strategies: bridge-building between fields. Trends Cogn. Sci. **22**(7), 651–665 (2018). https://doi.org/10.1016/j.tics.2018.04.003

18. Kirschenbaum, M., Palmer, D.W.: Perceptualization of particle swarm optimization. In: 2015 Swarm/Human Blended Intelligence Workshop (SHBI). IEEE (2015). https://doi.org/10.1109/shbi.2015.7321681
19. Krafft, P.M., et al.: Human collective intelligence as distributed Bayesian inference. arXiv preprint arXiv:1608.01987 (2016). https://arxiv.org/pdf/1608.01987.pdf
20. Krause, J., Ruxton, G.D., Krause, S.: Swarm intelligence in animals and humans. Trends in Ecol. Evol. 25(1), 28–34 (2010). https://doi.org/10.1016/j.tree.2009.06.016
21. Kurvers, R.H.J.M., Wolf, M., Naguib, M., Krause, J.: Self-organized flexible leadership promotes collective intelligence in human groups. Roy. Soc. Open Sci. 2(12), 150222 (2015). https://doi.org/10.1098/rsos.150222
22. Lepri, B., Staiano, J., Shmueli, E., Pianesi, F., Pentland, A.: The role of personality in shaping social networks and mediating behavioral change. User Model. User-Adap. Interact. 26(2–3), 143–175 (2016). https://doi.org/10.1007/s11257-016-9173-y
23. Lin, A.Y.M., Huynh, A., Lanckriet, G., Barrington, L.: Crowdsourcing the unknown: the satellite search for Genghis Khan. PLoS ONE 9(12), e114046 (2014). https://doi.org/10.1371/journal.pone.0114046
24. Mathews, N., Christensen, A.L., O'Grady, R., Mondada, F., Dorigo, M.: Mergeable nervous systems for robots. Nat. Commun. 8(439) (2017). https://doi.org/10.1038/s41467-017-00109-2
25. Mekler, E.D., Brühlmann, F., Tuch, A.N., Opwis, K.: Towards understanding the effects of individual gamification elements on intrinsic motivation and performance. Comput. Hum. Behav. 71, 525–534 (2017). https://doi.org/10.1016/j.chb.2015.08.048
26. Metcalf, L., Askay, D.A., Rosenberg, L.B.: Keeping humans in the loop: pooling knowledge through artificial swarm intelligence to improve business decision making. Calif. Manage. Rev. 61(4), 84–109 (2019)
27. Michel, O.: Cyberbotics Ltd., Webots^{TM}: professional mobile robot simulation. Int. J. Adv. Robot. Syst. 1(1), 40–43 (2004)
28. Millard, A.G., et al.: The Pi-puck extension board: a Raspberry Pi interface for the e-puck robot platform. In: 2017 IEEE/RSJ International Conference on Intelligent Robots and Systems (IROS), pp. 741–748. IEEE (2017)
29. Mondada, F., et al.: The e-puck, a robot designed for education in engineering. In: Proceedings of the 9th Conference on Autonomous Robot Systems and Competitions, vol. 1, pp. 59–65. IPCB: Instituto Politécnico de Castelo Branco (2009)
30. Montes de Oca, M.A., Ferrante, E., Scheidler, A., Pinciroli, C., Birattari, M., Dorigo, M.: Majority-rule opinion dynamics with differential latency: a mechanism for self-organized collective decision-making. Swarm Intell. 5(3–4), 305–327 (2011). https://doi.org/10.1007/s11721-011-0062-z
31. Moussaïd, M., Helbing, D., Garnier, S., Johansson, A., Combe, M., Theraulaz, G.: Experimental study of the behavioural mechanisms underlying self-organization in human crowds. Proc. Roy. Soc. B Biol. Sci. 276(1668), 2755–2762 (2009)
32. Moussaïd, M., et al.: Crowd behaviour during high-stress evacuations in an immersive virtual environment. J. Roy. Soc. Interface 13(122), 20160414 (2016). https://doi.org/10.1098/rsif.2016.0414
33. Mulders, D., De Bodt, C., Bjelland, J., Pentland, A., Verleysen, M., de Montjoye, Y.A.: Inference of node attributes from social network assortativity. Neural Comput. Appl. 1–21 (2019). https://doi.org/10.1007/s00521-018-03967-z

34. Nakamura, J., Csikszentmihalyi, M.: The concept of flow. Flow and the Foundations of Positive Psychology, pp. 239–263. Springer, Dordrecht (2014). https://doi.org/10.1007/978-94-017-9088-8_16

35. Pedersen, M.K., Rasmussen, N.R., Sherson, J.F., Basaiawmoit, R.V.: Leaderboard effects on player performance in a citizen science game. In: Proceedings of the 11th European Conference on Game Based Learning, vol. 531 (2017)

36. Pinciroli, C., Talamali, M.S., Reina, A., Marshall, J.A.R., Trianni, V.: Simulating Kilobots within ARGoS: models and experimental validation. In: Dorigo, M., Birattari, M., Blum, C., Christensen, A.L., Reina, A., Trianni, V. (eds.) ANTS 2018. LNCS, vol. 11172, pp. 176–187. Springer, Cham (2018). https://doi.org/10.1007/978-3-030-00533-7_14

37. Pinciroli, C., et al.: ARGoS: a modular, parallel, multi-engine simulator for multi-robot systems. Swarm Intell. **6**(4), 271–295 (2012). https://doi.org/10.1007/s11721-012-0072-5

38. Quigley, M., et al.: ROS: an open-source robot operating system. In: ICRA Workshop on Open Source Software, vol. 3, p. 5, Kobe, Japan (2009)

39. Quinn, A.J., Bederson, B.B.: Human computation: a survey and taxonomy of a growing field. In: Proceedings of the International Conference on Human Factors in Computing Systems (2011)

40. Rosenberg, L., Baltaxe, D., Pescetelli, N.: Crowds vs swarms, a comparison of intelligence. In: 2016 Swarm/Human Blended Intelligence Workshop (SHBI). IEEE (2016). https://doi.org/10.1109/shbi.2016.7780278

41. Rosenberg, L.B.: Human swarms, a real-time method for collective intelligence. In: 20/07/2015–24/07/2015. The MIT Press (2015). https://doi.org/10.7551/978-0-262-33027-5-ch117

42. Rubenstein, M., Cornejo, A., Nagpal, R.: Programmable self-assembly in a thousand-robot swarm. Science **345**(6198), 795–799 (2014). https://doi.org/10.1126/science.1254295

43. Sørensen, J.J.W.H., et al.: Exploring the quantum speed limit with computer games. Nature **532**(7598), 210–213 (2016). https://doi.org/10.1038/nature17620

44. Sørensen, J.J.W., et al.: Exploring the quantum speed limit with computer games. Nature **532**(7598), 210–213 (2016)

45. Thrash, T., et al.: Evaluation of control interfaces for desktop virtual environments. Presence Teleoperators Virtual Environ. **24**(4), 322–334 (2015). https://doi.org/10.1162/pres_a_00237

46. Valentini, G., Ferrante, E., Dorigo, M.: The best-of-n problem in robot swarms: formalization, state of the art, and novel perspectives. Front. Robot. AI **4** (2017). https://doi.org/10.3389/frobt.2017.00009

47. Valentini, G., Hamann, H., Dorigo, M.: Self-organized collective decision-making in a 100-robot swarm. In: Proceedings of the Twenty-Ninth AAAI Conference on Artificial Intelligence (AAAI 2015), pp. 4216–4217. AAAI Press (2015)

48. Vasile, C., Pavel, A., Buiu, C.: Integrating human swarm interaction in a distributed robotic control system. In: 2011 IEEE International Conference on Automation Science and Engineering, pp. 743–748. IEEE (2011)

49. Walker, P., Amraii, S.A., Chakraborty, N., Lewis, M., Sycara, K.: Human control of robot swarms with dynamic leaders. In: 2014 IEEE/RSJ International Conference on Intelligent Robots and Systems, pp. 1108–1113. IEEE (2014)

50. Wang, H., Sun, C.T.: Game reward systems: gaming experiences and social meanings. In: Proceedings of DiGRA 2011 Conference: Think Design Play (2012)

51. Werfel, J., Petersen, K., Nagpal, R.: Designing collective behavior in a termite-inspired robot construction team. Science **343**(6172), 754–758 (2014). https://doi.org/10.1126/science.1245842
52. Whiten, A., Hinde, R.A., Laland, K.N., Stringer, C.B.: Culture evolves. Philos. Trans. Roy. Soc. B Biol. Sci. **366**(1567), 938–948 (2011). https://doi.org/10.1098/rstb.2010.0372
53. Zhao, H., et al.: A networked desktop virtual reality setup for decision science and navigation experiments with multiple participants. J. Vis. Exp. **138**(e58155) (2018). https://doi.org/10.3791/58155

Memory Induced Aggregation
in Collective Foraging

Johannes Nauta(✉)[iD], Pieter Simoens[iD], and Yara Khaluf[iD]

Department of Information Technology–IDLab, Ghent University–imec,
Ghent, Belgium
johannes.nauta@ugent.be

Abstract. Foraging for resources is critical to the survival of many animal species. When resources are scarce, individuals can benefit from interactions, effectively parallelizing the search process. Moreover, communication between conspecifics can result in aggregation around salient patches, rich in resources. However, individual foragers often have short communication ranges relative to the scale of the environment. Hence, formation of a global, collective memory is difficult since information transfer between foragers is suppressed. Despite this limitation, individual motion can enhance information transfer, and thus enable formation of a collective memory. In this work, we study the effect of individual motion on the aggregation characteristics of a collective system of foragers during collective foraging. Using an agent-based model, we show that aggregation around salient patches can occur through formation of collective memory realized through local interactions and global displacement using Lévy walks. We show that the Lévy parameter that defines individual dynamics, and a decision parameter that defines the balance between exploration and exploitation, greatly influences the macroscopic aggregation characteristics. When individuals prefer exploration, global aggregation around a single patch occurs when explorative bouts are relatively short. In contrast, when individuals tend to exploit the collective memory, explorative bouts should be longer for global aggregation to occur. Local aggregation emerges when exploration is suppressed, regardless of the value of the decision parameter.

1 Introduction

Foraging is a key aspect in the survival of many animals, in both individual as well as collective systems. Besides its importance in ecology, foraging has been an inspiration source for engineers and researchers to design systems that can solve similar tasks in parallel. Therefore, designing foraging behaviors and understanding the underlying dynamics of natural foraging has always been of high interest. Here, we focus on analyzing the impact of individual behavior on aggregation within a collective system during a foraging task.

In foraging, the distribution over the available resources within the environment is unknown. Hence, foragers must resort to random searches. Whereas

© Springer Nature Switzerland AG 2020
M. Dorigo et al. (Eds.): ANTS 2020, LNCS 12421, pp. 176–189, 2020.
https://doi.org/10.1007/978-3-030-60376-2_14

individual random searches can be optimized over a wide variety of constraints and resource distributions [3, 14, 15, 38, 47, 48, 50, 51, 53], a collective system might benefit from interactions between individuals to enhance the efficiency of the search process [5, 13, 23, 34]. In many environments, resources are both sparse and patchy [19, 20, 24, 49], making locating the patches rich in resources the primary goal in the random search process. Hence, a collective system can benefit from the parallel search process to quickly locate patches. Moreover, attraction towards successful foragers can lead to aggregation within the patches [17, 28], thereby increasing foraging success within a collective system.

When individuals are capable of learning the spatial distribution over resources, the importance of the random search diminishes in favour of informed movement [7, 12, 15, 29, 30]. Thus, a trade-off between exploration and exploitation arises [1, 13, 18]. Especially in cases where resources are ephemeral, this trade-off is not trivially solved and requires constant adaptation of the collective system. Also, when resources are scarce, aggregation strategies can emerge [4, 45], indicating that specific circumstances increase the importance of interactions between foragers and can bring about collective behavior in otherwise individually acting foragers. Interestingly, more interactions do not necessarily equate to more efficient foraging. It has been shown that both an excess and a lack of information, stemming from large respectively small communication radii, result in less optimal searches when compared to intermediate communication ranges [27, 28]. In collective systems, the spread of information is a crucial aspect that influences the systems' ability to build global knowledge, and hence can enable the collective system to act as a single unit. Both interactions between foragers and individual motion influence this information spread substantially [22].

While many animals have evolved intricate means of communication over relatively long distances (e.g. [36]), a typical feature of swarms is their limited communication range [8, 16, 21, 40]. Previous studies on collective foraging have considered extremely large communication ranges [5, 13], hampering potential of application towards swarm robotics or animals with limited communication ranges. Other approaches have considered nests as aggregation sites wherein information is passed on to conspecifics, indicative of social insects such as ants [32, 43] and bees [46]. However, many animal groups do not act as a single unit, since behavior is often influenced by individual preferences, thus not necessitating cooperative behavior [10]. Additionally, from an engineering perspective, selective pressure can additionally decrease the benefit of nests since travels towards the nest for communication can incur loss of time or energy. Our approach introduces localized group formation, which avoids the need for a nesting site when communication ranges are limited.

In such localized groups, information can be easily shared even if communication ranges are limited. However, since formation of a global knowledge requires a system to be well-mixed, i.e. each individual has an equal probability to encounter any other individual in the system, motility patterns greatly influence the macroscopic behavior of the collective system. While there exists a vast body of work on motility patterns for individuals (e.g. [2,6,31,47]), previous studies regarding collective systems have either restricted motion to walks on a lattice [5,13] or have considered simplified dynamics [39,44], hence omitting the importance of more intricate motility patterns on the spread of information. Nevertheless, both target detection and the mixing properties of the collective system are highly dependent on the individual dynamics of the foragers, and thus should be included when studying collective foraging.

This work aims to highlight the importance of motility for transferring information, specifically in a foraging setting. We focus on the impact of individual motion on the macroscopic behavior of the collective system during collective foraging, where individuals have limited communication and perception ranges. More specifically, we study foraging within a patchy environment, and show that aggregation around salient resource rich patches occurs due to local interaction between foragers. To this end, we develop an agent-based dynamical model that enables formation of a collective memory based on the number of collective visits to a patch. In turn, the number of visits is directly influenced by the perceived quality of the patch, resulting in aggregation around the most salient patches. Moreover, we show that differences in individual motion result in vastly different macro behavior of the collective system. Depending on a decision parameter that controls how much a forager exploits or explores, we show that the type of motion heavily influences the aggregation properties. Global aggregation in exploitative foragers occurs when the explorative bouts are relatively long, while the opposite is true for explorative foragers. Hence, the impact of the individual motion depends on the decision parameter of the collective system, but can nonetheless result in global aggregation even when the communication ranges are limited.

2   Methods

2.1   Environment

Let us consider a two-dimensional $L \times L$ environment with periodic boundaries. The periodic boundaries reflect an environment much larger than the forager. It allows us to study the collective behavior in isolation of other, more invasive boundary effects. Within the environment, a total of M patches are distributed randomly, where each patch carries a weight $w_i \in [0, w_{max}]$, with $w_{max} < 1$. Similar to [13], the weight corresponds to the quality of the patch, and acts as the probability of staying on that patch whenever a decision is made by the forager. Hence, high quality patches ($w \approx 1$) encourage the forager to stay on that patch, in contrast to low quality patches ($w \approx 0$).

The patch sparsity greatly influences the macroscopic behavior of the collective system. In particular, when patch density is high, attraction to other successful foragers carries less significance, since more foragers are successful just by having a higher probability of patch detection. In contrast, sparse patch distributions have been shown to increase the benefits of communication between individuals [5]. Moreover, collective systems in such environments have been shown to display spatial aggregation [45], even evolving attraction towards conspecifics when famine was introduced to originally solitary foragers [4]. In this work, we aim to study aggregation during foraging, and thus focus on environments wherein patches are sparse.

2.2 Foragers

We consider a collective system consisting of N foragers with random initial positions. Our main focus is on the large-scale features of the collective foraging task, and hence we do not account for finite-size effects such as collision avoidance. This can be justified by reinstating that the scale of the environment is much larger than the individual scale, and that we can assume the patches to be large enough to contain all N foragers within a single patch. Foragers can detect patches within a detection radius R, and communicate with others within a communication radius $r > R$. Every forager has access to memory consisting of two components: (i) the location of patches detected by the forager, and (ii) the total number of visits (both individual and from others) to those particular patches. Both memory components are built through patch detection and communication with others. The resulting collective memory weights patches according to the total number of visits to that patch by the collective [13].

We discretize time, and let each forager move during a time step $t \rightarrow t+1$ according to the following rules:

(i) *Feeding.* If the forager is on a patch, it stays on that patch with a probability equal to the patch quality, i.e. the weight w_i. Note that $w_i = 0$ if the forager is not on a patch.

(ii) *Memory mode.* With probability $(1 - w_i)q$, the forager receives information from others in its communication radius r (if any), accumulates the memories and combines it with its own memory, and samples a goal patch. The probability of sampling a particular goal patch is proportional to the number of times that patch was visited by all foragers (including itself) within range r at time step t. A travel angle θ is computed through the relative position between the forager and the sampled goal state. Then, the forager travels along the computed angle with steps of fixed length ℓ_0, until patch detection, completing the step.

(iii) *Random walk mode.* With probability $(1 - w_i)(1 - q)$, the forager executes a random walk. Here, we consider each forager executing a Lévy walk, where walk distances are sampled from a truncated inverse power-law

$$p(\ell) = \begin{cases} Z\ell^{-\alpha} & \text{for } \ell_0 < \ell \leq L \\ 0 & \text{otherwise} \end{cases}, \tag{1}$$

with Z the normalization constant and α the Lévy parameter (see below). The orientation angle θ is sampled uniformly between 0 and 2π. The forager travels along the sampled angle with steps of fixed length ℓ_0, completing the walk after traversing the sampled distance. The walk is interrupted when the forager detects a patch. The random search thus ends either when the sampled distance is traversed, or when a patch has been detected.

The above rules are similar to the decision rules from [13], but includes both a shared memory component and a more intricate motility pattern, allowing us to study the effects of motility on information sharing, leading to aggregation in patchy environments. The decision parameter q defines the probability to exploit (collective) memory, where low values of q correspond to highly explorative foragers, and high values of q indicate foragers that are much more likely to exploit their (and their neighbors') memory. The memory mode results in aggregation around the most visited patches. Since the visits are directly influenced by the patch quality w_i, this means aggregation is most likely to occur on salient patches. Essentially, the information that foragers have stored in their memory is broadcasted continuously. However, only when a forager enters memory mode, this information is actually used to accumulate the collective memory. Hence, the transfer of memory between foragers is directional, meaning that the receiving forager does not influence the memory of neighboring conspecifics. Additionally note that memory accumulation makes the forager permanently adopt the memory of its neighbors. In principle, when the communication radius encapsulates all foragers, i.e. $r = r_{max} = L/\sqrt{2}$, each individual adopts the collective memory as its own. Furthermore, the computation of the travel angle only considers relative positions, and hence does not assume global knowledge. Lack of global knowledge is indeed an important characteristic of swarms with limited perception ranges [8, 16].

The random walk that the individuals execute is a Lévy walk, truncated to reflect the periodic boundary conditions of the environment. The lower truncation at ℓ_0 and the upper truncation at the environment size L ensure that walks occur within the relevant scales [35]. This type of walk results in statistically long *flights* to occur, where a flight corresponds to walking in the same direction for a specific distance. The Lévy walk captures different motility patterns, ranging from ballistic (straight line) motion at $\alpha \to 1$, and approximating Brownian motion for $\alpha \geq 3$. Intermediate values of α alternate long flights with local (Brownian-like) motion, displaying scale-free behavior typical of power-laws [9, 52]. These long bouts of straight line motion are responsible for the optimization of random searches for sparse resources, with a known optimum around $\alpha \approx 2$ (see e.g. [3, 48, 50] and references therein). For a more detailed description, we refer the interested reader to previous works on Lévy walks in a foraging setting [47, 52].

2.3   Measuring Aggregation

While there exist many ways in which one can define cohesion within a collective system of individuals, there is no single established metric that describes aggregation over a wide range of settings. We introduce several metrics that measure aggregation, where combining the results allows us to draw conclusions about the macro behavior of the collective system.

First, we measure the average distance of each individual to the center of mass of the collective

$$\Delta_{com} = \frac{C}{N} \sum_{i=1}^{N} ||\boldsymbol{x}_i - \boldsymbol{c}||, \tag{2}$$

where $||\boldsymbol{x}_i - \boldsymbol{c}||$ the Euclidean distance between the position vector of forager i and the position vector $\boldsymbol{c}$ of the center of mass of the system, i.e. the mean position of the collective[1]. Hence, we measure how well aggregated the collective is around their collective center. The constant $C = \mathbb{E}(\Delta_{com})$ is the (numerically computed) expected value[2] that acts as a normalization constants that ensures that $0 \leq \Delta_{com} \leq 1$. Without any aggregation, $\Delta_{com} = 1$, i.e. the collective system at any point in time just reflects a uniformly distributed set of foragers (i.e. $\Delta_{com} = \mathbb{E}(\Delta_{com})$). Global aggregation on a single patch has $\Delta_{com} = 0$, indicating that every forager occupies the same position. However, note that this measure does not capture situations where subsets of the collective system aggregate on multiple patches, i.e. local aggregation.

Next, we measure the average fraction of number of neighbors each individual has as

$$\langle n \rangle = \frac{1}{N-1} \sum_{i=1}^{N} n_i(r), \tag{3}$$

where $N-1$ denotes the maximum number of neighbors possible, and $n_i(r)$ the number of neighbors of forager i within a communication radius r. Hence, when r encapsulates the entire environment, resulting in a static, fully connected network, $\langle n \rangle = 1$ by definition. Thus, $\langle n \rangle$ captures the size of the proximity network of each forager. However, while foragers can be outside of each others communication radius, they can still be part of the same connected component if they share a common neighbor. A connected component is defined as a collection of foragers for which each member has at least one neighbor, within its communication radius r, that is also a member. Hence, we additionally measure the size of the *giant component* G, being the connected component with the largest numbers of members. Thus,

$$G = \max_{k} g_k(r), \tag{4}$$

where g_k is the size, i.e. the number of members, of connected component k. When the system aggregates globally, the size of the connected component should grow towards the size of the collective system, i.e. $G = N$. In contrast, when each individual has no direct neighbors, i.e. $g_k(r) \to 1$ when $r \to 0$, all connected components consist of a single forager and thus $G = 1$.

Finally, we measure the number of connected components K. When no individual has a neighbor, the total number of connected components is equal the size of the collective system $K = N$, whereas aggregation on a single patch results in a single connected component, thus $K = 1$.

---

[1] The masses of each forager are equal and hence can be omitted.

[2] Computation of this expected value assumes a uniform distribution with the center of mass located at the center of the environment $\boldsymbol{c} = (L/2, L/2)$.

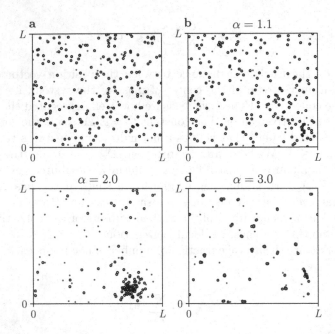

Fig. 1. Illustration of the dynamics for $N = 200$ foragers, with $q = 0.5$ and communication radius $r = 0.01L$. Other parameters are $L = 1000$, $M = 50$, and $w_{max} = 0.9$. (a) Initial positions of a single realization (seed). (b)–(d) Final positions after 10^6 steps for $\alpha = 1.1, 2.0, 3.0$. Black hollow circles indicate foragers, while small red circles are patch positions.

3 Results

We employ a Monte-Carlo (MC) scheme wherein the described dynamics are run after random initialization of the foragers and patches (see Fig. 1a). Initially, both $N = 200$ foragers, and $M = 50$ patches with $w_{max} = 0.9$, are uniformly distributed over the environment with size $L = 1000R$. Here, patches can be detected within a detection radius $R = 1$. We set the communication radius $r = 0.01L$, encouraging aggregation by ensuring that resources are sparse relative to the size of the collective system. The dynamics are run for $T = 10^6$ steps until equilibrium, in which the aforementioned measures are collected.

First, we discuss the MC scheme on the time dynamics of the aggregation. The initial positions (Fig. 1a), and the positions after the MC simulation (Fig. 1b–d) serve as an illustrative example. It highlights the influence of the Lévy parameter α on macroscopic behavior of the collective. Time dynamics of the measures are shown in Fig. 2. Low values $\alpha \to 1$, shown in Fig. 1b, do not result in any aggregation due to each individual executing long explorative bouts, both lowering patch detection probabilities as well as decreasing the frequency of memory mode usage. In this case, foragers are over-exploring and rarely exploit the collective component due to the suppression of communication between foragers.

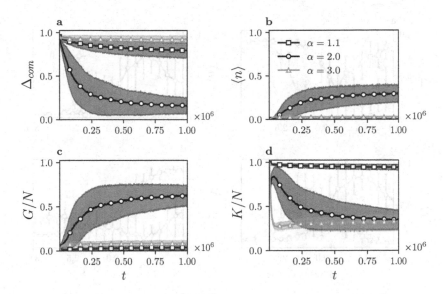

Fig. 2. Aggregation dynamics over time of measures of the collective with $q = 0.5$, and $\alpha = 2.0$. (a) Normalized mean distance towards center of mass. (b) The average fraction of neighbors (Eq. (3)). (c) Normalized number of the population within the giant component. (d) Normalized number of connected components within the collective. Shaded areas represent one standard deviation averaged over 100 realizations.

When $\alpha = 2.0$, we notice from Fig. 1c that global convergence around a salient patch occurs, since the size of the giant component grows to include over 60% of the total number of individuals while being restricted to a relative small area around the center of mass (see Fig. 2a,c). This is due to the alternation of long flights and local displacements when exploring. The long flights result in an increased coverage of the environment and result in patches being located more efficiently due to the optimization of the individual searches for patches [47,48]. The short, localized explorative bouts enables the individual foragers to enter the decision process more often, increasing the likelihood of exploiting memory. This stimulates communication between foragers, leading to formation of a collective memory which is depicted by global aggregation around a salient patch. The formation of a collective memory is additionally reflected in an increase in the number of neighbors $\langle n \rangle$ (see Fig. 2b), which indicates a well-mixed collective system wherein information is spread relatively well.

In contrast, when the individual motion approximates Brownian motion for $\alpha = 3.0$, they aggregate locally around detected salient patches. The dispersive qualities of the motion pattern lacks global displacements since heavy tails are suppressed. Hence, information about the current favored patch by each individual is unable to spread, hindering formation of a collective memory and resulting in the local aggregation depicted in Fig. 1d. Furthermore, the number

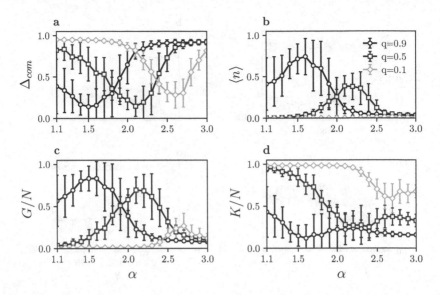

Fig. 3. Influence of Lévy parameter α and memory probability q on several metrics for foragers with limited communication radius $r = 0.01L$. Error bars indicate one standard deviation, averaged over 100 different realizations. (a) Normalized average distance towards the center of mass. (b) Average fraction of number of neighbors within communication radius r (see Eq. (3)). (c) Normalized size of the giant component (see Eq. (4)). (d) Normalized number of connected components.

of connected components decreases over time as individuals are more likely to adhere to a specific component over a long time due to the aforementioned lack of dispersive behavior (see also Fig. 2d). Surprisingly, the number of connected components is approximately equal to when $\alpha = 2.0$. This is because when α decreases, the length of explorative bouts of the foragers increase, and hence their tendency to leave a crowded patch increases (see the lone stragglers illustrated in Fig. 1c). Nonetheless, the fact that the average distance towards the center of mass, the number of neighbors and the size of the giant component do not drastically change over time, displays the lack of global consensus of the collective when their dispersivity is suppressed.

Next we extrapolate this example to capture the influence of both the Lévy parameter α and the exploitation probability q, on the aggregation dynamics of a collective system. Note that the detection radius $R = 1 \ll L$ is unchanged, and hence patch detection is difficult, incentivizing aggregation [13].

First, we notice that the value of α for which global aggregation is most prominent differs for each researched exploitation probability (see Fig. 3). When individuals are highly likely to exploit their own and their neighbors' memory, for $q = 0.9$, aggregation is most prominent for $\alpha \approx 1.5$. This is reflected in the minimization of Δ_{com} and the number of connected components K, and maximization of both $\langle n \rangle$ and the size of the giant component G. Note that the total number of connected components almost converges towards a fixed value when the heavy tail of the length of explorative bouts is suppressed for larger values of α. However, the accompanying increase in Δ_{com} illustrates that global aggregation does not occur at these values of α. Additionally, $\langle n \rangle$ and G decrease as α continues to increase, indicating that the number of connected components is minimized through the occurrence of local aggregation around salient patches. This result originates from the exploitative nature of the foragers, since they are much more likely to stay around a previously visited patch than to explore. This additionally limits the number of other foragers to communicate with, resulting in local aggregation.

When the collective is much more likely to explore instead of exploit their memory models, i.e. $q = 0.1$, we see that aggregation is less prominent, although it does still occur due to the likelihood of staying in a patch when the patch quality is high ($w \approx w_{max}$). Nonetheless, resulting from the memory accumulation of their local neighborhood, the increase in the number of neighbors is relatively small compared to when individuals tend more towards using their memory. This is moreover supported by the fact that foragers are simply less likely to decide to exploit their memory, and hence most foragers are continuously exploring. Therefore, when the length of the explorative bouts increases for low values of α, the memory component is suppressed and aggregation does not occur. Aggregation is strongest when the individual motion starts to resemble Brownian-like motion, for $\alpha \approx 2.6$. In this case, the smallest value of Δ_{com} coincides with the largest connected component. Similarly, $\langle n \rangle$ being largest coincides with the smallest number of connected components K, indicative of local aggregation around salient patches.

When individuals are equally likely to explore as to exploit, i.e. $q = 0.5$, we note that extreme values are obtained for $\alpha \approx 2.1$. The maximum number of neighbors is lower than when foragers exploit more for $q = 0.9$, due to the more explorative nature of individuals resulting in increased dispersive behavior. The minimum of Δ_{com} occurs at a higher value of α as q decreases. Together with the coinciding maxima of the size of the giant component, this indicates that individuals do not stray far from the center of mass during explorative bouts when $q = 0.5$. Effectively, as also illustrated in Fig. 1c, the majority of foragers stay close to the most salient patch. Distant patches are rarely visited by the collective system, since the collective memory effectively steers foragers back towards the center of mass located on the most salient patch. Similar to when $q = 0.9$, higher values of α for foragers with $q = 0.5$ do not display global aggregation, since Δ_{com} is maximized (Fig. 3a), and both $\langle n \rangle$ and G are minimized (Fig. 3b,c). Instead, local aggregation occurs, as depicted by the small number of connected components (see Fig. 3d) when α increases.

4   Conclusion

In this work, we have studied the aggregation dynamics of a collective system during a foraging task. By limiting the communication range between individuals, we isolated individual motion as the prime candidate for information transfer. Collective memory was formed as patches were visited using a Lévy walk for exploration. The memory strength towards a specific patch was related to the number of visits to the patch. During exploitation, the collective memory of nearby conspecifics was accumulated by the forager, and was used for informed motion towards a salient patch. We have shown that the proposed memory model can result in global aggregation around the most salient patches, even when communication ranges of individuals are limited. Furthermore, local aggregation occurs when the Lévy walk used for exploration tends more towards Brownian motion, $2 < \alpha < 3$. The value of α that results in the strongest aggregation depends on the value of the decision parameter q, which defines the probability of choosing to exploit the current available information when a decision is made. More exploitative foragers, $q = 0.9$, displayed the largest giant components around $\alpha \approx 1.5$, while simultaneously minimizing the distance towards the center of mass and the number of connected components, indicating global aggregation around a single salient patch. For explorative foragers, $q = 0.1$, global aggregation occurs at higher values of $\alpha \approx 2.6$, but is less pronounced. Balanced foragers with $q = 0.5$ display global aggregation around $\alpha \approx 2.1$, effectively displaying a middle ground between the two other extremes. Hence, we have shown that global aggregation can occur even when communication ranges are limited.

Whereas we have assumed that the probability to remain on a patch is proportional to the patch quality, this assumption implicitly states that patch quality is an objective measure. Realistically, this is simply not true, as the quality can be subjective or change over time depending on the needs of the system [11,25,41]. Hence, there is a dire need to research individual and collective assessment of patch quality. This should enable adaptation towards a priori unknown patches and allows a collective system to forage for the necessary resources.

Furthermore, in this study we have assumed that communication between individuals can only occur between neighboring foragers. However, studies have shown that animal groups can have different topologies within their social environment [33], such as scale free networks [26,37]. Indeed, when including long-range interactions, individual motion is perhaps less influential on the flow of information within the collective system. However, short communication ranges are a fundamental component in swarm robotics and cannot simply be omitted. Studying the importance of individual motion on different communication networks within collective systems is a topic for future study.

While we have studied the effects of individual motion and collective memory formation within the context of foraging, we believe that these results can be applied to more general concepts regarding spread of information within collective systems with limited communication ranges, such as robotic swarms [8] and active particles [42].

Acknowledgments. The authors would like to thank Ilja Rausch for useful discussions and providing invaluable resources specific to the domain.

References

1. Audibert, J.Y., Munos, R., Szepesvári, C.: Exploration-exploitation tradeoff using variance estimates in multi-armed bandits. Theoret. Comput. Sci. **410**(19), 1876–1902 (2009)
2. Bartumeus, F., Campos, D., Ryu, W.S., Lloret-Cabot, R., Méndez, V., Catalan, J.: Foraging success under uncertainty: search tradeoffs and optimal space use. Ecol. Lett. **19**(11), 1299–1313 (2016)
3. Bartumeus, F., da Luz, M.G.E., Viswanathan, G.M., Catalan, J.: Animal search strategies: a quantitative random-walk analysis. Ecology **86**(11), 3078–3087 (2005)
4. Bennati, S.: On the role of collective sensing and evolution in group formation. Swarm Intell. **12**(4), 267–282 (2018). https://doi.org/10.1007/s11721-018-0156-y
5. Bhattacharya, K., Vicsek, T.: Collective foraging in heterogeneous landscapes. J. Roy. Soc. Interface **11**(100), 20140674 (2014)
6. Boyer, D., Falcón-Cortés, A., Giuggioli, L., Majumdar, S.N.: Anderson-like localization transition of random walks with resetting. J. Stat. Mech. Theory Exp. **2019**(5), 053204 (2019)
7. Bracis, C., Gurarie, E., Van Moorter, B., Goodwin, R.A.: Memory effects on movement behavior in animal foraging. PloS One **10**(8), e0136057 (2015)
8. Brambilla, M., Ferrante, E., Birattari, M., Dorigo, M.: Swarm robotics: a review from the swarm engineering perspective. Swarm Intell. **7**(1), 1–41 (2013). https://doi.org/10.1007/s11721-012-0075-2
9. Clauset, A., Shalizi, C.R., Newman, M.E.J.: Power-law distributions in empirical data. SIAM Rev. **51**(4), 661–703 (2009)
10. Danchin, E., Giraldeau, L.A., Valone, T.J., Wagner, R.H.: Public information: from nosy neighbors to cultural evolution. Science **305**(5683), 487–491 (2004)
11. De Fine Licht, H.H., Boomsma, J.J.: Forage collection, substrate preparation, and diet composition in fungus-growing ants. Ecol. Entomol. **35**(3), 259–269 (2010)
12. Fagan, W.F., et al.: Spatial memory and animal movement. Ecol. Lett. **16**(10), 1316–1329 (2013)
13. Falcón-Cortés, A., Boyer, D., Ramos-Fernández, G.: Collective learning from individual experiences and information transfer during group foraging. J. Roy. Soc. Interface **16**(151), 20180803 (2019)
14. Faustino, C., Lyra, M., Raposo, E., Viswanathan, G., da Luz, M.: The universality class of random searches in critically scarce environments. EPL (Europhys. Lett.) **97**(5), 50005 (2012)
15. Ferreira, A., Raposo, E., Viswanathan, G., Da Luz, M.: The influence of the environment on Lévy random search efficiency: fractality and memory effects. Physica A **391**(11), 3234–3246 (2012)
16. Hamann, H.: Swarm Robotics: A Formal Approach. Springer, Cham (2018). https://doi.org/10.1007/978-3-319-74528-2
17. Haney, J.C., Fristrup, K.M., Lee, D.S.: Geometry of visual recruitment by seabirds to ephemeral foraging flocks. Ornis Scand. **23**, 49–62 (1992)
18. Katz, K., Naug, D.: Energetic state regulates the exploration-exploitation trade-off in honeybees. Behav. Ecol. **26**(4), 1045–1050 (2015)
19. Kéfi, S., et al.: Spatial vegetation patterns and imminent desertification in mediterranean arid ecosystems. Nature **449**(7159), 213 (2007)
20. Khaluf, Y., Ferrante, E., Simoens, P., Huepe, C.: Scale invariance in natural and artificial collective systems: a review. J. Roy. Soc. Interface **14**(136), 20170662 (2017)

21. Jimenez-Delgado, G., Balmaceda-Castro, N., Hernández-Palma, H., de la Hoz-Franco, E., García-Guiliany, J., Martinez-Ventura, J.: An integrated approach of multiple correspondences analysis (MCA) and fuzzy AHP method for occupational health and safety performance evaluation in the land cargo transportation. In: Duffy, V.G. (ed.) HCII 2019. LNCS, vol. 11581, pp. 433–457. Springer, Cham (2019). https://doi.org/10.1007/978-3-030-22216-1_32

22. Khaluf, Y., Simoens, P., Hamann, H.: The neglected pieces of designing collective decision-making processes. Front. Robot. AI **6**, 16 (2019)

23. Khaluf, Y., Van Havermaet, S., Simoens, P.: Collective Lévy walk for efficient exploration in unknown environments. In: Agre, G., van Genabith, J., Declerck, T. (eds.) AIMSA 2018. LNCS (LNAI), vol. 11089, pp. 260–264. Springer, Cham (2018). https://doi.org/10.1007/978-3-319-99344-7_24

24. Levin, S.A.: Multiple scales and the maintenance of biodiversity. Ecosystems **3**(6), 498–506 (2000). https://doi.org/10.1007/s100210000044

25. Lihoreau, M., et al.: Collective foraging in spatially complex nutritional environments. Philos. Trans. Roy. Soc. B **372**(1727), 20160238 (2017)

26. Lusseau, D., Newman, M.E.: Identifying the role that animals play in their social networks. Proc. R. Soc. Lond. B Biol. Sci. **271**(suppl_6), S477–S481 (2004)

27. Martínez-García, R., Calabrese, J.M., López, C.: Optimal search in interacting populations: Gaussian jumps versus Lévy flights. Phys. Rev. E **89**(3), 032718 (2014)

28. Martínez-García, R., Calabrese, J.M., Mueller, T., Olson, K.A., López, C.: Optimizing the search for resources by sharing information: Mongolian gazelles as a case study. Phys. Rev. Lett. **110**(24), 248106 (2013)

29. Menzel, R., et al.: Honey bees navigate according to a map-like spatial memory. Proc. Nat. Acad. Sci. **102**(8), 3040–3045 (2005)

30. Nauta, J., Khaluf, Y., Simoens, P.: Hybrid foraging in patchy environments using spatial memory. J. Roy. Soc. Interface **17**(166), 20200026 (2020)

31. Pemantle, R., et al.: A survey of random processes with reinforcement. Probab. Surv. **4**, 1–79 (2007)

32. Pinter-Wollman, N., et al.: Harvester ants use interactions to regulate forager activation and availability. Animal Behav. **86**(1), 197–207 (2013)

33. Pinter-Wollman, N., et al.: The dynamics of animal social networks: analytical, conceptual, and theoretical advances. Behav. Ecol. **25**(2), 242–255 (2014)

34. Pitcher, T., Magurran, A., Winfield, I.: Fish in larger shoals find food faster. Behav. Ecol. Sociobiol. **10**(2), 149–151 (1982). https://doi.org/10.1007/BF00300175

35. Pyke, G.H.: Understanding movements of organisms: it's time to abandon the Lévy foraging hypothesis. Methods Ecol. Evol. **6**(1), 1–16 (2015)

36. Ramos-Fernández, G.: Vocal communication in a fission-fusion society: do spider monkeys stay in touch with close associates? Int. J. Primatol. **26**(5), 1077–1092 (2005). https://doi.org/10.1007/s10764-005-6459-z

37. Ramos-Fernández, G., Boyer, D., Aureli, F., Vick, L.G.: Association networks in spider monkeys (Ateles geoffroyi). Behav. Ecol. Sociobiol. **63**(7), 999–1013 (2009). https://doi.org/10.1007/s00265-009-0719-4

38. Raposo, E.P., Buldyrev, S.V., da Luz, M.G.E., Santos, M.C., Stanley, H.E., Viswanathan, G.M.: Dynamical robustness of Lévy search strategies. Phys. Rev. Lett. **91**, 240601 (2003)

39. Rausch, I., Khaluf, Y., Simoens, P.: Scale-free features in collective robot foraging. Appl. Sci. **9**(13), 2667 (2019)

40. Rausch, I., Reina, A., Simoens, P., Khaluf, Y.: Coherent collective behaviour emerging from decentralised balancing of social feedback and noise. Swarm Intell. **13**(3–4), 321–345 (2019). https://doi.org/10.1007/s11721-019-00173-y

41. Rodrigues, M.A., et al.: Drosophila melanogaster larvae make nutritional choices that minimize developmental time. J. Insect Physiol. **81**, 69–80 (2015)
42. Romanczuk, P., Bär, M., Ebeling, W., Lindner, B., Schimansky-Geier, L.: Active Brownian particles. Eur. Phys. J. Spec. Top. **202**(1), 1–162 (2012). https://doi.org/10.1140/epjst/e2012-01529-y
43. Schafer, R.J., Holmes, S., Gordon, D.M.: Forager activation and food availability in harvester ants. Animal Behav. **71**(4), 815–822 (2006)
44. Talamali, M.S., Bose, T., Haire, M., Xu, X., Marshall, J.A., Reina, A.: Sophisticated collective foraging with minimalist agents: a swarm robotics test. Swarm Intell. **14**(1), 25–56 (2020). https://doi.org/10.1007/s11721-019-00176-9
45. Torney, C.J., Berdahl, A., Couzin, I.D.: Signalling and the evolution of cooperative foraging in dynamic environments. PLoS Comput. Biol. **7**(9), e1002194 (2011)
46. Visscher, P.K.: Group decision making in nest-site selection among social insects. Annu. Rev. Entomol. **52**(1), 255–275 (2007)
47. Viswanathan, G.M., Da Luz, M.G., Raposo, E.P., Stanley, H.E.: The Physics of Foraging: An Introduction to Random Searches and Biological Encounters. Cambridge University Press, Cambridge (2011)
48. Viswanathan, G.M., Buldyrev, S.V., Havlin, S., Da Luz, M., Raposo, E., Stanley, H.E.: Optimizing the success of random searches. Nature **401**(6756), 911 (1999)
49. Weimerskirch, H.: Are seabirds foraging for unpredictable resources? Deep Sea Res. Part II **54**(3), 211–223 (2007)
50. Wosniack, M.E., Santos, M.C., Raposo, E.P., Viswanathan, G.M., da Luz, M.G.E.: Robustness of optimal random searches in fragmented environments. Phys. Rev. E **91**, 052119 (2015)
51. Wosniack, M.E., Santos, M.C., Raposo, E.P., Viswanathan, G.M., da Luz, M.G.: The evolutionary origins of Lévy walk foraging. PLoS Comput. Biol. **13**(10), e1005774 (2017)
52. Zaburdaev, V., Denisov, S., Klafter, J.: Lévy walks. Rev. Mod. Phys. **87**(2), 483 (2015)
53. Zhao, K., et al.: Optimal Lévy-flight foraging in a finite landscape. J. Roy. Soc. Interface **12**(104), 20141158 (2015)

Modeling Pathfinding for Swarm Robotics

Sebastian Mai[(✉)][iD] and Sanaz Mostaghim[iD]

Faculty of Computer Science,
Otto von Guericke University Magdeburg, Magdeburg, Germany
{sebastian.mai,sanaz.mostaghim}@ovgu.de

Abstract. This paper presents a theoretical model for path planning in multi-robot navigation in swarm robotics. The plans for the paths are optimized using two objective functions, namely to maximize the safety distance between the agents and to minimize the mean time to complete a plan. The plans are designed for various vehicle models. The presented path planning model allows us to evaluate both decentralized and centralized planners. In this paper, we focus on decentralized planners and aim to find a set of Pareto-optimal plans, which enables us to investigate the fitness landscape of the problem. For solving the multi-objective problem, we design a modified version of NSGA-II algorithm with adapted operators to find sets of Pareto-optimal paths for several agents using various vehicle models and environments. Our experiments show that small problem instances can be solved well, while solving larger problems is not always possible due to the large complexity.

1 Introduction

Robot swarms and multi-robot systems in general become very popular and are used in various applications [2,14]. In most of such applications, navigation to certain positions and path planning for several robots within an environment pose a challenge for algorithm design. Currently robots rely on purely reactive behavior resulting in sub-optimal paths. The long term goal of our research is to extend the capabilities of the agents to plan future actions, taking into account the intend of their neighbors, to obtain more efficient behaviors. In order to optimally navigate within a moving robotic swarm, we need a framework for planning, evaluating and executing trajectories in a very dynamic environment.

In this paper we propose a model to represent trajectories for multiple robots within a swarm, respecting the kinematic constraints of the robots given by a specific vehicle model. We evaluate the plans by to objectives: Path length and safety radius. While the length of the paths indicate the time needed, the safety radius assesses how much (spatial) errors in plan-execution can be tolerated. In order to create a meaningful classification and comparison between future navigation behaviors we use multi-objective optimization to obtain a set of Pareto-optimal solutions that represent different trade-offs between both objectives. We can compare the obtained solutions either to the solution of another planning algorithm or

© Springer Nature Switzerland AG 2020
M. Dorigo et al. (Eds.): ANTS 2020, LNCS 12421, pp. 190–202, 2020.
https://doi.org/10.1007/978-3-030-60376-2_15

trajectories generated by a navigation policy in a simulation environment. Comparing to a set of Pareto-optimal solutions instead of a single solution allows us to see which trade-off a given method makes and whether its results are optimal with respect to a given trade-off. Thus, we gain more insights into the fitness landscape of the problem than by using single-objective optimization.

The remainder of this paper is structured as follows. In Sect. 2 we describe the state of the art planning algorithms for multi robot path planning. Our own model for describing plans and our approach to plan optimization is described in Sect. 3. The experiments we performed are described and evaluated in Sect. 4. The last section concludes the paper and outlines future research.

2   Related Works

In the following, we give an overview about different navigation and pathfinding algorithms for single and multi-robot scenarios. There is a long list of literature about such algorithms for one single robot e.g. [10,21]. New challenges appear once the single robot planning problem is extended to a multi-robot problem, where paths for multiple robots traveling through the same space are planned simultaneously. Collisions may occur when two robots' paths cross, but only if those robots travel through the crossing at the same time. This means a multi-robot planner must be aware of time, which is not necessarily the case for a single robot path planning. In addition, plans of multiple robots usually involve dependencies between the paths leading to the fact that solving a multi-robot planning problem is more complicated than solving multiple single robot problems. Furthermore, in single robot pathfinding we often assume that a solution must not contain a position/state twice. This assumption can help to bound the search space in single robot pathfinding, however, the assumption does not hold for the multi-robot case resulting in an infinitely large search space[1].

The Multi-Agent Pathfinding problem (MAPF) is a more general version of the problem modeled in this paper. The MAPF problem is described very well in a review by Stern et al. [16]. Many studies have been conducted around MAPF. The most common method to solve the problem is to use search-based solvers [8]. Surynek et al. [18] account for methods that convert the MAPF problem to a satisfiability (SAT) problem that can be solved by existing SAT solvers. In contrast to our method, this approach is more suited to problem instances like mazes, where it is difficult to find a valid solution. A new study by Wang and Rubenstein [19] aims to solve a special version of the MAPF problem by means of local interactions. Their approach is especially interesting because it could be considered to be a swarm approach to the MAPF problem. Our notion of safety/risk that is based on distance. In literature, multiple authors also consider the robustness of a plan. Here, robustness does not refer to robust optimization, but the property of a plan to tolerate errors in plan-execution. Atzmon et al. [1] recently published a work on robustness in MAPF. In contrast to our spacial model of

---

[1] A different technique to bound the search space is to plan only for a fixed time-frame, however, this means a new, problem-specific parameter needs to be introduced.

safety, the authors model robustness using temporal delays (k-robustness) and provide strategies to cope with those delays during plan execution. Olivierabreak et al. [12] show that robustness is a very important feature in a MAPF solution to a robotic context, as the inherent uncertainty in plan execution is likely to "break" plans with tight tolerances [12,17]. Street et al. [17] model this type of uncertainty as congestion and propose a congestion-aware planning algorithm. Concerning multiple objectives in path planning, we can refer to our previous work [20] where we establish multi-objective multi-agent pathfinding as a realistic, scalable benchmark for large-scale multi-objective optimization algorithms.

Usually, the MAPF problem is solved within a graph representation of the environment. In contrast to that, our model of the problem assumes a continuous space, where each robot has a position. Continuous space representations are also used by Krontiris et al. [9] and Čáp et al. [3]. Čáp et al. also use the notion of a minimal distance between robots [3].

In swarm robotics, navigation strategies usually are examined without taking a global perspective into account, in contrast to multi-robot systems. Silva and Nadia [15] present an approach using wave algorithms to gather sub-swarms that move in formations. In a recent article, Metoui et al. [11] use artificial potential fields and a decentralized architecture to solve a multi-robot navigation task. They use a kinematic model that is based on a continuous space representation of the environment. Several algorithms and various policies for collective behavior in multi-robot systems are listed by Rossi et al. [13].

3   Modeling Robot Navigation

In the following, we present our navigation model which encodes a set of trajectories M for k robots by a waypoint representation S. To decode a solution we use a vehicle model: a function $M = \mathcal{V}(S)$ that computes the paths from the waypoints. The vehicle model is an abstract representation of the kinematic constraints of the robots and is specific to the type of locomotion used by the robot. Figure 1 depicts a path $p(t)$ of a single robot that is represented by a sequence of $n = 4$ waypoints $W = (\boldsymbol{w}_1, \boldsymbol{w}_2, \cdots \boldsymbol{w}_n)$. A path $\boldsymbol{p}(t) : T \longrightarrow P$ for one robot is a function that maps the time step $t \in T$ to the pose $(x, y, \theta) \in P$ the robot occupies at time t, where the pose (x, y, θ) denotes the position and orientation of the robot. To compute $M = \mathcal{V}(S)$ we apply the vehicle model to each robot's sequence of waypoints independently: $p_i(t) = \mathcal{V}(W_i)$.

The quality of M is evaluated by two objective functions: **Risk** f_{R^*} and **Length** f_L. The proposed vehicle models and the objective functions are explained in the following sections.

3.1   Vehicle Models

Vehicle models provide a method to efficiently encode solution candidates that represent a robot's kinematic constraints by default. The vehicle model is used to compute a path-segment that connects two consecutive waypoints (w_i, w_{i+1}).

Fig. 1. Four waypoints (circles) with orientation (as indicated by dotted, blue arrows). The path (black line) was created by using the Bezier vehicle model. Start and end of the arrows indicate the control points of the Bezier curve for each segment of the path. (Color figure online)

The path from start w_0 to goal w_n is a concatenation of multiple segments. To show that our model is generic to different robot types we consider five vehicle models in this paper:

1. **Straight**: This vehicle model is the most simple one which connects the waypoints by linear movements and the angular components of the poses are ignored. During the movement θ points towards the direction of movement. The rotation to and from the start and goal poses happen without time delay. This is a theoretical model and does not represent any specific robotic system.
2. **Rotate Translate Rotate (RTR)**: The RTR model represents the motion of a robot able to turn in place, e.g. with a differential drive. It adds the rotational movement that is missing in the Straight model. The movement from the start position w_i consists of a rotation from θ_i to the orientation for straight movement θ_T (Rotate), straight movement to the next position (Translate) and rotation from θ_T to the goal orientation θ_{i+1} (Rotate).
3. **Dubins** vehicle model: This is a well known vehicle model for wheeled robots without the capability to turn in place or change directions. In Dubins vehicle model the robots always move forward. Dubins model is proven to compute the optimal path between two waypoints composed of straight lines and circle segments with fixed radius [7,10].
4. **Reeds-Shepp**: Similar to Dubins vehicle model, this model works by connecting straight lines and circle segments, but the robot is allowed to change directions (like a car when parking) [10].
5. **Bezier**: The last vehicle model uses Bezier curves to generate a path. This vehicle model does not correspond to a real-world kinematic model, and is used for comparison purposes. The vehicle model creates a Bezier curve with two control points associated to each path segment. The first control point is placed at a distance of B_d and angle θ from the waypoint w_i at the start of the segment, the second control point is placed at the same distance B_d and an angle $-\theta$ with respect to the waypoint at the end of the segment w_{i+1}. Figure 1 shows an example for a path with four waypoints that are used to generate three path segments.

3.2 Evaluating Plan Quality

In this section, we explain how we can evaluate a path in terms of two objectives: risk and length. **Risk** f_R^* is the objective that accounts for collisions between two agents and between agents and obstacles. To compute the risk objective we assume a circular robot that has to keep a safe distance towards obstacles and other robots at all times. Therefore, we consider a plan to be safer when the maximum distance to the closest agent or obstacle is bigger throughout each time step. In order to calculate the value for safety (f_R) for a plan for all agents, we consider the closest encounter of all agents with the environment d_E and the closest encounter of two agents towards each other d_A. d_E is calculated by finding the minimum distance d_o between all the paths $p(t) \in M$ and the obstacles in the environment. In an encounter between two agents the safety radius of both agents is affected, therefore the distance between two agents is counted with twice the weight, as the distance to the closest obstacle.

$$\text{Safety Radius: } f_R = \begin{cases} \min(\frac{1}{2}d_E, d_A) & \text{iff: } d_E > 0 \\ d_P & \text{iff: } d_E = 0 \end{cases} \tag{1}$$

$$\text{Environment Distance: } d_E = \min_{\forall p(t) \in M} d_o(p(t)) \tag{2}$$

$$\text{Inter-Agent Distance: } d_A = \min_{\forall i,j: i < j \leq k, \forall t} d(p_i(t), p_j(t)) \tag{3}$$

$$\text{Penalty Term: } d_P = -\frac{\sum_{i=1}^{k} \|\{t | \forall t : d_o(p_i(t)) = 0\}\|}{\sum_{i=1}^{k} L_i} \tag{4}$$

L_i is the length of a path p_i in terms of travel-time. Furthermore, we make some adjustments to the obtained objective value. In case the paths of the agents cross an obstacle, we assume that the value of the safety radius is not zero, but we use a negative penalty term d_P. This penalty helps the optimization algorithm to find obstacle-collision free paths more quickly. Our definition of safety radius f_R is a maximization objective. To convert f_R to a minimization function, we compute $f_R^* = 100 - f_R$, which represents the risk of a collision during plan execution. Additionally, we make the assumption that agents vanish as soon as they reach their goals. The value of f_R indicates the radius in which the circular robots can move and execute the plan without collisions, i.e. the small values result in critical plans. The safety radius helps the decision maker to select a solution from the Pareto front. Using the size of the robots, we can introduce a lower bound to the safety radius (f_R) of feasible solutions that can be included as a preference in multi-objective optimization.

The **length** f_L is the second objective. In MAPF, there are two possible ways to measure f_L: Makespan and Flow-time. Flow-time is the mean time of completion, while makespan is the time for the agent arriving latest. We use the flow-time objective (5), where L_i is the number of time steps in path i.

$$f_L = \frac{\sum_{i=1}^{k} L_i}{n} \tag{5}$$

3.3 Multi-objective Multi-path Planning

In this section, we explain how to perform the optimization algorithm to deal with the multi-objective problem of minimizing f_R and f_L. For this purpose, we modify the NSGA-II algorithm [6] to be able to solve our proposed problem.

As already mentioned, the encoded solution to the problem for k robots is stored in the set S and therefore, each individual in the NSGA-II algorithm represents a set of paths S_i which contains all k paths for k agents. For each waypoint in one path $p(t)$, we have (x, y, θ) at time t. In order to evaluate the quality of the solutions based on the above defined objective functions, we decode the solution S using a vehicle model $\mathcal{V}(S)$. We consider x and y to be defined in the space where the navigation is supposed to be defined and θ is set to a value in the range $[0, 2\pi]$. In case a waypoint is outside of the navigation area (e.g. after mutation), the path is penalized in the same way as if the robot moves through an obstacle and the angle is mapped to the correct domain.

In the following, we modify the genetic operators of the original NSGA-II algorithm to adapt to our own use-case. We consider two different types of **mutation**: A mutation for path smoothing and a Gaussian mutation on the encoded individuals. The mutation operator for path smoothing works as follows: One waypoint w_i in the waypoint vector is selected randomly and the path between the adjacent waypoints is computed as defined by the vehicle model $p_i(t) = \mathcal{V}((w_{i-1}, w_{i+1}))$. The selected waypoint is then placed randomly on the path $w_i := p_i(t \in (0, L_{p_i}))$ between the two adjacent waypoints. The Gaussian mutation adds $\mathcal{N}(0, \sigma_m)(d_{max} - d_{min})$ to each variable, where d_{min}, d_{max} are the minimum and maximum value the variable can assume within the domain.

As for the **crossover** operation, we use a two-point crossover on the paths of one randomly selected agent in two different solutions S_i, S_j. For this purpose, we first select a random agent A with the waypoints W_{iA} in solution S_i and W_{jA} in solution S_j. W_{iA} and W_{jA} are both n-tuples of poses that encode the path of the agent A in both solutions. We select two random cut-off points u and w with $0 < u < w < n$ and swap all poses with an index[2] between u and w between the two solutions. Thus, only the path of one agent is changed during crossover and a (sub-)sequence of the waypoints is swapped between solutions.

The solution of this multi-objective optimization algorithm is an approximation set of Pareto-optimal solutions, from which we need to select one possible solution. In order to reduce the number of alternatives and enforce a fast convergence, we introduce a constraint on the f_R objective as a preference for feasible solutions. We set a parameter f_R^{min}, because the safety radius must cover the footprint of the robot. When the safety radius is too small a collision will happen even with perfect plan execution. By introducing this constraint, only solutions are valid in which two robots never get closer than f_R^{min}. This constraint affects the environmental selection of the algorithm. In case there are several solutions with $f_R > f_R^{min}$, we only select feasible solutions. In case there are not enough feasible solutions, we use the solutions with the best values of f_R.

---

[2] Start- and goal configuration are fixed and never affected by the crossover operation.

4   Experiments

To validate our proposed approach, we perform several experiments on three maps of the environment (Cross, Bar and Empty) as shown in Fig. 3. We use the Python DEAP [4] framework to implement the problem and the modified NSGA-II algorithm [6] as described in Sect. 3. In all the experiments, the agents start from pre-defined fixed positions and are supposed to reach pre-defined goals. We set the following parameters for the mutation operators $\sigma_m = 0.01$, $p_g = 0.66$, $p_s = 0.33$, a mutation rate $p_{mut} = 0.8$ and a crossover rate $p_{cross} = 0.4$. The crossover rate is low compared to the usual settings for NSGA-II [6]. However, this delivers the best value for our specific use-case according to our preliminary experiments. f_R^{min} is part of the problem setting, we used a value of $f_R^{min} = 5$. All the experiments are performed for 31 independent runs[3] over 400 generations for each setting of the algorithm. In order to measure the quality of the obtained solutions of the modified NSGA-II, we record the hypervolume values [5] with respect to the reference point $h_{ref} = (f_R^* = 100, f_L = 400)$.

4.1   Evaluation

The first part of the evaluation is dedicated to the general properties of the optimization problem. The goal is to understand the movements and trajectories of robots with various vehicle models. Since the two vehicle models Straight and Bezier are theoretical models which do not represent any robotic system, we use them as baseline for comparisons. We study robots with differential drive represented by the RTR model and more restricted robots with non zero steering radius (e.g. cars) represented by the Reeds-Shepp model. Dubins vehicle model is a special case of the Reeds-Shepp model, where the robots are not allowed to change directions.

Figure 2 shows the obtained approximated solutions (combined non-dominated sets over 31 runs) for the five vehicle models. We fix the number of robots to 3. In all environments, we observe that, as expected, the theoretical Straight model obtains the best solutions. For each category of vehicle models, we observe that only Reeds-Shepp can reach very close to the theoretical models. An important detail is that the RTR model is always outperformed by the Straight model, because the cost of orientation changes is neglected by the Straight model. Considering additional restriction of the movement direction leads to deterioration in terms of both objectives. This can be observed by comparing the solutions of Reeds-Shepp and Dubins vehicle models, where Reeds-Shepp outperforms the Dubins vehicle.

Since Dubins vehicle represents most of the existing robotic systems and due to the space limit in this paper, we select Dubins vehicle models for further experiments[4]. Figure 3 shows two selected solutions generated by the algorithm

---

[3] We did exactly 31 runs, so the median fitness (and commonly used quantiles) correspond to a specific run.

[4] More vehicle models and source code: http://www.ci.ovgu.de/Research/Codes.html.

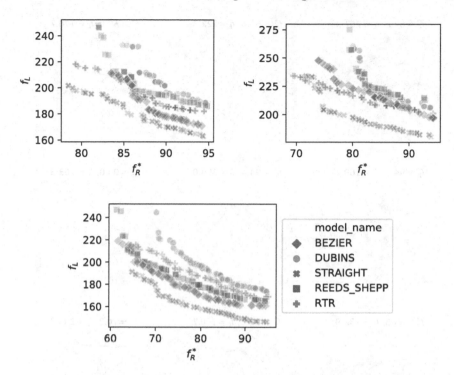

Fig. 2. Obtained solutions using various vehicle models in terms of f_L and f_R^* for three environments Cross (top left), Bar (top right) and Empty (bottom).

for Dubins vehicle model with $k = 3$ and 7 agents. These solutions represent the extreme points on the Pareto-front, i.e. each is the best in terms of one of the objectives. Our hypothesis is that solutions that perform well in the f_L objective, perform poorly in the f_R^* objective and vice versa. Hence, solutions with short, direct paths have a lower safety radius and solutions with a larger safety radius always have long paths. In Fig. 3 we can observe this trade-off. The solutions in the top row (best f_R^*) achieve low risk by keeping a larger distance to both the obstacles and other agents, while the solutions in the bottom row go very close to obstacles and agents. This holds for both $k = 3$ and $k = 7$. These agents shorten the path to their goal. Looking at the results with $k = 3$, we observe that the blue agent in the top left image takes a really long detour, it passes the gap between the obstacles only after the other agents moved through. However, the paths in the top middle- and right image are still rather direct and do not contain obvious detours.

In the following experiments, we analyse the impact of the number of agents and waypoints on the quality of the obtained paths. Table 1 shows the results in terms of HV for the three environments. We mark the best median hypervolume of the runs for each setting in bold font. We observe that the best result is reached by using one to three waypoints, indicating that adding more waypoints increases

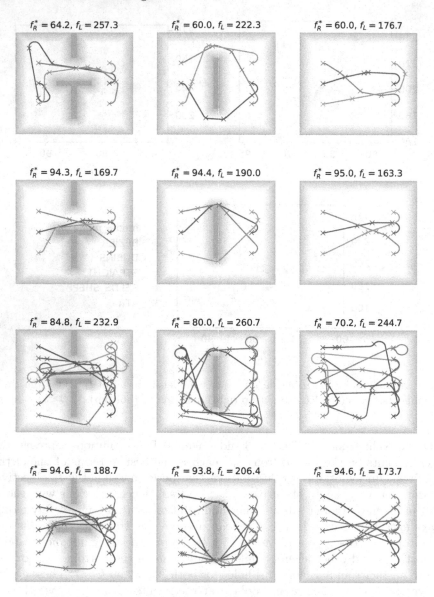

Fig. 3. Paths for $k = 3$ and 7 agents using Dubins vehicle model with $n = 3$ waypoints (denoted by ×) in environments: Cross, Bar, Empty (left to right).

the problem difficulty. Additionally, the solution with one waypoint is rarely the best solution, especially in the environments without obstacle – one waypoint is often not sufficient to express a good path. Overall, we conclude that the number of waypoints n directly scales the size of the search-space, without changing the underlying problem. In theory, when adding waypoints, the performance should

Table 1. Median HV and IQR for all runs at generation 400 with Dubins model.

env.	k / n	2	3	4	5	7	9	11
bar	1	7508(28)	7332(31)	**5953**(2211)	3417(1221)	1526(545)	749(923)	121(302)
	2	**7910**(11)	**7702**(40)	5635(1940)	4048(868)	2551(715)	**1590**(536)	**871**(1107)
	3	7889(28)	7692(1299)	5399(1919)	**4124**(1783)	**2573**(435)	1558(618)	671(917)
	4	7863(347)	6387(2029)	5051(1143)	3607(666)	2144(617)	844(1452)	0(619)
	5	7794(314)	5994(1916)	4253(912)	3188(1046)	1445(1722)	253(1016)	0(420)
cross	1	7674(644)	4685(79)	3979(1370)	2641(1728)	931(722)	312(722)	0(130)
	2	**9148**(27)	5055(479)	**4492**(425)	**3678**(562)	2050(457)	**1289**(1747)	0(961)
	3	9138(10)	**5730**(1014)	4472(492)	3541(642)	**2115**(720)	1183(1114)	**394**(866)
	4	9107(26)	5357(742)	4393(918)	3376(815)	1880(756)	494(1046)	0(0)
	5	9002(2888)	4971(954)	3662(1067)	2678(1316)	624(1527)	0(0)	0(0)
empty	1	9301(24)	**9261**(5)	**7538**(23)	8275(907)	4708(914)	3404(620)	2148(460)
	2	**9313**(24)	9260(13)	7525(104)	**8488**(633)	**5082**(552)	**3671**(649)	**2421**(435)
	3	9297(27)	9242(23)	7466(189)	8171(805)	4912(784)	3293(554)	2315(359)
	4	9283(31)	9226(45)	7207(210)	7613(880)	4572(827)	2997(451)	1859(301)
	5	9266(36)	9050(201)	7079(258)	6776(1083)	3934(571)	2298(549)	1358(475)

always stay the same or improve the result of the optimization, because the new waypoint can be on the path (resulting in the same path). However, the new waypoint can also help to express a solution that can not be represented by fewer waypoints. In the following, we study the influence of the number of waypoints in more detail. Figure 4 shows the Pareto-front for a different number of waypoints and 7 agents. The front computed with $n = 1$ does not achieve the full diversity, $n = 3$ clearly outperforms the front for $n = 1$ in convergence and diversity. The front for $n = 5$ has few very good solutions, but often is not able to converge far enough.

One remaining parameter to study concerns the number of agents k which is most important for the difficulty of the problem as can be observed in Figure 3. Referring to the results in Table 1, we can clearly see that the HV decreases as more agents are added to the problem. There are two reasons for this: (1) The HV of the true Pareto-front decreases when adding more agents. The reason concerns the safety radius objective f_R. As more agents are added to the same working-area in the environment f_R is impaired as there is less space per agent. (2) The algorithm is not able to find the correct solution due to increased task difficulty. Additionally, we observe that for the Empty environment, there are not large changes in HV when comparing the results for $k = 2$ and 3 agents (the same holds for 4 and 5 agents). We believe that this is an artifact of the distances in our start-configuration. In the Cross environment which has the most number of obstacles, we observe that the algorithm does often fail to converge for $k = 11$ agents. This illustrates that our algorithm is able to solve the problem for instances with few agents, but certainly not for tens our even hundreds of

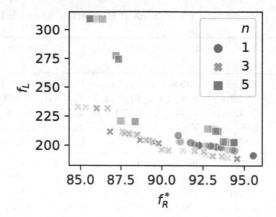

Fig. 4. Combined Pareto-fronts for Dubins vehicle using 7 agents.

agents. In addition, we believe that this could be due to the specified environment (size as well as start and goal positions for the agents). It is important to note that with eleven agents and three waypoints there are already $11 \cdot 3 \cdot 3 = 99$ variables to optimize, resulting in a high dimensional search-space. Overall, the results show that the algorithm is capable of finding solutions for small problem instances (i.e. $k \leq 10$).

5   Conclusion

In this paper, we proposed a mathematical model to represent and evaluate movement plans for a swarm of robots. In addition, we implemented an optimization algorithm based on NSGA-II [6] that is capable of computing a Pareto-optimal set of plans for a given navigation scenario.

Our experiments show that the algorithm is able to find valid plans, unless the environment gets too complex or the number of variables is raised. Our proposed model and the algorithm generalize well to different robot-types and vehicle models and can yield insight into the objective space of multi-robot path planning problems. Different navigation-policies that use our model can now be compared on how they are situated in the objective space. In the future, we also plan to expand and apply the model proposed in this paper. We will examine various start and goal positions and more complex problems with different sizes. In addition, we aim to implement the plan execution in a swarm of real robots (similar to [12]), to see how our method can deal with the uncertainty that is present in real robots.

References

1. Atzmon, D., Stern, R., Felner, A., Wagner, G., Barták, R., Zhou, N.F.: Robust multi-agent path finding and executing. J. Artif. Intell. Res. **67**, 549–579 (2020). https://doi.org/10.1613/jair.1.11734

2. Bayindir, L.: A review of swarm robotics tasks. Neurocomputing **172**, 292–321 (2016). https://doi.org/10.1016/j.neucom.2015.05.116
3. Čáp, M., Novák, P., Vokřínek, J., Pěchouček, M.: Multi-agent RRT*: sampling-based cooperative pathfinding. In: 12th International Conference on Autonomous Agents and Multiagent Systems 2013, AAMAS 2013 2, pp. 1263–1264 (2013)
4. De Rainville, F.M., Fortin, F.A., Gardner, M.A., Parizeau, M., Gagné, C.: DEAP: a python framework for evolutionary algorithms. In: Proceedings of the 14th Annual Conference Companion on Genetic and Evolutionary Computation, GECCO 2012, New York, pp. 85–92. ACM (2012). https://doi.org/10.1145/2330784.2330799
5. Deb, K.: Multi-Objective Optimization using Evolutionary Algorithms. Wiley, Hoboken (2001)
6. Deb, K., Pratap, A., Agarwal, S., Meyarivan, T.: A fast and elitist multiobjective genetic algorithm: NSGA-II. IEEE Trans. Evol. Comput. **6**(2), 182–197 (2002). https://doi.org/10.1109/4235.996017
7. Dubins, L.E.: On curves of minimal length with a constraint on average curvature, and with prescribed initial and terminal positions and tangents. Am. J. Math. **79**(3), 497–516 (1957)
8. Felner, A., et al.: Search-based optimal solvers for the multi-agent pathfinding problem: summary and challenges. In: Tenth Annual Symposium on Combinatorial Search (SoCS), pp. 29–37 (2017). https://www.aaai.org/ocs/index.php/SOCS/SOCS17/paper/view/15781
9. Krontiris, A., Sajid, Q., Bekris, K.E.: Towards using discrete multiagent pathfinding to address continuous problems. In: AAAI Workshop - Technical Report WS-12-10, pp. 26–31 (2012)
10. LaValle, S.M.: Planning algorithms (2006). https://doi.org/10.1017/CBO9780511546877
11. Metoui, F., Bouaanid, B., Abdelkrim, M.N.: Path planning for a multi robot system with decentralized control architecture. Stud. Syst. Decis. Control **270**, 229–259 (2020). https://doi.org/10.1007/978-981-15-1819-5_12
12. Oliveira, G.M.B., et al.: A cellular automata-based path-planning for a cooperative and decentralized team of robots. In: IEEE congress on Evolutionary Computation (CEC), pp. 739–746. IEEE (2019). https://doi.org/10.1109/cec.2019.8790205
13. Rossi, F., Bandyopadhyay, S., Wolf, M., Pavone, M.: Review of multi-agent algorithms for collective behavior: a structural taxonomy. IFAC-PapersOnLine **51**(12), 112–117 (2018). https://doi.org/10.1016/j.ifacol.2018.07.097
14. Schranz, M., Umlauft, M., Sende, M., Elmenreich, W.: Swarm robotic behaviors and current applications. Front. Robot. AI **7** (2020). https://doi.org/10.3389/frobt.2020.00036
15. Silva, L., Nedjah, N.: Efficient strategy for collective navigation control in swarm robotics. Procedia Comput. Sci. **80**, 814–823 (2016). https://doi.org/10.1016/j.procs.2016.05.371
16. Stern, R., et al.: Multi-agent pathfinding: definitions, variants, and benchmarks. ArXiv Preprint, June 2019. https://arxiv.org/abs/1906.08291
17. Street, C., Lacerda, B., Mühlig, M., Hawes, N.: Multi-robot planning under uncertainty with congestion-aware models. In: Proceedings of the 19th International Conference on Autonomous Agents and Multiagent Systems, p. 9 (2020)
18. Surynek, P., Felner, A., Stern, R., Boyarski, E.: An empirical comparison of the hardness of multi-agent path finding under the makespan and the sum of costs objectives. In: Proceedings of the 9th Annual Symposium on Combinatorial Search, SoCS 2016 (SoCS), pp. 145–146, January 2016

19. Wang, H., Rubenstein, M.: Walk, stop, count, and swap: decentralized multi-agent path finding with theoretical guarantees. IEEE Robot. Autom. Lett. 5(2), 1119–1126 (2020). https://doi.org/10.1109/LRA.2020.2967317
20. Weise, J., Mai, S., Zille, H., Mostaghim, S.: On the scalable multi-objective multi-agent pathfinding problem. In: Accepted at Congress on Evolutionary Computing CEC 2020 (2020)
21. Zafar, M.N., Mohanta, J.C.: Methodology for path planning and optimization of mobile robots: a review. Procedia Comput. Sci. **133**, 141–152 (2018). https://doi.org/10.1016/j.procs.2018.07.018

Motion Dynamics of Foragers in Honey Bee Colonies

Fernando Wario[1]([⊠]), Benjamin Wild[2], David Dormagen[2], Tim Landgraf[2], and Vito Trianni[1]

[1] ISTC, National Research Council, Rome, Italy
{fernando.wario,vito.trianni}@istc.cnr.it
[2] Department of Mathematics and Computer Science, Freie Universität Berlin, Berlin, Germany
{b.w,david.dormagen,tim.landgraf}@fu-berlin.de

Abstract. Information transfer among foragers is key for efficient allocation of work and adaptive responses within a honey bee colony. For information to spread quickly, foragers trying to recruit nestmates via the waggle dance (dancers) must reach as many other non-dancing foragers (followers) as possible. Forager bees may have different drives that influence their motion patterns. For instance, dancer bees need to widely cover the dance floor to recruit nestmates, the more broadly, the higher the food source profitability. Followers may instead move more erratically in the hope of meeting a dance. Overall, a good mixing of individuals is necessary to have flexibility at the level of the colony behavior and optimally respond to changing environmental conditions. We aim to determine the motion pattern that precedes communication events, exploiting a data-driven computational model. To this end, real observation data are used to define nest features such as the dance floor location, shape and size, as well as the foragers' population size and density distribution. All these characteristics highly correlate with the bees walking pattern and determine the efficiency of information transfer among bees. A simulation environment is deployed to test different mobility patterns and evaluate the adherence with available real-world data. Additionally, we determine under what conditions information transfer is most efficient and effective. Owing to the simulation results, we identify the most plausible mobility pattern to represent the available observations.

1 Introduction

Honey bee colonies, along with ant and termite colonies, are the best-known examples of superorganisms, social groups made up of members of the same species which display signs of self-organization and collective intelligence [16,22]. The honey bee foraging behavior has been intensely studied by the scientific community. Nevertheless, despite the general mechanisms underlying self-organization during foraging activities being well understood [11,17], there is still much to learn about the effects of individual differences among bees and how

© Springer Nature Switzerland AG 2020
M. Dorigo et al. (Eds.): ANTS 2020, LNCS 12421, pp. 203–215, 2020.
https://doi.org/10.1007/978-3-030-60376-2_16

such differences impact the overall behaviors. In particular, information transfer among foragers is a key aspect, as it determines how the colony flexibly modulates the workload and adapts to external contingencies and internal demands. For information transfer, a good mixing of individuals is necessary, and this is supported by the ability of workers to move and meet other workers carrying valuable information.

The goal of this study is to understand how the motion patterns of foragers influence information exchanges. More specifically, we focus on waggle dances, whereby foragers recruit nestmates to valuable food patches. We aim at identifying the features of the motion pattern followed by bees before dancing and following behaviors. To this end, real observational data are used to define fundamental environmental properties (comb surface characteristics) such as dance floor location, shape and size, as well as colony features such as forager population size and density distribution, all characteristics that highly correlate with the bees motion pattern [14]. Then, a simulation environment is deployed to test different mobility patterns for forager bees within the hive. To determine the mobility pattern of simulated bees, we assume that foragers may or may not take into account the detailed characteristics of the dance floor. A correlated random walk model [1,6] follows the assumption that only an approximate location for the dance floor is known to the bees, which is modelled as a location bias toward which bees turn with a fixed probability. Conversely, a random waypoint model [2]) follows the assumption that the location and dimensions of the dance floor are known, as the model postulates that displacements are determined by randomly sampling target locations within the relevant areas. From simulations, we also obtain the interaction rate among foragers, which shows under what conditions information transfer is most effective and efficient. On such a basis, can we shed light on the most plausible assumptions by matching real-world observations with the simulations resulting from different mobility patterns? Answering this question will provide interesting hypotheses for further studying the information transfer abilities among forager bees, and will also suggest design principles for the efficient implementation of swarm robotic systems.

In the following, we describe the methodology used to model the environmental properties of the hive and to test the selected mobility patterns in our simulation environment (see Sect. 2). Then, we present and evaluate the results obtained from simulations in Sect. 3. Finally, in Sect. 4 we discuss the plausibility of the different mobility patterns being presented, and propose as a follow up to this work a detailed comparison between our simulation results and real-world data at the single trajectory level.

2   Methodology

As mentioned above, this study is grounded on real-world data, which were obtained using the BeesBook System [18], an experimental system that allows tracking marked bees within an observation hive during weeks. The system is highly reliable, localizing markers with a 98% recall at 99% precision and decoding more than 98% of the markers correctly [3,21]. A BeesBook dataset consists of a list of registers detailing the position and identity of each marked bee—once detected—for each video frame during the full extension of the experimental season. Additionally, through the Waggle Dance Detector module [19], the Bees-Book system provides a record of dance activity that enumerates all detected dances, including duration and location on the honeycomb surface.

From this dataset, we obtained valuable information such as the spatial distribution of foragers and of the dances they performed, as well as the average speed of foragers. This information was used to define models for the dance floor and the density distribution of foragers on the honeycomb surface. This process is explained in the following section.

2.1 Data Preprocessing

The dataset used for the analysis was collected in 2016 during the months of July and August in Berlin, Germany. From this dataset, we considered 12 days between August the 8th and August the 19th, 2016. To focus our attention on the spatial distribution of foragers and dances, we first analyzed the local weather conditions—solar radiation, temperature and rainfall—during the experiment dates. These are known to impact foraging and, consequently, dancing activity [5,8]. From the analysis, we decided to limit the observation time window between 10:00 and 16:00 UTC+2, which covers the most favorable conditions for foraging and accordingly recorded the most relevant dance activity (see the top left panel in Fig. 1). Overall, we considered a dataset with a total of 72 h, that shows fairly similar activity across all days and hours.

2.2 Foragers and Dances Distribution Models

Once obtained the dataset for the analysis, we extracted the information about the distribution of foragers and dances over the comb, in order to obtain an empirical model. In the case of the dance distribution, we observed that, within the considered time window, the dance rate was substantially homogeneous (see top-right panel in Fig. 1). Hence, we focused on the spatial distribution only. We began by dividing the surface of the comb in 21×37 cells of ($1\,cm^2$) surface (the dimensions of the comb used during the recording season was $(21 \times 37\,cm)$). Then, we computed the total number of dances for each cell during a day, as well as over the full temporal extension of the dataset. As the differences in the spatial distribution across days were negligible, we focused on the cumulative distribution over all days. This cumulative distribution was then normalized to represent the probability of a dance occurring in each of the defined cells. Finally, we fitted a 2D Gaussian function using non-linear least squares method to define a model for the dance floor (Gaussian centred at $\mu_d = (8.17, 12.42)$ with standard deviation $\sigma_d = (4.92, 3.96)$ and with standard deviation errors of $[5.28e-2, 4.25e-2, 7.53e-2, 6.02e-2]$, see the inset in the left panel of Fig. 2)

For the foragers' spatial distribution, first foragers were identified based on their social interaction patterns and spatial distributions in the nest [20], then

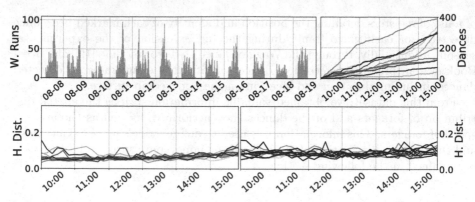

Fig. 1. Top left: Distribution of dances over multiple days. The number of waggle runs detected by the BeesBook system over a time interval of 1 min is displayed throughout the different days. Note that each dance consists of multiple waggle runs. Top right: Cumulative number of dances over different days. The cumulative distribution is approximately linear over the different days, indicating a constant rate of dances, although this rate differs from day to day. Bottom left: Hellinger distance between consecutive empirical density distributions during a day. Density distributions are computed over time intervals of 10 min, and compared to the previous time interval to show variations over time. Bottom right: Hellinger distance between empirical density distribution over 10-min intervals, and the overall model obtained from the whole dataset.

a grid over the image of the comb surface was defined. This grid, however, comprised of 46×70 cells, following the original structure of the BeesBook data-set. All the spatial parameters obtained for the model were later scaled properly. We then extracted the time series of the positions of all foragers, splitting the time series of each day into intervals of 10 min, and computing the cumulative distribution of foragers on the grid within each interval. These empirical distributions computed over these intervals were used to evaluate their homogeneity over time. To this end, we computed the similarity between distributions by means of the Hellinger distance [10]. The analysis shows that foragers density is fairly consistent during the selected window of time (see bottom-left panel in Fig. 1).

Similar to the dance distribution, once we validated the temporal homogeneity of the foragers' density, we computed the normalized-cumulative distribution over the full temporal extension of the dataset. Finally, to obtain a parametric model of the forager density, we fitted a 2D Gaussian function to the cumulative distribution using non-linear least squares method (Gaussian centred at $\mu_p = (8.73, 11.07)$ cm, with standard deviation $\sigma_p = (9.99, 5.60)$ cm and with standard deviation errors of $[7.62\mathrm{e}{-}2, 4.29\mathrm{e}{-}2, 1.20\mathrm{e}{-}2, 6.14\mathrm{e}{-}2]$, see Fig. 2 left). We computed the Hellinger distance between model and distributions of foragers over time to verify that the model was providing a good representation over and across different days (see bottom-right panel in Fig. 1). This analysis revealed that the model grasps sufficiently well the empirical density distribution of foragers extracted from the data across the full observation period.

Overall, the Gaussian models match reasonably well the real-world data, as shown in Fig. 2 left. In particular, the model for the density distribution of foragers matches visibly well to real-world data, also confirmed by the small Hellinger distance of 3.59e−2. Conversely, the Gaussian model for dances has a worse match to the observation data, mainly due to the absence of parts of the dances, as the observation camera for detecting dances was covering only the bottom-left section of the honeycomb close to the entrance. It is interesting to notice that the distribution of dances is similar to the density distribution, but shifted towards the entrance to the honeycomb. We hypothesize that such a shift stems from the fact that dances are executed right after the forager has returned from a foraging trip. Hence, they could be performed closer to the entrance as the forager trajectory starts there.

2.3   Multi-agent Simulations

We have built a simulation environment prepared to progressively incorporate the features derived from the real-world observations, and to test the effect of different mobility patterns. The virtual arena is customized after the dimensions of the honeycombs used during the experimental seasons (21×37 cm). Since our study focuses on mobility patterns that precede and follow dance communication activity, we only simulate the behavior of forager agents. For each simulation, we consider a forager population of 200 agents, that corresponds to the average number of foragers observed during the experimental season. We divided the forager population into two groups, dancers and followers. According to the literature [7], between 5% and 10% of the colony population engage in foraging activities, depending on the colony size and the resources available in the vicinity of the hive. For the colony studied during the experimental season and for the days we consider to define our models, around 35% of the colony population was identified as being part of the forager class. For simplicity, we consider in our simulations 20% of the forager population (or 7% of the colony population) as dancers (hence, at any time, we count 40 dancers and 160 followers). While both dancers and followers adopt the same mobility pattern and move over the comb surface at the same average speed (fixed to 5 mm/s in compatibility to the observation data), they display different behaviors concerning dance communication. Dancer agents are the only ones that can switch from move to dance state, during which they stay still in place and broadcast their known foraging site. The foraging site is not relevant for the present study, hence it is fixed and identical for all dancers. In order to reproduce the uniform distribution of dances over time observed on the experimental data, dancer agents stop and dance with a fixed probability per unit time p_d, which can be tuned to reproduce the rate observed experimentally in a given day. Considering that the simulation is advanced by one step every 0.25 s, to obtain in average 400 dances in a day we set $p_d = 1.16e-4$. Each dance event lasts a fixed amount of time (3 s).

Followers, on the other hand, continuously patrol their vicinity in search of dancing agents (<u>move</u> state). When they come close enough to an agent actively dancing (within 1 body length, i.e., 2 cm), they switch to the <u>follow</u> state, also standing still in place until the dancer ends its broadcast. If the interaction between dancer and follower(s) lasts long enough (>0.25 s), the communication is considered successful and the follower acquires knowledge of the foraging location communicated by the dancer.

The simulations employ the Gaussian model for the density distribution of foragers to determine the mobility pattern of the bees. At initialization, dancers are positioned at the bottom-left corner of the arena, which corresponds to the entrance to the hive, while followers are initialized at random positions on the comb surface following the density distribution model. For each mobility pattern, dancer agents evaluate at each step whether to dance or not, while followers stop only when they perceive a dance in their proximity, as specified above. After performing a dance, the agent is removed from the arena and a new dancer is introduced at the entrance. In this way, we mimic the behavior of dancer bees that leave the hive after unloading and communicating the foraging source to their nestmates [4,15]. Also, in this way we want to test the hypothesis that the dance distribution is shifted due to a bias in the starting position of the dancers' trajectories.

In this paper, we report the results for two different mobility patterns adapted to the density distribution model: a random waypoint model (RWM) and a biased correlated random walk (CRW). The former uses the foragers density model as a probability distribution function to draw intermediate location goals. The latter uses the estimated center of the dance floor as a bias for the random walk. The details are provided below. Other mobility patterns like Lévy walks [1] could be considered, which are however less suited for a constrained space like the beehive.

Random Waypoint Model (RWM). This mobility pattern allows agents to explore the whole arena by choosing a random destination and moving straight until the destination is reached. In our simulations, the choice of the new destination is proportional to the empirical density distribution of foragers. More specifically, each new destination is drawn randomly exploiting the 2D Gaussian model we obtained from real-world data. Whenever agents stop—to dance or to follow a dance—they lose memory of their previous destination, and a new one is drawn when motion is resumed. This mobility pattern assumes that foragers have some knowledge of their location over the honeycomb—i.e., a map—that they exploit to choose where to move next.

Biased Correlated Random Walk (CRW). A correlated random walk is the simplest mobility pattern that can be imagined for the bees, as well as for many biological and artificial systems [1,6,9]. With this model, agents alternate straight walks and random turns. In this simulations, the duration of the straight walks is sampled from a folded normal distribution $N(0, \sigma_w)$, with $\sigma_w = 0.75$ s.

The turning angle is instead drawn from a wrapped Cauchy distribution, characterized by the following probability density function:

$$f_C(\theta; \mu, \rho) = \frac{1 - \rho^2}{2\pi \left(1 + \rho^2 - 2\rho \cos\left(\theta - \mu\right)\right)}. \tag{1}$$

where μ represents the average and ρ the skewness of the distribution. In our simulation, $\mu = 0$ implies that the turning angle is correlated with the current direction of motion, while the parameter ρ is varied to control the degree of correlation of the random walk, obtaining different levels of persistence in moving towards a given direction. Considering that $0 \leq \rho < 1$, smaller values lead to a more uniform distribution, hence less correlated walks, while higher values of ρ correspond to a skewed distribution, hence highly correlated walks.

The *location bias parameter* β is used to calibrate the agents' bias to move towards the center of the foragers' density distribution. At every turning event, the agent evaluates whether to draw a new random angle or to orientate towards the center of the foragers' density model, based on the probability β. In the latter case, a Gaussian noise $N(0, \sigma_\beta)$ is also added to the rotation angle, with $\sigma_\beta = 0.2\pi$. This value has been empirically tuned to account for imprecision in the rotation towards the center of the dance floor.

While moving, agents can reach the borders of the arena. Since only one side of the comb is simulated, the arena is considered to be bounded, and when agents come across one of the borders during their motion, they stop and change direction moving away from the border towards the center of the forager density distribution. Additionally, whenever agents stop to dance or to follow a dance, they lose memory of the previous direction of motion, and they chose a new orientation uniformly-random as soon as they resume motion.

3   Results

We performed extensive simulations to understand the effect of the mobility pattern on (i) the density distribution of agents during simulations, (ii) the distribution of dances by simulated agents and (iii) the ability to transfer information between dancers and followers. In all simulations, dancers and followers employ the same mobility pattern. We implemented a total of thirteen different scenarios: one with RWM calibrated with the empirical density distribution of foragers, and 12 with the CRW by varying the parameters $\rho \in \{0, 0.3, 0.6, 0.9\}$ and $\beta \in \{0.01, 0.05, 0.1\}$. For each scenario, we ran 100 simulations, each one for $T = 28800$ time steps, equivalent to 2 h of colony activity. Similar to what was done with the real-world observation data, we divide the arena in a grid to compute the spatial distribution of foragers and dances. We also record which dancers and followers interact during dance communication events (dance partners) to analyze the information transfer and the level of mixing in the population.

The density distribution of all forager agents under the RWM mobility pattern is shown in Fig. 2 right. The correspondence with the Gaussian model calibrated on the empirical density distribution of foragers is remarkable, as also testified by the small Hellinger distance between the empirical density distribution and the simulations, which averages to 0.028 (see Fig. 4). This is somewhat expected given that the RWM exploits the full Gaussian model of the empirical density distribution to determine target destinations; hence movements are constrained within the areas with higher observed density. The distribution of dances shows a pattern similar to the forager density, with a negligible shift towards the entrance. This is because the RWM is characterized by a quick diffusion towards the area in which target destinations are sampled, hence it is not impacted significantly by the initial position of the (dancer) agents.

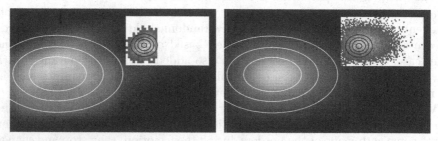

Fig. 2. Left: Overall empirical density distribution of foragers computed on the whole dataset. The background heatmap represents the empirical density distribution obtained from data. The white isolines represent the Gaussian model fitted on the data (centred at $\mu_p = (8.73, 11.07)$ cm, with standard deviation $\sigma_p = (9.99, 5.60)$ cm). Inset: overall dance distribution obtained from data. The black isolines correspond to the Gaussian model fitted on these data (center at $\mu_d = (8.17, 12.42)$ with standard deviation $\sigma_d = (4.92, 3.96)$). Note that the observation camera for dance events covers only the bottom-left part of the honeycomb, hence data points on the right part are missing. Right: density distribution of foragers obtained from simulations using the RWM. The heatmap corresponds to the distribution, while the white isolines correspond to the Gaussian model estimated from the real-world data. Inset: distribution of dances obtained from simulations. The black isolines correspond to the Gaussian model estimated from the data.

When the CRW mobility pattern is employed for dancers and followers, the interplay between persistence in motion and bias to return to the dance floor strongly determines the spatial distribution of the agents, as shown in Fig. 3. Specifically, the larger the location bias β, the narrower the dispersion of agents around the center of the density distribution model. Indeed, when the agents orientate towards the dance floor center with higher probability, they remain clustered and do not diffuse much across the honeycomb. The correlation coefficient ρ instead determines how much an agent would persist in a chosen direction. Generally speaking, higher values of ρ correspond to larger diffusion. This is particularly visible when the location bias β is small (left column in Fig. 3), but has the opposite effect with a strong location bias. Indeed, if an agent frequently reorients towards the center, a high persistence will contribute to move to it even during the subsequent walks, while a small persistence would make agents quickly bend in a completely different direction, hence reducing the impact of the location bias. By comparing the Hellinger distance obtained over the 100 runs shown in Fig. 4, we observe that the best values are for an intermediate level of β, while we observe that ρ has smaller effects, with opposite trends for small or high values of β. Compared to the RWM, the CRW density distributions are slightly worse, but not much difference is observable for $\beta = 0.05$.

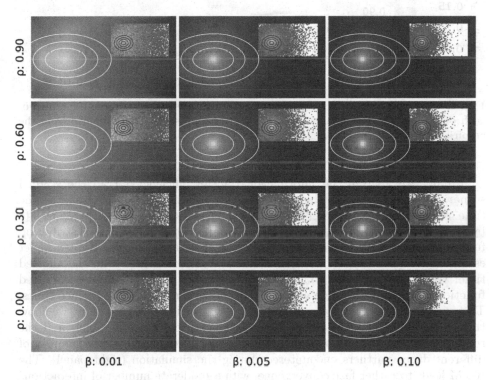

Fig. 3. Density distribution of foragers for each combination of β and ρ tested with simulations. The insets represent the distribution of dances for the same combination of parameters.

The insets in Fig. 3 show the dance distribution obtained with simulations when the CRW mobility pattern is employed, and compare to the Gaussian fit obtained from real-world data (black). The shift towards the entrance is remarkable especially for low location bias ($\beta = 0.01$) and for small CRW persistence ($\rho = 0$). The former entails that movements are not frequently oriented toward the center, the latter entails a small diffusion of the agents. Hence, dancers do not reach the dance floor quickly and dances are mostly performed close to the entrance. Conversely, when β is high, dancers quickly stabilize their motion around the final distribution, and the shift towards the entrance is less visible, in a similar way to what observed with the RWM. Intermediate values of the location bias correspond to the best qualitative match between the dance distribution observed in simulation and the model obtained from real data.

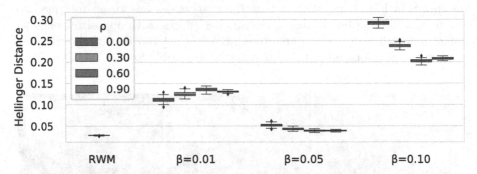

Fig. 4. The Hellinger distance computed between the density distribution obtained in each of the 100 simulation runs, compared to the Gaussian model obtained from the empirical density distribution.

Finally, we analyze the information transfer efficiency for all the studied scenarios (see Fig. 5). To this end, we compute the convergence time as the time required for all followers to obtain information about the foraging site by attending to one dance, at least. We compute the cumulative distribution function of the convergence times across the 100 runs using the Kaplan-Meier estimation [12], censoring those runs that do not converge within the allotted time. We fit a Weibull distribution on the estimated function and use the fitted function to compute average and standard deviation of the convergence times. The average values are shown in the left panel of Fig. 5. Additionally, we show the average number of followers for each dance event (middle panel) and the redundancy of information received by a forager, computed as the number of different dance partners encountered during the simulation (right panel). The RWM leads to rather fast convergence, with a moderate number of interactions and mild redundancy (see the red arrow on the colorbars in Fig. 5). Concerning the CRW, it is possible to note that, the higher the location bias β, the faster the transfer of information between dancers and followers. This is because the foragers are compact around the center of the dance floor, and interactions are numerous (higher number of followers per dance event) but also very redundant (higher number of partners per follower). Also in this case, the CRW persistence ρ has opposite effects depending on the location bias β, as the two parameters concur in determining the diffusion of agents away from the dance floor.

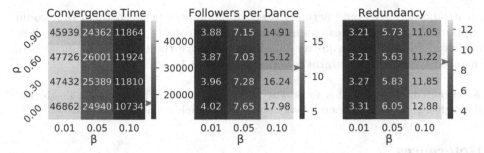

Fig. 5. Information transfer efficiency and effectiveness with different mobility patterns. Each panel represents the average over 100 runs. The matrices show the results for CRW. The results for the RWM are indicated by a red triangle on the colorbar. Left: Average convergence time. Centre: Average number of followers for each dance event. Right: Redundancy of information received, computed as the average number of different partners recorded for each follower agent.

4   Discussion and Conclusions

The comparison made between the empirical density distribution and the simulated one allows us to speculate about the plausibility of the mobility pattern we have implemented. The RWM is clearly the one that produces the best match. However, its implementation would entail that bees are precisely aware of their location over the honeycomb, relying on a kind of map to choose the next location to move to. Additionally, the negligible shift towards the entrance observed in the distribution of dances also makes less plausible the RWM. In fact, to obtain a better match, dancer bees should employ a different map than followers bees, but as each forager can take both roles, it would be difficult to imagine that the employed map changes according to the role. On the contrary, the CRW model is based on much more parsimonious assumptions. Here, we assume only that bees can re-orient towards the dance floor—with noise—with a certain probability. Having a sense of the direction of the dance floor is a much less cognitively-demanding ability than a complete map of the honeycomb. Additionally, the CRW provides a better match for the distribution of dances, for the same parameters that minimize the difference between empirical and simulated density distribution of foragers (i.e., $\beta = 0.05$). Finally, the ability to transfer information among foragers—as observed with the CRW—also suggests that intermediate levels of location bias allow to satisfactory deal with the trade-off between convergence time and redundancy.

Future work will attempt to confirm the above discussion by looking at the detailed bees trajectories available from the BeesBook system. By looking at the real trajectories, a data-driven model can be made to determine what type of motion foragers perform, and how this is impacted by the local density of bees. Most importantly, we want to differentiate the trajectories that precede a dance communication event, to distinguish between dancers, followers and "idle" foragers, in order to understand how the behavioral state of foragers impacts on its motion. We could actually observe differences between them that could support better techniques to spread information rapidly within the swarm. The gathered knowledge can be very useful to improve the design of artificial bee-inspired systems (e.g., swarm robotics systems [9,13]).

Acknowledgements. This work was partially supported by CONACYT, Mexico, through a grant for postdoctoral stay abroad, scholarship holder No. 272227.

References

1. Bartumeus, F., Da Luz, M.G., Viswanathan, G.M., Catalan, J.: Animal search strategies: a quantitative random-walk analysis. Ecology **86**(11), 3078–3087 (2005). https://doi.org/10.1890/04-1806
2. Bettstetter, C., Hartenstein, H., Pérez-Costa, X.: Stochastic properties of the random waypoint mobility model. Wirel. Netw. **10**(5), 555–567 (2004). https://doi.org/10.1023/B:WINE.0000036458.88990.e5
3. Boenisch, F., Rosemann, B.M., Wild, B., Wario, F., Dormagen, D., Landgraf, T.: Tracking all members of a honey bee colony over their lifetime. Front. Robot. AI **5**, 1–10 (2018). https://doi.org/10.3389/FROBT.2018.00035. https://arxiv.org/abs/1802.03192
4. Camazine, S., Sneyd, J.: A model of collective nectar source selection by honey bees: self-organization through simple rules. J. Theoret. Biol. **149**(4), 547–571 (1991). https://doi.org/10.1016/S0022-5193(05)80098-0
5. Clarke, D., Robert, D.: Predictive modelling of honey bee foraging activity using local weather conditions. Apidologie **49**(3), 386–396 (2018). https://doi.org/10.1007/s13592-018-0565-3
6. Codling, E.A., Plank, M.J., Benhamou, S.: Random walk models in biology. J. R. Soc. Interface **5**(25), 813–834 (2008). https://doi.org/10.1098/rsif.2008.0014
7. Danka, R.G., Gary, N.E.: Estimating foraging populations of honey bees (Hymenoptera: Apidae) from individual colonies. J. Econ. Entomol. **80**(2), 544–547 (1987). https://doi.org/10.1093/jee/80.2.544
8. Devillers, J., Doré, J.C., Tisseur, M., Cluzeau, S., Maurin, G.: Modelling the flight activity of Apis mellifera at the hive entrance. Comput. Electron. Agric. **42**(2), 87–109 (2004). https://doi.org/10.1016/S0168-1699(03)00102-9
9. Dimidov, C., Oriolo, G., Trianni, V.: Random walks in swarm robotics: an experiment with kilobots. In: Dorigo, M., et al. (eds.) ANTS 2016. LNCS, vol. 9882, pp. 185–196. Springer, Cham (2016). https://doi.org/10.1007/978-3-319-44427-7_16
10. Hellinger, E.: Neue Begründung der Theorie quadratischer Formen von unendlichvielen Veränderlichen. Journal für die reine und angewandte Mathematik **136**, 210–271 (1909). http://eudml.org/doc/149313
11. Johnson, B.R.: Division of labor in honeybees: form, function, and proximate mechanisms, January 2010. https://doi.org/10.1007/s00265-009-0874-7
12. Kaplan, E.L., Meier, P.: Nonparametric estimation from incomplete observations. J. Am. Stat. Assoc. **53**(282), 457–481 (1958). https://doi.org/10.1080/01621459.1958.10501452
13. Miletitch, R., Dorigo, M., Trianni, V.: Balancing exploitation of renewable resources by a robot swarm. Swarm Intell. **12**(4), 307–326 (2018). https://doi.org/10.1007/s11721-018-0159-8

14. Ortis, G., Frizzera, D., Seffin, E., Annoscia, D., Nazzi, F.: Honeybees use various criteria to select the site for performing the waggle dances on the comb. Behav. Ecol. Sociobiol. **73**(5), 1–9 (2019). https://doi.org/10.1007/s00265-019-2677-9

15. Seeley, T.D.: Social foraging by honeybees: how colonies allocate foragers among patches of flowers. Behav. Ecol. Sociobiol. **19**(5), 343–354 (1986). https://doi.org/10.1007/BF00295707

16. Seeley, T.D.: The honey bee colony as a superorganism. Am. Sci. **77**(6), 546–553 (1989)

17. Seeley, T.D.: Honeybee Democracy. Princeton University Press, Princeton (2010)

18. Wario, F., Wild, B., Couvillon, M.J., Rojas, R., Landgraf, T.: Automatic methods for long-term tracking and the detection and decoding of communication dances in honeybees. Front. Ecol. Evol. **3**, 1–14 (2015). https://doi.org/10.3389/fevo.2015.00103

19. Wario, F., Wild, B., Rojas, R., Landgraf, T.: Automatic detection and decoding of honey bee waggle dances. PLoS ONE **12**(12), 1–16 (2017). https://doi.org/10.1371/journal.pone.0188626. http://arxiv.org/abs/1708.06590

20. Wild, B., et al.: Social networks predict the life and death of honey bees. bioRxiv (2020). https://doi.org/10.1101/2020.05.06.076943. https://www.biorxiv.org/content/early/2020/05/06/2020.05.06.076943

21. Wild, B., Sixt, L., Landgraf, T.: Automatic localization and decoding of honeybee markers using deep convolutional neural networks, February 2018. http://arxiv.org/abs/1802.04557

22. Wilson, D.S., Sober, E.: Reviving the superorganism. J. Theoret. Biol. **136**(3), 337–356 (1989). https://doi.org/10.1016/S0022-5193(89)80169-9

Multi-robot Coverage Using Self-organized Networks for Central Coordination

Aryo Jamshidpey$^{(\boxtimes)}$ ⓘ, Weixu Zhu ⓘ, Mostafa Wahby ⓘ, Michael Allwright ⓘ, Mary Katherine Heinrich ⓘ, and Marco Dorigo ⓘ

IRIDIA, Université Libre de Bruxelles, Brussels, Belgium
{aryo.jamshidpey,weixu.zhu,mostafa.wahby,michael.allwright,
mary.katherine.heinrich,mdorigo}@ulb.ac.be

Abstract. We propose an approach to multi-robot coverage that combines aspects of centralized and decentralized control, based on the existing 'mergeable nervous systems' concept. In our approach, robots self-organize a dynamic ad-hoc communication network for distributed asymmetric control, enabling a degree of central coordination. In the coverage task, simulated ground robots coordinate with UAVs to explore an arena as uniformly as possible. Compared to strictly centralized and decentralized approaches, we test our approach in terms of coverage percentage, coverage uniformity, scalability, and fault tolerance.

1 Introduction

Multi-robot coverage control targets the systematic, uniform observation of a physical area or terrain. A widely studied approach is coverage path planning, in which the motion of robots is often centrally planned and coordinated, sometimes with prior knowledge of the size and shape of the environment [4]. Centralized approaches to coverage path planning have high performance, but are limited in terms of scalability and fault tolerance, due to a lack of redundancy that results in single points of failure and communication bottlenecks. Self-organized approaches to coverage, by contrast, are typically scalable and fault-tolerant, but are slow and inefficient compared to centralized approaches (e.g., [10]).

We propose a novel approach to multi-robot coverage control that seeks to combine aspects of centralized and decentralized approaches. Our approach is based on the existing concept of 'mergeable nervous systems' (MNS) [15], where robots assemble and physically connect, and temporarily yield control of their sensors and actuators to a single brain robot. In our prior work [25], we have extended the MNS concept to strictly wireless communication, rather than making use of physical connections. In our approach, robots establish asymmetric control over a dynamic ad-hoc communication network that is established and managed exclusively through self-organization. In this way, a self-organized network is used to implement some degree of central coordination, combining aspects of centralized and decentralized control. In this paper, we apply our

© Springer Nature Switzerland AG 2020
M. Dorigo et al. (Eds.): ANTS 2020, LNCS 12421, pp. 216–228, 2020.
https://doi.org/10.1007/978-3-030-60376-2_17

MNS approach to the task of multi-robot coverage. We also define comparable centralized and decentralized approaches to the considered coverage task and compare their performance with our hybrid solution. Increased decentralization in multi-robot systems typically involves increased parallelization and redundancy, such that a group of robots governed by centralized control is likely to be faster and more efficient than those governed by decentralized control. Therefore, we would expect a centralized approach to outperform the MNS approach, which in turn should outperform a decentralized approach. We test this using experiments in simulation. While decentralization might cause a decrease in efficiency and speed, it also provides desirable features such as increased scalability and fault tolerance. We therefore test the MNS approach to assess how well these features typical of decentralization have been preserved in our approach. Specifically, we assess scalability in terms of robot–robot communication and interference, and fault tolerance in terms of performance after robot failures.

1.1   Related Work

Coverage control has been widely studied in sensor networks (e.g., [6,23]), and has also been studied for search and exploration tasks with single robots and multi-robot systems. In the task of single robot coverage, the robot should gather information about the environment as efficiently as possible [7,11]. The overall time for a coverage task can be decreased by using multiple robots, but multi-robot approaches require solutions to efficient coordination. In centralized approaches, multi-robot coverage control is often approached as a path planning problem [4,5,24] making use of optimization or learning techniques. These approaches sometimes incorporate aspects of decentralized control. For instance, in [20], decentralized path planning relies on reinforcement and imitation learning through a centralized planner. In [14], robots use decentralized control to initially spread out in the environment, and then use a centralized approach for online learning of a density function.

In decentralized approaches, solutions to spatial coordination include leaving markings during exploration (e.g., artificial pheromones [12]), or maintaining communication (e.g., via line-of-sight [18]). Connectivity maintenance has been investigated in [13], seeking to maximize coverage and minimize communication overhead, and also has been investigated in [22], using a Voronoi tessellation approach to add fault tolerance. Connectivity maintenance during task parallelization has also been studied—using a distributed navigation controller and a global layer for task scheduling, in [16] it is shown that an hybrid centralized/decentralized approach can maintain connectivity in a scenario in which robots are deployed towards certain task-specific locations in the environment. Efficiency is also a key challenge for decentralized approaches, as they are prone to redundancy. In [10], a large number of robots perform coverage by simple collision avoidance, but full coverage is not guaranteed and efficiency is low, as robots frequently revisit areas. In a similar approach that reduces repeated coverage [8], robots leave markings during exploration; in another, a pheromone-based approach is used to achieve coverage efficiency [21]. Similar to pheromone-based approaches, activated beacons are used in [1] to guide coverage in a swarm of UAVs. Finally, coverage has also been studied in a heterogeneous swarm of robots with different sensing capabilities [19].

2   Methods

We investigate the applicability of the 'mergeable nervous systems' (MNS) [15] concept to the task of multi-robot coverage control. As the MNS concept combines aspects of centralized and decentralized control, we compare the performance of our approach to that of fully centralized control (i.e., all robots are controlled using global communication by a single robot with a global view) or fully decentralized control (i.e., all robots are controlled independently). In the decentralized approach, robots explore the environment by means of a random walk without any centralized coordination. In the MNS and centralized approaches, robots maintain a target formation while exploring the environment in a coordinated way. In the centralized approach, all robots are given motion instructions by one robot, whose identity as the central coordinating entity (i.e., the *master*) is predetermined and static. In our MNS approach, robots form a self-organized communication network—specifically, a directed rooted tree, where each link connects a parent robot to a child robot. One robot in the MNS is dynamically assigned the role of the *brain*, through self-organization (for the details of this process, see our prior work [25]). The robots use the network to receive motion instructions from their respective parents in the communication topology—except for the brain robot, which defines its own motion.

In this section, we describe the methods for our experiments. First, we define the coverage task. Second, we define the two motion behaviors that the three approaches can utilize during the coverage task. One is collision avoidance that is performed by robots independently, and is used in all three approaches. The other is perimeter following, which directs the motion of one robot, in both the approaches where robots are coordinated. The perimeter-following behavior is used by the brain robot in the MNS approach, and by the master robot in the centralized approach. Third, we define the target formations that robots maintain in the MNS and centralized approaches. Fourth, we give the implementation details of the three approaches. Overall, we keep the implementation details of the centralized and decentralized approaches as similar as possible to those of the MNS approach, to facilitate direct comparability. Finally, we describe the details of our simulation setup and the types of experiments conducted.

2.1   Coverage Task

We define the coverage task as uniform environment exploration—the robots should collectively visit every portion of the environment, and spend equal time visiting each portion. The environment is an enclosed square arena with randomly distributed small obstacles. The portions of the arena that need to be visited are the cells of a 16×16 overlay grid (i.e., 256 cells of equal size).

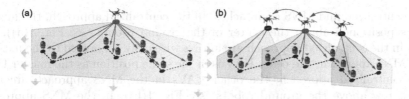

Fig. 1. Communication network topologies, and robot positions in the formations (centralized and MNS approaches). Red arrows show network connections. Light blue zones indicate approximate UAV field of view. Dashed black lines are not connections, but help to visualize the zigzag line that ground robots form. (a) Centralized approach. Connections are predetermined and static. The UAV is the master. (b) MNS approach. Dark blue arrows indicate UAV interchangeability. Network connections are self-organized; the UAV at the center is the brain. (Color figure online)

The environment is explored by differential drive ground robots, which are capable of detecting obstacles and other ground robots. In the MNS and centralized approaches, the ground robots are accompanied by camera-equipped UAVs that send motion instructions to other robots (both ground robots and other UAVs).

In all approaches, ground robots independently avoid obstacles and other ground robots. They are equipped with a ring of short-range proximity sensors. If a robot senses an object in a direction within 60° of its heading, it performs collision avoidance: it turns right if it senses objects only to the left of its heading; otherwise, it turns left. When not avoiding collisions, a robot follows its default motion behavior in the respective control approach.

In the centralized and MNS approaches, one robot—the master UAV and brain UAV respectively—is equipped with a simple motion controller to follow the arena perimeter. This controller moves the UAV forward in a straight line unless it detects a boundary, in which case it turns 90° to the left, and then moves forward again. This results in a counter-clockwise motion around the arena. As this perimeter-following behavior is deterministic, one loop of the master or brain UAV around the perimeter always takes the same amount of time, in both the centralized and MNS approaches. A master or brain UAV begins this perimeter-following behavior after the ground robots have established the target formation from their randomly distributed starting positions. The master or brain UAV then continues the behavior at a constant speed, irrespective of the speed of the other robots, until experiment termination. Given the size and shape of the target formation and the arena, this simple counter-clockwise path is sufficient to enable coverage.

In the centralized and MNS approaches, robots establish a target formation from randomly distributed starting positions, and then maintain that formation during coverage. In the target formation used here, ground robots are positioned in a zigzag line (see Fig. 1). The zigzag line formation is selected to reduce the occurrence of robot-robot collisions, compared to a straight line formation with smaller gaps between robots. The target formation of ground robots is identical in the centralized and MNS approaches. In the centralized approach, the master UAV is positioned above the center of the ground robots (see Fig. 1(a)). All robots in the centralized approach are always wirelessly connected to the master. In the MNS approach, the brain UAV is in the same position as the master UAV of the centralized approach; the other UAVs in the MNS approach are in a straight row above the ground robots (see Fig. 1(b)). In the MNS approach, the communication network topology is a caterpillar tree—i.e., a tree in which all inner nodes are on one central path, to which each leaf node is connected (see Fig. 1(b)). The centralized and MNS approaches use one and three UAVs, respectively, and each use nine ground robots. The decentralized approach uses nine ground robots, which perform a random walk without UAV guidance.

2.2   Approaches to Multi-robot Coverage

In the centralized approach, the default motion behavior for all robots is directly controlled by the *master* UAV that acts as central coordinating entity. The master UAV can directly communicate with all robots constantly, and can always see all robots and the full environment, regardless of its position. At the beginning of an experiment, the master sends all robots motion instructions, to move them into the target formation. The master then uses its perimeter-following behavior to follow the arena perimeter, while simultaneously sending all robots motion instructions, to maintain the target formation (relative to the master UAV).

In the decentralized approach, the default motion behavior for all ground robots is simply forward motion. At initiation, the robots are distributed randomly and begin moving in random directions. They only change direction as a result of collision avoidance. Due to the density of obstacles in the environment, collision avoidance is sufficient to change the robots' directions frequently enough for environment exploration.

In our MNS approach, the default motion behavior for non-brain robots is received from parents in the communication network. Our MNS approach is based on the existing concept of 'mergeable nervous systems' [15], for physically connected robots that we have extended to wireless connections in prior work [25]. In this approach, a heterogeneous swarm of UAVs and ground robots forms a target communication network topology through a self-organized process, and then uses this network to pass motion instructions between neighbors, moving robots into positions and orientations that match a given target formation. Please refer to [25] for details of the process by which the MNS is established and maintained. In the approach, a UAV can establish links with ground robots in its field of view, and can establish links with other UAVs when there is a shared ground robot in both their fields of view. In the experiments here, robots initially use the MNS process to establish the communication network and target formation. Then, the UAV that has become the brain (one of the three UAVs) begins to follow the arena perimeter. As the brain moves, it sends motion instructions to each of its children, which subsequently send motion instructions to their own children, thereby moving the whole formation.

2.3 Experiment Setup

The experiments are conducted using the ARGoS multi-robot simulator [17], with robot models implemented using an extension [2,3]. The 4×4 m^2 arena is fully enclosed, with its bottom-left corner at $(0, 0)$ of the coordinate frame. Static $4 \times 4 \times 2$ cm^3 obstacles are positioned randomly in the 3.7×3.7 m^2 center of the arena (with uniform distribution). The arena has a 16×16 overlay grid (with 0.25×0.25 m^2 cells). The UAV model has a maximum speed of 7.4 cm/s, and is equipped with a downward-facing camera. In the MNS approach, each UAV views a 1.5×1.75 m^2 rectangular ground area, at the 1.5 m flight altitude used in the experiments. Collectively, the three UAVs in the default MNS formation have a 1.5×2.75 m^2 view. By contrast, the UAV in the centralized approach has a full view of the arena at all times. The ground robot model has an average speed of 6.8 cm/s, and is equipped with a ring of 12 outward-facing proximity sensors with a 5.0 cm range. Ground robots are topped with fiducial markers encoding unique IDs, which the UAVs use to detect the relative positions and orientations of the ground robots. In our setup, UAVs are unable to detect obstacles. In the MNS approach, the communication range for UAVs and ground robots is 1 m. In the centralized approach, the master UAV has unlimited communication with all ground robots. The mechanical bodies of UAVs and ground robots are represented by simple 2.5 cm radius cylinders. In all approaches, if a ground robot reaches the arena boundary, its normal motion behavior is temporarily overridden—it turns to a random direction in the 180° range facing away from the boundary, then drives straight forward.

3 Results

In this section, we give the results of our experiments testing performance, scalability, and fault tolerance. In all experiments, we record robot positions. In all three approaches, ground robots initially face random directions and are positioned randomly in a 1.0×1.25 m^2 rectangular area against the southern arena boundary, following a uniform distribution. In the centralized and MNS approaches, the UAVs are positioned above the ground robots, near the southern boundary. Once the formation is established, the master or brain UAV has the southern boundary in view, and therefore turns left to start following the arena perimeter. For the centralized and MNS approaches, we define a *round* as one complete loop around the arena perimeter.

3.1 Performance

The performance experiments compare the three approaches in terms of *coverage percentage* (i.e., the percentage of grid cells visited by at least one ground robot) and *coverage uniformity* (i.e., the uniformity of the total time robots spend in each grid cell), and in terms of the time and energy expended (according to potential consumption rates). We test the performance of the three approaches (centralized, MNS, decentralized), with three different numbers of obstacles (100, 200, 300)—in total nine performance experiments.

Real-world energy consumption of UAVs and ground robots can vary considerably. Therefore, we test five possible ratios of UAV-to-ground-robot energy consumption $\{0.5, 1, 2, 3, 4\}$, with the ground robot consumption rate fixed at 30 units per step. Ratios over 1 represent scenarios with small simple ground robots and powerful UAVs, and ratios of 1 and 0.5 represent scenarios with large complex ground robots (e.g., quadruped robots) and minimal lightweight UAVs. Experiments testing the MNS approach terminate at step 4710, at the completion of one round; others terminate when the robots have consumed the same total energy as the MNS approach, under the same energy ratio. For example, under energy ratio 0.5, if the MNS approach has consumed 100 energy units at step 4710, then 100 energy units is the energy budget for the other approaches under that ratio. For each experiment type, we execute 10 experiment runs for each energy ratio termination time.

Table 1. The coverage percentage results of the performance experiments.

		MNS			Centralized		Decentralized	
Ratio	Energy: /10^6 Units	Time (steps)	#Obstacles : Coverage percentage	Time (steps)	#Obstacles : Coverage Percentage	Time (steps)	#Obstacles : Coverage percentage
0.5	1.48365	4710	100 : 96.9% 200 : 95.7% 300 : 92.2%	5206	100 : 98.4% 200 : 97.7% 300 : 96.1%	5495	100 : 86.7% 200 : 80.5% 300 : 76.2%
1	1.6956	4710	100 : 96.9% 200 : 95.7% 300 : 92.2%	5652	100 : 98.8% 200 : 98.1% 300 : 96.7%	6280	100 : 90.2% 200 : 84.8% 300 : 80.1%
2	2.1195	4710	100 : 96.9% 200 : 95.7% 300 : 92.2%	6423	100 : 98.8% 200 : 98.1% 300 : 97.7%	7850	100 : 94.9% 200 : 92.2% 300 : 87.1%
3	2.5434	4710	100 : 96.9% 200 : 95.7% 300 : 92.2%	7065	100 : 98.8% 200 : 98.1% 300 : 97.7%	9420	100 : 97.2% 200 : 95.3% 300 : 91.8%
4	2.9673	4710	100 : 96.9% 200 : 95.7% 300 : 92.2%	7609	100 : 99.2% 200 : 98.4% 300 : 98.1%	10990	100 : 98.4% 200 : 97.3% 300 : 94.5%

Coverage Percentage. We compare the coverage percentage of the three approaches under equal energy expenditure. Table 1 shows that the centralized approach outperforms the other approaches for all energy ratios and obstacle densities, although its performance is only slightly better than that of the MNS approach. When the energy ratio is less than 3, the MNS approach outperforms the decentralized approach for all obstacle densities. If time expenditure is considered, Table 1 shows that, for energy ratios of 3 and 4, the decentralized approach takes more than twice as much time as the MNS approach and achieves only slightly better coverage percentage. Because of the high UAV energy cost in these cases, the decentralized approach is allowed much longer exploration time than the MNS approach. Table 1 also shows that coverage percentage becomes lower for all approaches as obstacle density increases. Figure 2 shows the coverage percentage over time for the three approaches, in two obstacle setups at energy ratio 4. We report results only for one energy ratio because the graphs are similar for all energy ratios—the only difference being in the change in the performance gaps between the three approaches. We have chosen to report energy ratio 4 because it is the worst energy ratio for the MNS approach; the gap between the MNS approach and the better-performing centralized approach is largest in this ratio, and the gap between the MNS approach and the worse-performing decentralized is smallest in this ratio. Figure 2 therefore shows that, in all cases, the MNS approach substantially outperforms the decentralized approach, and the centralized approach slightly outperforms the MNS approach. Energy ratio 4 bears the worst performance for the MNS approach because the MNS uses three UAVs, as opposed to one UAV or no UAVs.

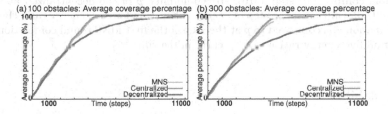

Fig. 2. Average coverage percentage for MNS, centralized, and decentralized approaches, with an energy ratio of 4. (a) 100 obstacles setup; (b) 300 obstacles setup.

Coverage Uniformity. For coverage uniformity, we assess the centralized, MNS and decentralized approaches at the timestep of the first complete MNS round, and additionally assess the decentralized experiments at energy exhaustion. For each run, $v_i \in \mathbf{v}$ is defined as the total time spent by all robots in cell i. The coverage uniformity p is the norm of $\mathbf{v}$, calculated as follows:

$$p = \sum_{i=1}^{256} \sqrt{|v_i - M(\mathbf{v})|}, \tag{1}$$

where $M(\mathbf{v})$ is the median of $\mathbf{v}$. The smaller the value of p, the more uniformity between cells; the most uniform case is $p = 0$. Figure 3(a) shows the coverage uniformity p of all three approaches, at the step of the first MNS round completion (step 4710). The centralized approach is the most uniform (i.e., the smallest p, on average). Figure 3(a) shows that, in terms of coverage uniformity, the MNS approach substantially outperforms the decentralized approach, and the centralized approach slightly outperforms the MNS approach. While the MNS and centralized approaches have approximately similar uniformity in later rounds (i.e., later in time), the uniformity of the decentralized approach becomes steadily worse over time, as shown in Fig. 3(b). The worsening of uniformity over time, in the decentralized approach, is most pronounced in the highest obstacle density. Figure 3(a) also shows that, at the completion of the first MNS round, the uniformity of all three approaches worsens as obstacle density increases.

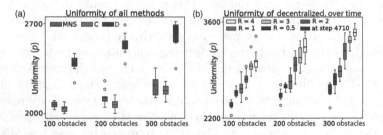

Fig. 3. Coverage uniformity (lowest p is the most uniform). (a) Uniformity p of all approaches (MNS; centralized, C; and decentralized, D), at the timestep of the first MNS round completion (step 4710). (b) Uniformity p of the decentralized approach, over time. Uniformity p at the step of energy exhaustion for each energy ratio (R)—i.e., p at termination—is compared to p at the step of the first MNS round completion (step 4710) for all five energy ratios—i.e., p early in the run.

3.2 Scalability

The scalability and fault tolerance experiments test whether the MNS approach displays features that would typically be observed in decentralized robot systems (cf. [9]). We evaluate scalability in the MNS approach in terms of communication (i.e., the number of messages exchanged) and interference (i.e., the number of robot-robot collisions). The scalability experiments are conducted in an arena without obstacles, with three different swarm sizes that are arranged in the same type of target formation as the default (caterpillar tree, zigzag line—see Fig. 1), with 10 runs per swarm size. The sizes are: 1) two UAVs, four ground robots; 2) four UAVs, eight ground robots; 3) six UAVs, twelve ground robots. Figure 4(a) shows that the number of messages increases linearly with increasing swarm size, as robots communicate only with their neighbors in the network. Figure 4(b) shows that the number of collisions increases steadily at the beginning of the experiments, as the MNS formation is being established. Once the MNS is established and begins to explore the environment, no further robot–robot collisions are observed.

3.3 Fault Tolerance

The ability of the MNS approach to replace a robot after failure, or to repair a broken network connection, has been demonstrated in [25]. In this paper, we investigate the fault tolerance of the MNS approach when ground robots fail and cannot be replaced or repaired, evaluated according to connectivity and coverage percentage. When a ground robot fails, its network link(s) are broken, and it can no longer move or communicate with any robots. We use the default MNS formation (see Fig. 1) and the setup with 100 obstacles, and impose failure at step 400. We test failure of the following numbers of ground robots (10 runs each), out of 12 total robots in the swarm: 1, 3, 5, 7, 8. We assess the impact of the failures on coverage percentage results. As UAV-to-UAV connections are established indirectly when mutually viewing ground robots, we also assess the impact of the ground robot failures on parent connectivity (i.e., whether the brain UAV maintains communication with the other UAVs, which are parent nodes in the communication network). Figure 4(c) shows that coverage percentage decreases as the number of failures increases. Figure 4(d) shows that parent connectivity is maintained in all cases of 5, 3, or 1 failure(s). In cases of 7 or 8 failures—in which more than half of the swarm fails—connectivity is maintained in 70% and 50% of runs, respectively.

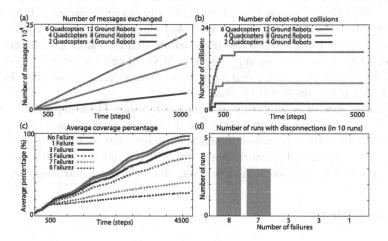

Fig. 4. (a–b) Scalability. (a) Number of messages exchanged in the MNS over time. (b) Number of robot-robot collisions over time. (c–d) Fault tolerance. (c) Coverage percentage over time, in MNSs with varying number of failing ground robots. (d) Number of runs that suffer a brain–parent disconnection, according to number of ground robot failures (out of 10 total runs for each number of failures).

4   Discussion

In terms of coverage percentage and coverage uniformity, the MNS approach substantially outperforms the decentralized approach. The decentralized approach requires approximately twice as much time as the MNS approach to reach similar coverage percentage, when the energy consumption ratio is at least 3 (see Table 1). If the ratio is less than 3, the decentralized approach never reaches the coverage percentage of the MNS approach, due to exhaustion of the energy budget. The MNS approach also achieves better coverage uniformity than the decentralized approach; coverage uniformity in the decentralized approaches worsens over time. The lower performance of the decentralized approach is due to the uneven distribution of robots that occurs during a random walk. As expected, the centralized approach outperforms the other two approaches in coverage percentage and coverage uniformity. However, the performance difference between the centralized and MNS approaches is relatively small, compared to that between the MNS and decentralized.

The scalability of the MNS approach, in terms of number of messages exchanged and robot-robot collisions, is good (see Fig. 4(a,b)). The number of messages increases linearly with increasing swarm size, and no robot-robot collisions are observed after the MNS is established, in any swarm size. The MNS approach is also fault-tolerant, in terms of connectivity and coverage performance. Substantial drops in performance only occur when more than half of the swarm fails. Connectivity in the MNS approach might be an advantage over the decentralized approach in consensus achievement tasks (e.g. collective decision making or collective sensing). As the MNS approach recovers from brain failure (see [25]), it also has an advantage over centralized approaches, in which all robots fail if the master UAV fails.

A possible direction for future development of our MNS approach would be to adapt the target formation on the fly. In this case, if failures are detected, the MNS could switch to a new formation shape that is better suited to the remaining swarm size. In future work, we will extend our MNS approach to apply it to tasks such as collective sensing, or localization and mapping.

5   Conclusions

We have presented an MNS approach to multi-robot coverage, and tested its coverage performance against strictly centralized and strictly decentralized approaches. Our results indicate that the MNS approach significantly outperforms the decentralized approach, but is slightly outperformed by the centralized approach. We have also tested the MNS approach for its performance in terms of scalabilty and fault tolerance—two features that are difficult to obtain with a centralized approach. Our results show that the MNS approach scales linearly in terms of inter-robot communication, and that its performance and connectivity are robust to failures if less than 50% of the ground robots fail. Overall, the results demonstrate that the MNS approach successfully combines aspects of centralized and decentralized control in a coverage task, as it achieves high performance (similar to centralized approaches), and achieves scalability and fault tolerance (similar to decentralized approaches).

Acknowledgements. This work is partially supported by the Program of Concerted Research Actions (ARC) of the Université libre de Bruxelles; by the Ontario Trillium Scholarship Program through the University of Ottawa and the Government of Ontario, Canada; by the Office of Naval Research Global (Award N62909-19-1-2024); by the European Union's Horizon 2020 research and innovation programme under the Marie Skłodowska-Curie grant agreement No 846009; and by the China Scholarship Council (grant number 201706270186). Mary Katherine Heinrich and Marco Dorigo acknowledge support from the Belgian F.R.S.-FNRS, of which they are a Postdoctoral Researcher and a Research Director respectively.

References

1. Albani, D., Nardi, D., Trianni, V.: Field coverage and weed mapping by UAV swarms. In: 2017 IEEE/RSJ International Conference on Intelligent Robots and Systems (IROS), pp. 4319–4325. IEEE (2017)
2. Allwright, M., Bhalla, N., Pinciroli, C., Dorigo, M.: ARGoS plug-ins for experiments in autonomous construction. Technical reports, TR/IRIDIA/2018-007, IRIDIA, Université Libre de Bruxelles, Brussels, Belgium (2018)
3. Allwright, M., Bhalla, N., Pinciroli, C., Dorigo, M.: Simulating multi-robot construction in ARGoS. In: Dorigo, M., Birattari, M., Blum, C., Christensen, A.L., Reina, A., Trianni, V. (eds.) ANTS 2018. LNCS, vol. 11172, pp. 188–200. Springer, Cham (2018). https://doi.org/10.1007/978-3-030-00533-7_15
4. Almadhoun, R., Taha, T., Seneviratne, L., Zweiri, Y.: A survey on multi-robot coverage path planning for model reconstruction and mapping. SN Appl. Sci. 1(8), 1–24 (2019). https://doi.org/10.1007/s42452-019-0872-y
5. Dewangan, R.K., Shukla, A., Godfrey, W.W.: Survey on prioritized multi robot path planning. In: IEEE International Conference on Smart Technologies and Management for Computing, Communication, Controls, Energy and Materials (ICSTM), pp. 423–428. IEEE (2017)
6. Dieber, B., Micheloni, C., Rinner, B.: Resource-aware coverage and task assignment in visual sensor networks. IEEE Trans. Circuits Syst. Video Technol. 21(10), 1424–1437 (2011)
7. Galceran, E., Carreras, M.: A survey on coverage path planning for robotics. Robot. Auton. Syst. 61(12), 1258–1276 (2013)
8. Ge, S.S., Fua, C.H.: Complete multi-robot coverage of unknown environments with minimum repeated coverage. In: Proceedings of the 2005 IEEE International Conference on Robotics and Automation, pp. 715–720. IEEE (2005)
9. Hamann, H.: Swarm Robotics: A Formal Approach. Springer, Cham (2018). https://doi.org/10.1007/978-3-319-74528-2
10. Ichikawa, S., Hara, F.: Characteristics of object-searching and object-fetching behaviors of multi-robot system using local communication. In: Proceedings of 1999 IEEE International Conference on Systems, Man, and Cybernetics, vol. 4, pp. 775–781. IEEE (1999)
11. Juliá, M., Gil, A., Reinoso, O.: A comparison of path planning strategies for autonomous exploration and mapping of unknown environments. Auton. Robots 33(4), 427–444 (2012)

12. Koenig, S., Liu, Y.: Terrain coverage with ant robots: a simulation study. In: Proceedings of the Fifth International Conference on Autonomous Agents, pp. 600–607. Association for Computing Machinery (2001)
13. Laouici, Z., Mami, M.A., Khelfi, M.F.: Cooperative approach for an optimal area coverage and connectivity in multi-robot systems. In: 2015 International Conference on Advanced Robotics (ICAR), pp. 176–181. IEEE (2015)
14. Luo, W., Sycara, K.: Adaptive sampling and online learning in multi-robot sensor coverage with mixture of gaussian processes. In: 2018 IEEE International Conference on Robotics and Automation (ICRA), pp. 6359–6364. IEEE (2018)
15. Mathews, N., Christensen, A.L., O'Grady, R., Mondada, F., Dorigo, M.: Mergeable nervous systems for robots. Nat. Commun. **8**, 439 (2017)
16. Panerati, J., Gianoli, L., Pinciroli, C., Shabah, A., Nicolescu, G., Beltrame, G.: From swarms to stars: task coverage in robot swarms with connectivity constraints. In: 2018 IEEE International Conference on Robotics and Automation (ICRA), pp. 7674–7681. IEEE (2018)
17. Pinciroli, C., et al.: ARGoS: a modular, parallel, multi-engine simulator for multi-robot systems. Swarm Intell. **6**(4), 271–295 (2012)
18. Rekleitis, I., Lee-Shue, V., New, A.P., Choset, H.: Limited communication, multi-robot team based coverage. In: IEEE International Conference on Robotics and Automation, 2004, Proceedings. ICRA 2004, vol. 4, pp. 3462–3468. IEEE (2004)
19. Santos, M., Egerstedt, M.: Coverage control for multi-robot teams with heterogeneous sensing capabilities using limited communications. In: 2018 IEEE/RSJ International Conference on Intelligent Robots and Systems (IROS), pp. 5313–5319. IEEE (2018)
20. Sartoretti, G., et al.: PRIMAL: pathfinding via reinforcement and imitation multi-agent learning. IEEE Robot. Autom. Lett. **4**(3), 2378–2385 (2019)
21. Schroeder, A., Ramakrishnan, S., Manish, K., Trease, B.: Efficient spatial coverage by a robot swarm based on an ant foraging model and the lévy distribution. Swarm Intel. **11**(1), 39–69 (2017)
22. Siligardi, L., et al.: Robust area coverage with connectivity maintenance. In: 2019 International Conference on Robotics and Automation (ICRA), pp. 2202–2208. IEEE (2019)
23. Wang, X., Han, S., Wu, Y., Wang, X.: Coverage and energy consumption control in mobile heterogeneous wireless sensor networks. IEEE Trans. Autom. Control **58**(4), 975–988 (2012)
24. Zafar, M.N., Mohanta, J.C.: Methodology for path planning and optimization of mobile robots: a review. Procedia Comput. Sci. **133**, 141–152 (2018)
25. Zhu, W., Allwright, M., Heinrich, M.K., Oğuz, S., Christensen, A.L., Dorigo, M.: Formation control of UAVs and mobile robots using self-organized communication topologies. In: Swarm Intelligence - Proceedings of ANTS 2020 - Twelfth International Conference. Lecture Notes in Computer Science. Springer, Heidelberg (2020)

Robot Distancing: Planar Construction with Lanes

Andrew Vardy$^{(\boxtimes)}$ (iD)

Department of Computer Science, Department of Electrical and Computer
Engineering, Memorial University of Newfoundland, St. John's, Canada
av@mun.ca

Abstract. We propose a solution to the problem of spatial interference between robots engaged in a planar construction task. Planar construction entails a swarm of robots pushing objects into a desired two-dimensional configuration. This problem is related to object clustering and sorting as well as collective construction approaches such as wall-building. In previous work we found robots were highly susceptible to collisions among themselves and with the boundary of the environment. Often these collisions led to deadlock and a dramatic reduction in task performance. To address these problems the solution proposed here subdivides the work area into lanes. Each robot determines its own lane and applies a novel control law to stay within it while nudging objects inwards towards the goal region. We show results using a realistic simulation environment. These results indicate that subdividing the arena into lanes can produce mild performance increases while being highly effective at keeping the robots separated. We also show that the introduction of lanes increases robustness to unforeseen obstacles in the environment.

1 Introduction

In this paper we address a problem observed in many studies of swarm robotics and multi-robot systems in general. That problem is spatial interference between robots. Researchers in swarm robotics have observed that when varying the number of robots within a fixed space, performance tends to initially increase as robots are added but then begins to level off and decrease as the robots get in each other's way [10]. There is likely no universal solution to this problem, but it remains worthwhile to attempt to increase the band of superlinear performance with increasing numbers of robots. We propose an approach to mitigate spatial interference by subdividing the environment into lanes for a swarm of robots engaged in a planar construction task. We previously defined planar construction as 'the gathering of ambient objects into some desired shape' [35]. Put simply, the task here is to gather objects into a linear region at the center of the arena—as if building a wall.

Planar construction can be considered a sub-area of collective robotic construction (recently reviewed in [24]). It involves the formation of a desired two-dimensional structure from ambient objects in the environment, requiring

© Springer Nature Switzerland AG 2020
M. Dorigo et al. (Eds.): ANTS 2020, LNCS 12421, pp. 229–242, 2020.
https://doi.org/10.1007/978-3-030-60376-2_18

a combination of capabilities: discovering objects, transporting them towards the growing structure, and manipulating the structure into a desired shape. There is a strong relationship with work on foraging [15,23,28], clustering objects of a single type [1,7,12,16,17,34], and sorting objects of different types [19–21,36,37]. The shape constructed here is linear and can be considered a wall. Several other groups have studied the formation of walls by a set of distributed agents [5,13,29,30].

While there has been a wide variety of work on subdividing space for other purposes (e.g. exploration [39]) relatively little research has gone into subdividing space for the purpose of collecting, clustering or manipulating groups of objects. Schneider-Fontán and Matarić demonstrated perhaps the earliest work in this vein [27]. They divided a workspace into equal-sized rectangular regions with one robot assigned to each territory. They addressed the problem of a reduction in workforce due to robot failures by dividing the overall height of the workspace by the number of active robots. Their experiments showed that an increase in the number of robots caused a reduction in performance, likely due to inaccuracies in the way that objects were handed-off from one territory to the next. Others have explored the notion of territories formed by the robots themselves [26] but not in the context of transporting objects. Perhaps the closest recent example involved subdividing space between an active work area and a waiting area for robots engaged in a collection and deposit task [18].

Our previous work on the problem of planar construction has focused on *orbital construction*, a parsimonious control algorithm that can gather objects into a shape defined by a scalar field [32]. We previously proposed a somewhat more complex algorithm for planar construction which relies upon a set of distinct landmarks to specify the shape of the desired structure [33]. In addition, we have investigated the use of reinforcement learning for the planar construction problem [31] and the allocation of roles to robots [11]. In our most recent work, we provided a realization of the orbital construction algorithm on physical robots [35]. This was partially successful but we were surprised at the level of spatial interference between robots and between the robots and the boundary of the environment. Often there would be collisions near the border which resulted in deadlock situations. These collisions would often attract other robots and task performance would plummet.

This experience with the negative effects of spatial interference motivates us to resolve this problem. The main solution proposed in this paper is to subdivide the work area into lanes that surround a goal region. We present results in simulation showing that the introduction of lanes can have a minor positive impact on performance and a substantial benefit in terms of increasing the distance maintained between robots. When robots operate in close proximity there is always some non-zero probability that they will collide and that this collision will result in reduced task performance. By maintaining separation into lanes we do not ensure high task performance but we do decrease the probability of collision-induced failures.

In Sect. 2 we provide some discussion on the top-down specification of a desired shape and how this relates to the swarm robotics credo of purely local sensing. Section 3 describes our algorithm and methodology. Section 4 presents experimental results using a simulation environment. Finally, Sect. 5 presents some discussion on our results and areas for future work.

2   The Template Perspective

While we are interested in robotic systems with self-organizing properties, it is likely that most practical applications will involve some level of top-down task specification. For a construction task, the user may want to specify the location and constraints on the design. For a task involving area coverage such as cleaning or painting the user would want to specify the area to be treated. For a task involving organization of items (e.g. clustering or sorting) the user may want to specify the area of operation as well as the sites where items are to be collected. Inspired by its usage in models of nest construction in wasps and ants [2] we use the term *template* as a structure or pattern that controls the outcome of an otherwise self-organized process.

The template may be expressed through pheromones which guide a construction process, with termite nest construction being perhaps the longest-studied example [4]. Computational studies have shown the generation of tunnels and domes by agents that deposit and sense various pheromones [14]. In Ladley and Bullock's model the stationary queen termite exudes a template pheromone that causes agents to deposit material at a characteristic distance from the queen, eventually leading to the constructing of a "royal chamber". In this model, the static template pheromone's influence combines with the influence of a *cement* pheromone that acts as a positive feedback mechanism, encouraging an agent to deposit material near other recently-deposited material.

For implementation on physical robots it is necessary to make the template accessible to the robots. One possibility is to use landmarks, which could be physical or virtual [33]. Most recently we defined a scalar field as the template and projected it onto a television screen that was the robots' working surface [35]. Phototransistors on the bottom of the robots would sense the local light value and use this to guide their movements. This strategy allowed us to incorporate the user-specified template of the shape while adhering to the swarm robotics principle of purely local sensing. However, this approach has several practical limitations. For example, it is clearly infeasible to install a large screen to act as a floor in real-world applications.

After exploring several different mechanisms for defining a template and making it accessible to the underlying robots we believe it is prudent to use a more generic physical setup that can model various types of concrete implementations. In this setup we have an overhead camera observing a set of robots on a tabletop. The robots have AprilTag markers for pose detection in the overhead image [22]. A desktop computer extracts these markers as well as markers or color codes we attach to other objects in the arena. We have developed a working system that takes this approach for marine swarm robotics research [9]. This system allows us to either make use of global information or simulate purely local sensors. Thus, we can explore a variety of solutions which either adhere (or do not adhere) to the swarm robotics convention of using purely local information [3].

It could be argued that whatever the physical realization, we are violating the conventions of swarm robotics by making a global template available to all robots. However, in the algorithm described below only a local sampling of information from the global template is utilized. The *structpath* data structure used in Werfel et al.'s well-known work on termite-inspired collective construction [38] is similar in this sense—robots use knowledge of their position to look up local information that modulates their behavior. In both Werfel et al.'s work and ours, this local information provides guidance to the robots on allowable motions while also serving to specify a desired shape to be constructed.

3   Methodology

3.1   Physical System Modeled

In this paper we only report results from simulation, but we are developing a physical system consists of 12 robots based on the Pololu 3pi robot, a small circular robot (9.5 cm in diameter) with differential-drive kinematics[1]. These robots operate on a table with a stadium-shaped boundary[2] to retain both objects and robots. The objects the robots operate on are cubes with a side length of 3 cm. We refer to these cubes as *pucks*. The robots have a middle plate that serves as a plow, pushing pucks inwards as the robots circumnavigate the arena along clockwise orbits. Our simulation model is shown in Fig. 1.

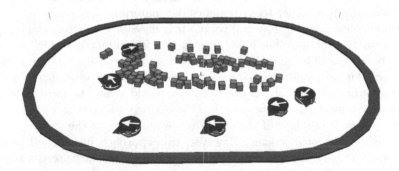

Fig. 1. Screenshot of the CoppeliaSim simulation in progress.

---

[1] https://www.pololu.com/docs/0J21.

[2] A stadium consists of a rectangular region in the middle with semicircular ends.

3.2   Generating the Template and Lanes

The template used to specify the desired linear shape as well as to specify lanes is the distance transform. This is a well-known image processing operation where some pixels, designated as sources, are set to zero. All other pixels are then set to contain the distance to the closest source. Sophisticated methods exist to compute the distance transform efficiently, even on complex meshes [6].

To obtain the distance image (the result of the distance transform), the image from the overhead camera is first processed using simple color-based thresholding to yield a binary image with a value of 0 for the stadium-shaped boundary. This image is flood-filled with 0's outside the boundary, yielding the image shown in Fig. 2(a). Figure 2(b) shows the distance image and Fig. 2(c) shows the gradient computed from the distance image. It is important to note that the distance image and gradient need only be computed once.

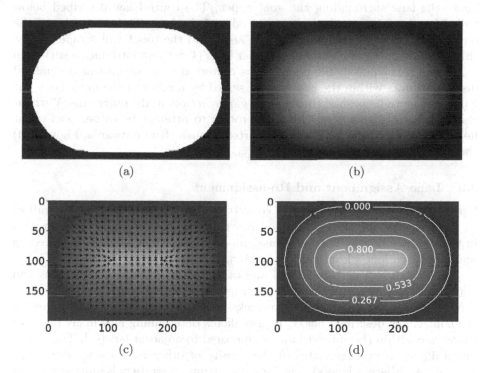

(a)   (b)

(c)   (d)

Fig. 2. Stages in the processing of an image from the overhead camera to yield the distance transform and its gradient. (a) Binary image set to 0 for the boundary and outside the boundary. (b) Distance transform. (c) Gradient of the distance transform. (d) Contours of the distance transform for $N_{lanes} = 3$.

The distance image will be denoted $D_{i,j}$ where i is the column index and j is the row index. We normalize this image by dividing by the maximum distance, $\max(D_{i,j})$, yielding $\tilde{D}_{i,j} \in [0,1]$.

We will refer to various levels, denoted as τ, which are scalar values defining a particular contour line of the distance image. The goal region is the ultimate destination for all pucks in our task and is defined as all (i, j) such that $\tilde{D}_{i,j} \geq \tau_{goal}$. In our experiment $\tau_{goal} = 0.8$. This value can be seen as the innermost contour line in Fig. 2(d).

Let N_{lanes} be the number of lanes chosen. Each lane is indexed by an integer $k \in [0, \ldots, N_{lanes} - 1]$. A pair of levels is used to define each lane: $(\tau_{low}(k), \tau_{high}(k))$. These are defined as follows for each k:

$$\tau_{low}(k) = \tau_{goal} \frac{k}{N_{lanes}} \tag{1}$$

$$\tau_{high}(k) = \tau_{goal} \frac{k+1}{N_{lanes}} \tag{2}$$

$k = 0$ indicates the outermost lane while $k = N_{lanes} - 1$ is the innermost lane—the lane surrounding the goal region. The control law described below will drive the robot to skirt the outer edge of the lane. The outer edge of the outermost lane is the boundary of the arena and the robot will scrape against this, pushing pucks inwards. For all inner lanes ($k > 0$) we introduce a small gap w_{puck} which is equal to the puck radius converted to its equivalent normalized distance. Pucks within the gap are not sensed by a robot in the inner lane, but will continue to be pushed through the gap by robots in the outer lane. Without the gap there is a tendency for inner robots to attempt to gather pucks that lie partially out of the lane and inadvertently push them outwards. Figure 2(d) shows the contours generated when $N_{lanes} = 3$.

3.3   Lane Assignment and Re-assignment

Upon initialization, each robot will convert its pose (x, y, θ) to the integer indices (i, j) which describe the column and row corresponding to the robot's position in the distance image. The initial assignment of the robot's lane is obtained by searching each lane's level pair $(\tau_{low}(k), \tau_{high}(k))$ for the one containing $\tilde{D}_{i,j}$.

Each robot can perceive pucks and other robots within a fixed radius and determine their pose and therefore their level in the distance image. This allows a robot to restrict its attention to pucks within its lane, which is required by the control law described below. It also affords determining how many pucks or robots are within the current lane as compared to adjacent lane(s)[3]. The robots maintain short-term estimates of the density of other robots seen within the current and adjacent lane(s). The lane re-assignment strategy is quite simple—if fewer robots are observed in an adjacent lane, the robot switches to that lane.

3.4   Control Law

In previous work on planar construction we used a very parsimonious controller which required sensing the scalar field at three positions in front of the robot,

---

[3] The innermost and outermost lanes have one adjacent lane, while other lanes have two adjacent lanes.

as well as sensing the presence of pucks in the left visual field [11,31,32,35]. This controller was effectively "computation-free" in the sense introduced in [8] meaning that it was a very simple reactive controller. This controller has worked very well in simulation but exhibits oscillations on physical robots which can lead the robot to get stuck on the environment's boundary.

We present here a novel proportional control law to guide each robot within its lane, while diverting outwards to gather pucks. In the absence of pucks, the control law drives the robot to follow the inside edge of the robot's lane, indexed by k. This inside edge is the contour corresponding to $\tau_{high}(k)$. If a puck within the robot's lane (and ahead of the robot) is visible, then we can compute the puck's position in the distance image as (p, q). The corresponding value in the distance image is $\tilde{D}_{p,q}$. We add a small quantity to this corresponding to the robot's radius in normalized distance units, denoted w_{robot}. The purpose of this addition is to guide the center of the robot to the point where the plow can meet the puck. We can then compute a target level that defines the contour the robot will strive to maintain,

$$\tau_{target} = \min(\tau_{high}(k), \tilde{D}_{p,q} + w_{robot}) \tag{3}$$

The gradient computed from the distance image is used to define the robot's desired angle. Whereas the distance image is two-dimensional, the gradient produces a 2-vector for every position in the distance image. We denote this vector as $\nabla \tilde{D}_{i,j} = [u\ v]^T$. The angle of the gradient relative to the robot is

$$\gamma = \text{atan2}(v, u) - \theta, \tag{4}$$

where θ is the robot's heading. If we were only concerned with moving orthogonal to the gradient in a clockwise orbit, the ideal value of γ would be $\frac{\pi}{2}$. In this case we could simply use the difference between $\frac{\pi}{2}$ and γ as an error signal to achieve such an orbiting controller. However, such an approach would not maintain the robot at the target level τ_{target}.

One simple error signal for controlling the shape of the orbit would be $\tau_{target} - \tilde{D}_{i,j}$ where (i, j) corresponds to the robot's current position. However, to combine the error signal for distance with the error signal for γ it is preferable to use a normalized quantity such as the following,

$$\epsilon = \frac{\tau_{target} - \min(\tilde{D}_{i,j}, 2\tau_{target})}{\tau_{target}}. \tag{5}$$

where $\epsilon \in [-1, 1]$. Whereas ϵ represents an error signal in distance, ξ represents an angular error that the controller acts to minimize,

$$\xi = \left(\frac{\pi}{2} - K\epsilon\right) \ominus \gamma \tag{6}$$

The operator $\ominus$ represents taking the smallest signed difference between angles. K is a constant that governs how aggressive the controller is in correcting for distance, versus correcting to be orthogonal to the gradient. In our experiments we use the value $K = 0.9$. The final step is to compute the robot's forward speed ν and angular speed ω in such a way as to minimize ξ. The following simple approach achieves this,

$$\nu = \cos \xi \qquad (7)$$
$$\omega = \sin \xi \qquad (8)$$

3.5   Simulation Environment and Metrics

In our previous work we used a custom two-dimensional simulator capable of very fast execution [31,35]. This simulator quickly resolved collisions, but not in the most realistic way. In particular, collisions between the robots and between the robots and the boundary were not predicted to be problematic. However, in transitioning to physical robots these collisions were found to be more sustained and to have more serious consequences than predicted [35]. For these reasons we have switched to the slower but more realistic simulator CoppeliaSim [25] for the experiments conducted below.

The main performance metric referred to below is the proportion of pucks in the goal region. We also compute both the average and the minimum distance between robots. These are calculated by considering all possible pairs of robots, without consideration for whether the robots lie within the same lane or not.

4   Experimental Results

Figure 3 shows the performance of the system operating on 75 pucks using 1, 2, or 3 lanes. The three rows in the figure correspond to measurements of the proportion of robots in the goal region (top row), average distance between robots (middle row), and minimum distance between robots (bottom row). The two columns correspond to 6 robots (left) and 12 robots (right).

We consider the 1-lane case to be a benchmark for comparison since all conditions are the same except for the subdivision into lanes. Considering the proportion of pucks successfully delivered to the goal region we see a different time evolution for 2 or 3 lanes compared with 1 lane, but not a large difference in performance. If steady-state performance is desired, it appears that 2 lanes would be best for these conditions for either 6 or 12 robots.

Considering average and minimum distances between robots (middle and bottom rows of Fig. 3, respectively) a much bigger difference between the number of lanes used is observed. This difference is larger for the case of 12 robots. The plots show 95% confidence intervals as shaded regions and we can conclude that when a pair of such intervals do not overlap that the underlying difference is statistically significant at the $p = 0.05$ level. If we focus on the end of each simulation run, we can say that that the introduction of lanes significantly increases the average distance between robots for 6 robots, and significantly increases both the average and minimum distance between robots for the 12 robot condition.

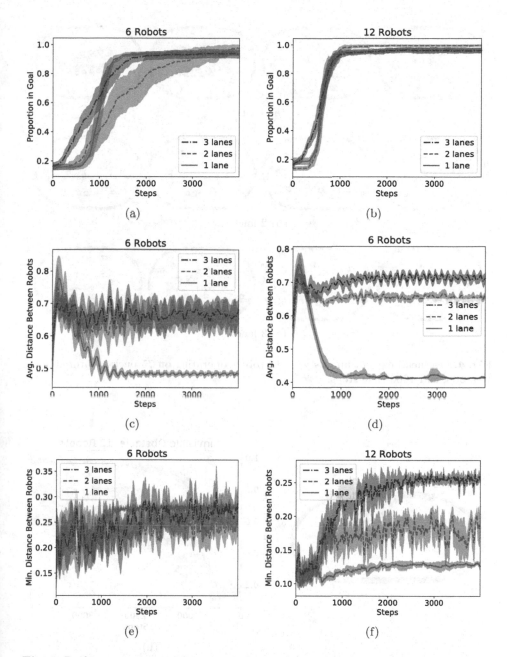

Fig. 3. Performance measured in terms of proportion of pucks in the goal region ((**a**) and (**b**)), average distance between robots ((**c**) and (**d**)), and minimum distance between robots ((**e**) and (**f**)). Each trace is an average of 10 trials. Shaded regions with matched colors represent 95% confidence intervals.

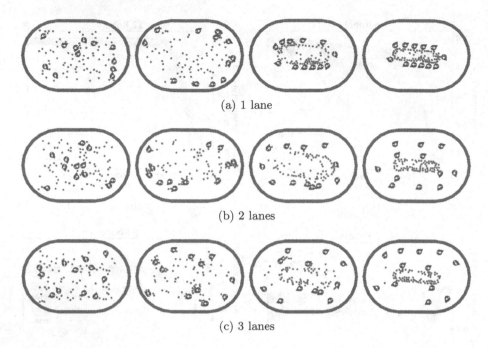

(a) 1 lane

(b) 2 lanes

(c) 3 lanes

Fig. 4. Overhead camera images with 12 robots operating on 75 pucks captured at 0, 100, 500, and 1000 time steps.

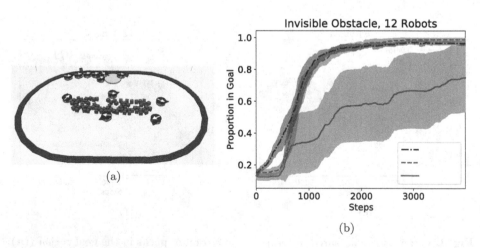

Fig. 5. Impact of an invisible obstacle. **(a)** Screenshot of the invisible obstacle reducing performance in the 2-lane configuration. **(b)** Proportion of pucks moved to the goal by 12 robots in the presence of the invisible obstacle.

Figure 4 shows the time evolution of overhead camera images captured by CoppeliaSim with 12 robots. These are samples from the first of 10 trials ran to produce the plots in Fig. 3. Figure 4(a) shows how with a single lane the robots tend to form a connected chain. The two lanes are discernible in Fig. 4(b) but in Fig. 4(c) we see that by time step 1000 (rightmost image) all robots have converged to the inner or outer lane, leaving none in the middle lane. This is a weakness of our lane re-assignment strategy since the middle lane has been vacated yet two pucks remain there. In fact, those 2 pucks remain outside of the goal region until the end of the run. It is this phenomenon which yields reduced steady-state performance for the 3-lane condition, as is evident in Fig. 3(b).

We are also interested in the susceptibility of the system to various disturbances. One such disturbance is an unforeseen obstacle or blockage. We inserted a low cylinder near the boundary of the arena which is not perceived by the robots. This cylinder could represent an irregularity in the floor that was invisible to the robots yet could hinder their progress. Figure 5 shows a screenshot and a plot averaged over 10 trials. It is clear that the single-lane solution is much more strongly affected by the cylinder.

5 Conclusions and Future Work

We have presented a strategy for resolving spatial interference between robots engaged in a planar construction task. Subdividing the work area into lanes implicitly divides the task among the robots in corresponding lanes. We cannot argue that the gains in performance are large or statistically significant, but it is clear that introducing multiple lanes increases both the average and minimum distance between robots, particularly for higher numbers of robots. Also, the use of lanes reduces the system's susceptibility to localized disturbances such as unforeseen obstacles.

However, much work remains to validate and improve the concepts introduced here. A hardware realization is necessary and we have paved the way for this by modeling our current fleet of robots and their operating environment. One clear weakness identified was the lane re-assignment strategy employed which led to the middle lane being prematurely vacated as shown in Fig. 4(c). We have used a direct comparison of the count of robots in the current lane versus adjacent lanes, which is certainly not the only possibility. The other cue to consider is the number of pucks observed, either in absolute terms or in comparison to adjacent lanes. We have previously tested the use of response thresholds in a similar role and will likely revisit this approach [11]. We are also interested in studying how our broader strategy for planar construction can be applied to more complex shapes.

Acknowledgments. Funding provided from the Natural Sciences and Engineering Research Council of Canada (NSERC) under Discovery Grant RGPIN-2017-06321. Thanks also to the constructive feedback of the ANTS 2020 reviewers who helped to clarify the contents presented here.

References

1. Beckers, R., Holland, O., Deneubourg, J.L.: From local actions to global tasks: stigmergy and collective robotics. In: Artificial Life IV, pp. 181–189. MIT Press, Cambridge (1994)
2. Bonabeau, E., Dorigo, M., Theraulaz, G.: Swarm Intelligence: From Natural to Artificial Systems. Oxford University Press, New York (1999)
3. Brambilla, M., Ferrante, E., Birattari, M., Dorigo, M.: Swarm robotics: a review from the swarm engineering perspective. Swarm Intell. **7**(1), 1–41 (2013)
4. Bruinsma, O.H.: An analysis of building behaviour of the termite Macrotermes subhyalinus (Rambur). Ph.D. thesis (1979)
5. Crabbe, F.L., Dyer, M.G.: Second-order networks for wall-building agents. In: International Joint Conference on Neural Networks (1999). IJCNN 1999, vol. 3, pp. 2178–2183. IEEE (1999)
6. Crane, K., Weischedel, C., Wardetzky, M.: The heat method for distance computation. Commun. ACM **60**(11), 90–99 (2017)
7. Deneubourg, J.L., Goss, S., Franks, N., Sendova-Franks, A., Detrain, C., Chrétien, L.: The dynamics of collective sorting robot-like ants and ant-like robots. In: First International Confernce on the Simulation of Adaptive Behaviour, pp. 356–363. MIT Press, Cambridge (1990)
8. Gauci, M., Chen, J., Li, W., Dodd, T.J., Groß, R.: Self-organised aggregation without computation. Int. J. Robot. Res. **33**(9), 1145–1161 (2014). https://doi.org/10.1177/0278364914525244
9. Gregory, C., Vardy, A.: microUSV: a low-cost platform for indoor marine swarm robotics research. HardwareX (2020). https://doi.org/10.1016/j.ohx.2020.e00105
10. Hamann, H.: Superlinear scalability in parallel computing and multi-robot systems: shared resources, collaboration, and network topology. In: Berekovic, M., Buchty, R., Hamann, H., Koch, D., Pionteck, T. (eds.) ARCS 2018. LNCS, vol. 10793, pp. 31–42. Springer, Cham (2018). https://doi.org/10.1007/978-3-319-77610-1_3
11. Ibrahim, D., Vardy, A.: Adaptive task allocation for planar construction using response threshold model. In: Martín-Vide, C., Pond, G., Vega-Rodríguez, M.A. (eds.) Theory and Practice of Natural Computing, pp. 173–183. Springer, Cham (2019). https://doi.org/10.1007/978-3-030-34500-6_12
12. Kazadi, S., Abdul-Khaliq, A., Goodman, R.: On the convergence of puck clustering systems. Robot. Auton. Syst. **38**(2), 93–117 (2002)
13. Kazadi, S., Wigglesworth, J., Grosz, A., Lim, A., Vitullo, D.: Swarm-mediated cluster-based construction. Complex Syst. **15**(2), 157 (2004)
14. Ladley, D., Bullock, S.: The role of logistic constraints in termite construction of chambers and tunnels. J. Theoret. Biol. **234**(4), 551 (2005). https://doi.org/10.1016/j.jtbi.2004.12.012
15. Lein, A., Vaughan, R.T.: Adaptive multi-robot bucket brigade foraging. In: Proceedings of the Eleventh International Conference on Artificial Life (ALife XI), August 2008
16. Maris, M., Boeckhorst, R.: Exploiting physical constraints: heap formation through behavioral error in a group of robots. In: IEEE/RSJ International Conference on Robots and Systems (IROS), vol. 3, pp. 1655–1660. IEEE Xplore (1996)
17. Martinoli, A., Ijspeert, A., Gambardella, L.: A probabilistic model for understanding and comparing collective aggregation mechanisms. In: Floreano, D., Nicoud, J.D.., Mondada, F. (eds.) Advances in Artificial Life. Proceedings of the 5th European Conference on Artificial Life (ECAL). Springer, Heidelberg (1999). https://doi.org/10.1007/3-540-48304-7_77

18. Mayya, S., Pierpaoli, P., Egerstedt, M.: Voluntary retreat for decentralized inter-ference reduction in robot swarms. In: 2019 International Conference on Robotics and Automation (ICRA), pp. 9667–9673. IEEE (2019)
19. Melhuish, C., Holland, O., Hoddell, S.: Collective sorting and segregation in robots with minimal sensing. In: 5th International Conference on the Simulation of Adaptive Behaviour. MIT Press, Cambridge (1998)
20. Melhuish, C., Sendova-Franks, A.B., Scholes, S., Horsfield, I., Welsby, F.: Ant-inspired sorting by robots: the importance of initial clustering. J. R. Soc. Interface **3**(7), 235–242 (2006)
21. Melhuish, C., Wilson, M., Sendova-Franks, A.: Patch sorting: multi-object clustering using minimalist robots. In: Kelemen, J., Sosík, P. (eds.) ECAL 2001. LNCS (LNAI), vol. 2159, pp. 543–552. Springer, Heidelberg (2001). https://doi.org/10.1007/3-540-44811-X_62
22. Olson, E.: AprilTag: a robust and flexible visual fiducial system. In: Proceedings of the IEEE International Conference on Robotics and Automation (ICRA), pp. 3400–3407. IEEE Xplore (May 2011)
23. Østergaard, E.H., Sukhatme, G.S., Matarić, M.J.: Emergent bucket brigading - a simple mechanism for improving performance in multi-robot constrained-space foraging tasks. In: In Autonomous Agents, pp. 2219–2223 (2001)
24. Petersen, K.H., Napp, N., Stuart-Smith, R., Rus, D., Kovac, M.: A review of collective robotic construction. Sci. Robot. **4**(28) (2019). https://doi.org/10.1126/scirobotics.aau8479. http://robotics.sciencemag.org/content/4/28/eaau8479
25. Rohmer, E., Singh, S.P.N., Freese, M.: CoppeliaSim (formerly V-REP): a versatile and scalable robot simulation framework. In: Proceedings of the International Conference on Intelligent Robots and Systems (IROS) (2013). http://www.coppeliarobotics.com
26. Schmolke, A., Mallot, H.: Territory formation in mobile robots. In: Artificial Life VIII, pp. 256–269 (2002)
27. Schneider-Fontán, M., Matarić, M.: Territorial multi-robot task division. IEEE Trans. Robot. Autom. **14**(5), 815–822 (1998)
28. Shell, D., Matarić, M.: On foraging strategies for large scale multi robot systems. In: IEEE/RSJ International Conference on Intelligent Robots and Systems (2006)
29. Soleymani, T., Trianni, V., Bonani, M., Mondada, F., Dorigo, M., et al.: An autonomous construction system with deformable pockets. Technical report, IRIDIA Technical Report Series, January, 2014 002. IRIDIA, Université Libre de Bruxelles, Brussels (2014)
30. Stewart, R.L., Russell, R.A.: A distributed feedback mechanism to regulate wall construction by a robotic swarm. Adapt. Behav. **14**(1), 21–51 (2006)
31. Strickland, C., Churchill, D., Vardy, A.: A reinforcement learning approach to multi-robot planar construction. In: IEEE International Symposium on Multi-Robot and Multi-Agent Systems (2019)
32. Vardy, A.: Orbital construction: swarms of simple robots building enclosures. In: 2018 IEEE 3rd International Workshops on Foundations and Applications of Self* Systems (FAS* W), pp. 147–153 (2018)
33. Vardy, A.: Landmark-guided shape formation by a swarm of robots. In: Correll, N., Schwager, M., Otte, M. (eds.) Distributed Autonomous Robotic Systems. SPAR, vol. 9, pp. 371–383. Springer, Cham (2019). https://doi.org/10.1007/978-3-030-05816-6_26
34. Vardy, A., Vorobyev, G., Banzhaf, W.: Cache consensus: rapid object sorting by a robotic swarm. Swarm Intell. **8**(1), 61–87 (2014). http://www.cs.mun.ca/av/supp/si12

35. Vardy, A., Ibrahim, D.S.: A swarm of simple robots constructing planar shapes. arXiv preprint arXiv:2004.13888 (2020)
36. Verret, S., Zhang, H., Meng, M.Q.H.: Collective sorting with local communication. In: IEEE/RSJ International Conference on Robots and Systems (IROS). IEEE Xplore (2004)
37. Wang, T., Zhang, H.: Multi-robot collective sorting with local sensing. In: IEEE Intelligent Automation Conference (IAC) (2003)
38. Werfel, J., Petersen, K., Nagpal, R.: Designing collective behavior in a termite-inspired robot construction team. Science **343**(6172), 754–758 (2014)
39. Wurm, K.M., Stachniss, C., Burgard, W.: Coordinated multi-robot exploration using a segmentation of the environment. In: 2008 IEEE/RSJ International Conference on Intelligent Robots and Systems, pp. 1160–1165 (2008)

The Pi-puck Ecosystem: Hardware and Software Support for the e-puck and e-puck2

Jacob M. Allen[1] , Russell Joyce[1] , Alan G. Millard[2(✉)] , and Ian Gray[1]

[1] Department of Computer Science, University of York, York, UK
{jma542,russell.joyce,ian.gray}@york.ac.uk
[2] Lincoln Centre for Autonomous Systems, University of Lincoln, Lincoln, UK
amillard@lincoln.ac.uk

Abstract. This paper presents a hardware revision of the Pi-puck extension board that now includes support for the e-puck2. This Raspberry Pi interface for the e-puck robot provides a feature rich experimentation platform suitable for multi-robot and swarm robotics research. We also present a new expansion board that features a 9-DOF IMU and XBee interface for increased functionality. We detail the revised Pi-puck hardware and software ecosystem, including ROS support that now allows mobile robotics algorithms and utilities developed by the ROS community to be leveraged by swarm robotics researchers. We also present the results of an illustrative multi-robot mapping experiment using new long-range Time-of-Flight distance sensor modules, to demonstrate the ease-of-use and efficacy of this new Pi-puck ecosystem.

1 Introduction

The e-puck robot platform [15] is widely-used for mobile robotics research, and is a popular choice for swarm robotics due to its size and commercial availability. Three hardware revisions of the original e-puck (v1.1–1.3) have been released commercially by GCtronic and EPFL since it was first developed in 2004, followed by the release of the e-puck2 in 2018. Our Pi-puck[1] extension board allows a Raspberry Pi single-board computer to be interfaced with an e-puck or e-puck2 to enhance its capabilities. It features a range of augmentations over the base e-puck design, including greater computational power, and increased communication, sensing and interfacing abilities. The first prototype design of the Pi-puck extension board was created by the York Robotics Laboratory (YRL) at the University of York, and was published in 2017 [12], before the release of the e-puck2. The latest version of the hardware was developed as a collaboration between YRL and GCtronic to support both the e-puck and e-puck2, and is available to purchase from GCtronic and its distributors.

Nedjah and Junior [17] argue that there is an urgent need to standardise many aspects of swarm robotics research, so that faster progress can be made

---

[1] https://www.york.ac.uk/robot-lab/pi-puck.

© Springer Nature Switzerland AG 2020
M. Dorigo et al. (Eds.): ANTS 2020, LNCS 12421, pp. 243–255, 2020.
https://doi.org/10.1007/978-3-030-60376-2_19

towards real-world applications. In particular, they call for standardisation of hardware and software – the Pi-puck aims to provide a common hardware and software ecosystem for researchers that wish to run embedded Linux and associated software on e-puck robots. The board was designed to replace the now deprecated Linux extension board developed by the Bristol Robotics Laboratory [9], and the Gumstix Overo COM turret [2]. The recently published Xpuck [7] is an e-puck extension that is similar in spirit to the Pi-puck – extending the e-puck with a powerful ODROID-XU4 single-board computer through custom hardware. This greatly enhances the robot's computational capabilities, but comes at the cost of size (the form factor of the XU4 is similar to that of the Raspberry Pi 3 or 4) and power consumption, necessitating the use of an auxiliary battery. The Xpuck was also developed prior to the release of the e-puck2, and its communication with the robot relies on an SPI bus that is not present on the e-puck2's expansion connector. In contrast, the Pi-puck has been designed around the Raspberry Pi Zero to provide a modest compute upgrade while minimising size and energy usage, and primarily uses I^2C for communication with the base robot, which is compatible with both the e-puck and e-puck2.

Nedjah and Junior [17] also encourage the use of Robot Operating System (ROS) [21] to facilitate standardisation. Although ROS has become the *de facto* standard for robotics middleware in single-robot and multi-robot studies, the swarm robotics research community has generally been reluctant to adopt it. This can partly be attributed to the fact that many swarm hardware platforms are microcontroller-based, so cannot run ROS on-board [24], however ROS integration can still be achieved via wireless communication and the `rosserial` interface – see the Mona [25] and HeRo [22] swarm platforms. Additionally, the ROS communication model is inherently centralised, which is antithetical to the philosophy of many swarm algorithms.

Rudimentary ROS support was implemented for the previous version of the Pi-puck [12], and the software infrastructure has now been updated to provide ROS Melodic support for the latest hardware revision, opening the door to a large body of existing work developed by the ROS community. This paper discusses the ROS drivers and ecosystem developed for the Pi-puck platform, and how they can be leveraged by other swarm robotics researchers. We recognise that the use of ROS may not be appropriate in some cases, and Pi-puck users may instead opt for lighter-weight software frameworks designed specifically for swarm robotics research such as Buzz [19], OpenSwarm [24], or SwarmTalk [26]. For resource-constrained experiments, the Pi-puck could be programmed to work with these frameworks instead of ROS.

2   Hardware Changes

There have been several major changes to the design of the Pi-puck hardware since it was first published in 2017 [12], which add new features, implement support for the e-puck2, and improve the stability of the platform for large-scale production. Many of these changes were made after consultation with members of

Fig. 1. *Left*: Pi-puck on e-puck2, with six Time-of-Flight distance sensor modules. *Centre:* Pi-puck board with YRL Expansion Board, XBee, and OLED display. *Right*: Pi-puck on e-puck1, with expansion board and e-ink pHAT showing an ArUco tag.

the swarm robotics community, and with GCtronic as the primary manufacturer of the e-puck robot. Figure 1 shows the latest version of the Pi-puck extension board connected to an e-puck robot, along with a further expansion board and attached hardware (detailed in Sect. 2.1), as well as six Time-of-Flight (ToF) distance sensor modules.

A full block diagram detailing the hardware of the Pi-puck platform is shown in Fig. 2, which includes the components added to the robot on the extension board itself, as well as the major communication buses between the Raspberry Pi, extension board hardware, and the base e-puck robot. The Raspberry Pi Zero WH is specifically supported, due to its wide availability, low cost, minimal power consumption, integrated wireless capabilities, and small physical footprint. However, it is feasible that other Raspberry Pi and compatible boards could be used with the Pi-puck if additional mechanical support and wiring were added.

Raspberry Pi Support. The first fundamental change is in the mounting of the Raspberry Pi board, which is now face-down. This allows Raspberry Pi boards with pre-soldered headers to be used, which are easier to acquire in large quantities. A small micro-USB shim has been added to allow USB communication between the Raspberry Pi and Pi-puck extension board, and an integrated USB hub allows up to three devices to utilise this connection simultaneously. The Raspberry Pi UART is also accessible through a micro-USB port on the Pi-puck board, via a USB-UART converter, allowing a text console on the Raspberry Pi to be accessed without additional hardware.

Sensor Modules. Six 4-pin sockets are provided around the edge of the robot for connecting optional I^2C sensor modules, allowing for a range of flexible options for experimentation. The mapping application detailed in Sect. 5 uses custom-designed, open-source[2] distance sensor boards based around the

---

[2] https://github.com/yorkrobotlab/pi-puck-tof-sensor.

VL53L1X ToF laser-ranging module (4 m range) – similar to the VL53L0X distance sensor on the front of the e-puck2 (with a 2 m range).

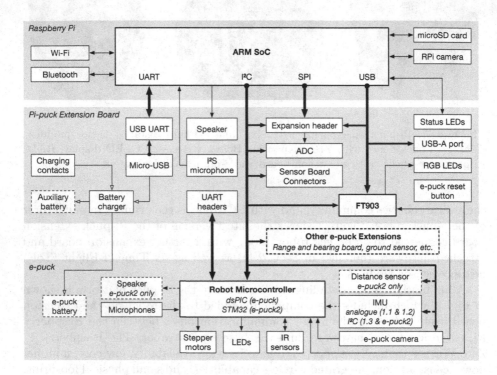

Fig. 2. Overview of interfaces between the Raspberry Pi, Pi-puck extension board, and e-puck hardware. Arrows show master to slave (I^2C/SPI), host to device (USB), or data/power direction. Dashed lines indicate hardware and connections that are optional or not available on all e-puck revisions.

Battery Power. The Pi-puck extension board can be powered from either the standard e-puck battery, an auxiliary battery connected through a JST-PH socket, or from both batteries simultaneously. When powered from an 1800 mAh e-puck battery, an idling Pi-puck with no expansion hardware will drain its battery in approximately 5 h. An active Pi-puck that is performing a simple obstacle avoidance algorithm controlled from the Raspberry Pi, while fitted with the extensions shown in Fig. 1 (including an XBee, OLED display and six ToF sensors), has been measured to last around 1.5 h. Both of these times could be increased significantly by attaching an auxiliary battery to augment the power provided by the e-puck battery.

The Pi-puck has two battery charging circuits on-board, to allow for charging each of the two batteries independently. Batteries can be charged either by connecting a 5 V power supply to the sprung charging contacts on the front of the robot (e.g. via an external charging wall), or through the integrated micro-USB socket. Additionally, both battery voltages can be measured in real-time from the Raspberry Pi using an on-board analogue-to-digital converter (ADC), allowing automatic shutdown or recharging behaviours to be triggered in software.

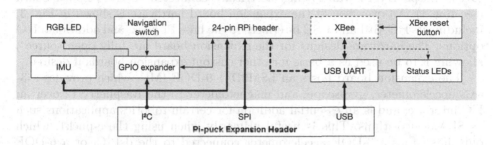

Fig. 3. Overview of interfaces between the Pi-puck and YRL Expansion Board. Arrows show master to slave (I^2C/SPI), host to device (USB), or data direction. Dashed lines indicate hardware and connections that are optional.

Camera Interface. An FTDI FT903 MCU is used to convert the e-puck's parallel camera interface to a USB Video Device Class (UVC) peripheral to make it accessible from the standard Linux kernel and applications on the Raspberry Pi. Like the Xpuck [7], the Pi-puck can capture 640 × 480 resolution video at 15 frames per second, enabling improved on-board image processing over the resource-constrained e-puck microcontroller. The Raspberry Pi configures the camera sensor via I^2C, then the FT903 creates the UVC device and streams the image. The FT903's firmware can easily be updated over USB using the Device Firmware Upgrade (DFU) standard, and its UART interface is broken out to header pins, to allow for customisation of the microcontroller firmware and support for potential future e-puck camera sensor design changes.

Additional I/O. The Raspberry Pi is connected directly to an I^2S microphone on the Pi-puck extension board, as well as an audio amplifier and speaker. The Pi-puck board has three additional RGB LEDs that can be individually controlled over the I^2C interface, via the FT903 microcontroller. A vertical USB-A port allows the connection of an additional USB device to the Raspberry Pi, through the on-board USB hub, and a further USB channel is attached to a general-purpose expansion header, which also breaks out the I^2C and SPI buses along with power and other signals. Both e-puck1 and e-puck2 UART interfaces are broken-out to pins on the Pi-puck board, to allow for easier debugging of the robot's microcontroller firmware using modern 3.3 V signals.

2.1 YRL Expansion Board

To complement the base Pi-puck platform, we have developed an additional board that connects to the expansion header on the top of the robot. This board has a similar form-factor to the Raspberry Pi Zero, and is physically mounted to the Raspberry Pi while being electrically connected to the Pi-puck board (see Fig. 1). An overview of the expansion board hardware is shown in Fig. 3, including its I^2C, SPI and USB interfaces to the base Pi-puck, and additional I/O options. The hardware designs for the expansion board are fully open source[3], allowing it to be used as a basis for other custom expansion boards if desired.

The expansion board hosts an LSM9DS1 9-DOF IMU, which provides a 3-axis accelerometer, gyroscope, and magnetometer to the Raspberry Pi over an I^2C interface, and is an essential addition for certain robotics applications such as SLAM algorithms. This is useful primarily when using the e-puck1, which only has either a 3-DOF accelerometer connected to the dsPIC, or a 6-DOF I^2C accelerometer and gyroscope (depending on the specific hardware revision), compared to the e-puck2 which has a built-in MPU-9250 9-DOF IMU.

The expansion board also provides a socket for an XBee radio module, accessible through a USB-UART interface to the Raspberry Pi, and LEDs for showing the radio status. Using a generic set of headers for the XBee interface allows multiple generations and specifications of XBee modules to be used, as long as they comply to the standard pinout. The Pi-puck's XBee interface enables peer-to-peer, point-to-point, or mesh networking between robots, and can be used for transmitting data with higher bandwidth than infrared transceivers, as well as estimating the distance of neighbouring robots by measuring received signal strength (see Fig. 4). In addition to robot-to-robot communication, the Pi-puck's XBee module could also be integrated with XBee-enabled experimental infrastructure like the IRIDIA Task Abstraction Module [1].

One benefit of the Pi-puck is the ability to leverage the Raspberry Pi ecosystem, which is taken further by the expansion board's 24-pin Raspberry Pi compatible header, allowing the robot to be extended with a large variety of existing hardware. Figure 1 shows the Pi-puck with the off-the-shelf Inky pHAT e-ink display from Pimoroni [18], which can be used in a swarm context for very low power dynamic agent identification through ArUco tags [3], and additional human-swarm interaction possibilities [14], as well as enabling simpler integration with tools like ARDebug [13]. This is achieved easily though using existing software packages with minimal modification, and without the need for custom Linux kernel drivers or firmware.

3 Software Ecosystem

This section describes the software ecosystem that supports the Pi-puck extension board hardware, including our customised Linux distribution, e-puck microcontroller firmware, and software interfaces to other e-puck extensions.

---

[3] https://github.com/yorkrobotlab/pi-puck-expansion-board.

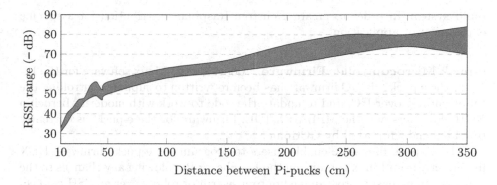

Fig. 4. Reported range of XBee packet RSSI (Received Signal Strength Indicator) values with varying transmission distance between a pair of Pi-pucks.

Linux Distribution. There are currently two main Linux distributions for the Pi-puck – one supported by GCtronic[4] (included on the Pi-puck's microSD card when purchased from them), and one created by YRL, which is detailed in this paper. This allow us to provide support for different features of the platform, while targeting different users and alternative approaches for packaging software.

The YRL Pi-puck software distribution offers a foundation for research and education, with a focus on open-source packages that are easy to modify, build and distribute. The core of the distribution is supplied as a set of Debian packages that are hosted in a package repository online[5], and are available in source format for modification if desired[6]. These packages cover the full Linux set-up of the Pi-puck hardware, as well as providing utilities for controlling and programming various devices on the robot. Distributing this software via Debian packages allows for easy installation on any Debian-based Linux distribution, automatic resolution of any dependencies, and straightforward updates.

To accelerate the initial configuration of each robot, the Pi-puck Debian packages are built into the Pi-puck Raspbian microSD card image[7], which is created using the `pi-gen` tool from the Raspberry Pi Foundation. This image is based on the standard Raspberry Pi Foundation Raspbian Buster image, but with additional build stages added to include the Pi-puck packages, and to modify the default Linux configuration to better support a swarm of robots. This image is supplied both in source form and as a file system image that can be directly copied onto a microSD card for use with the Raspberry Pi. Additional files for assisting with the deployment of a swarm of robots are included on the FAT32 `boot` partition of the SD card, allowing users to configure parameters such as Wi-Fi connection details and a unique robot hostname before the image is first booted. Users can also add additional packages and files into the `pi-gen`

---

[4] https://www.gctronic.com/doc/index.php?title=Pi-puck.

[5] https://www.cs.york.ac.uk/pi-puck/.

[6] https://github.com/yorkrobotlab/pi-puck-packages.

[7] https://github.com/yorkrobotlab/pi-gen.

build system, in order to create a custom Raspbian distribution for a specific experiment or application.

e-puck Microcontroller Firmware. Alongside the Linux software for the Pi-puck, the e-puck1 dsPIC firmware has been re-written to support controlling the robot entirely over I^2C, and to update the code to work with modern Microchip XC16 compilers and the MPLAB X IDE. Firmware for the e-puck2 is currently provided and supported by GCtronic.

The Pi-puck hardware enables users to program the e-puck1 firmware HEX file directly from Linux running on the Raspberry Pi, allowing any changes to the firmware to be easily programmed onto a swarm of robots over an SSH connection. We provide a new dsPIC bootloader firmware that must be programmed to each e-puck once using a standard PIC in-circuit debugger to enable this feature, after which all subsequent programming can be done directly from the Raspberry Pi, using provided programming scripts. Due to hardware differences, the same method of firmware programming does not work with the e-puck2's microcontrollers, however they could be programmed from the Raspberry Pi using Bluetooth, Wi-Fi or USB (with the use of an additional cable) instead.

Other e-puck Extensions. The Pi-puck hardware allows the Raspberry Pi to communicate with any devices on the e-puck's I^2C bus, including other e-puck extension boards, such as the range and bearing turret [5] and ground sensors module. Python software has been written to demonstrate how to interface with the range and bearing turret directly from the Raspberry Pi (with both the standard and DEMIURGE firmware[8]), without requiring any input from the e-puck's microcontroller, and is provided as an example in the Pi-puck software repository. Python code for communicating with the e-puck ground sensors extension from the Raspberry Pi is also provided, allowing for this to be included in high-level robot control applications.

4   Robot Operating System (ROS) Support

The base e-puck1 is able to provide limited ROS support via Bluetooth and an external computer, and the e-puck2 extends this functionality with the option of using Wi-Fi instead of Bluetooth [4]. The Pi-puck allows ROS nodes to be executed directly on the robot instead, thanks to the Raspberry Pi's embedded Linux operating system. ROS integration for the Pi-puck is implemented by the `pi_puck_driver` package, which contains a series of Python nodes[9] that support the features listed in this section. The Pi-puck ROS repository has been created for ROS Melodic (supported until May 2023), and has been tested on Raspbian Buster on the Raspberry Pi.

---

[8] https://github.com/demiurge-project/argos3-epuck/tree/master/erab_firmware.
[9] https://github.com/yorkrobotlab/pi-puck-ros.

Motors. The motors node communicates with a robot controller by subscribing to wheel speeds and publishing the step counts of the e-puck's stepper motors. This node interfaces with the e-puck's microcontroller via I^2C to request step counts and set the speed of each wheel independently, allowing for precise movement and turning. The motors node also publishes nav_msgs/Odometry messages, containing the robot's pose and linear/angular velocity, estimated from the motor step counts using dead reckoning.

A base_controller node is also provided to convert geometry_msgs/Twist messages (containing the desired linear and angular velocity of the robot) into control signals for the motors. This allows control of the robot to be abstracted away from specifying individual wheel speeds, and the implementation automatically scales the wheel speeds to account for combined linear and angular velocities exceeding physical limits.

Sensors. The short_range_ir node interfaces with the e-puck's analogue IR transceivers, and publishes their readings as eight separate sensor_msgs/Range messages (reflected IR, offset by ambient IR), so they can be used as proximity sensors. The raw IR readings are mapped to distances between 5 mm and 40 mm by applying a logarithmic least error fit to experimentally-measured sensor data. The long_range_ir node similarly interfaces with the Pi-puck's optional long-range ToF distance sensors (via STMicro's pre-built closed-source driver), and publishes their readings as up to six separate sensor_msgs/Range messages (depending on the number of sensor modules installed).

Many existing ROS SLAM packages (such as GMapping) require distance data to be presented as sensor_msgs/LaserScan messages rather than the point cloud that is obtained from the individual distance and ToF sensors. To solve this, a transform is provided that re-exposes a sensor_msgs/Range message as a sensor_msgs/LaserScan of three points, denoting the edges and centre of the field of view of the range sensor. Additionally, for sensor measurements to be mapped correctly onto the world, the reference frames for the sensors and the Pi-puck itself must be broadcast as a series of transforms, either statically or dynamically depending on whether the reference frame can move in relation to other reference frames. A Unified Robot Description Format (URDF) model is provided for this purpose, which contains a list of reference frames, and a list of static transforms between those reference frames. This also allows the Pi-puck sensor readings to be visualised on a 3D model in RViz.

Power, IMU, and OLED. The power node interfaces with the Pi-puck's ADC to obtain the voltages of the e-puck and auxiliary batteries, and publishes them, along with metadata such as whether the battery is currently charging, as sensor_msgs/BatteryState messages.

Additional support is provided for features of the YRL Expansion Board, such as the LSM9DS1 IMU and accessories using the 24-pin header. The imu node interfaces with the IMU on the expansion board, and publishes the accelerometer and gyroscope readings as `sensor_msgs/Imu` messages. Magnetometer readings are published as `sensor_msgs/MagneticField` messages, and the robot's pose is subsequently derived from these two message types using Madgwick's IMU and AHRS (Attitude and Heading Reference System) algorithm [10]. Calibration scripts are also provided in order to account for magnetic interference from the robot's motors and speaker. The `oled` node interfaces with the optional Adafruit PiOLED display, and subscribes to `std_msgs/String` and `sensor_msgs/Image` messages published by other nodes, allowing text and images to be displayed (e.g. robot ID, or status).

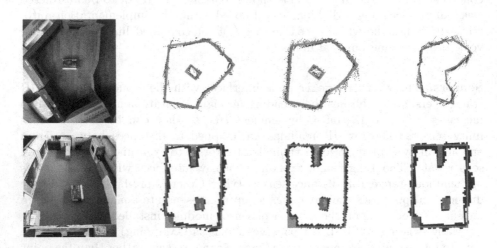

Fig. 5. Results of single-robot and multi-robot mapping experiments. *From left to right*: test arenas, single-robot mapping with obstacle avoidance, single-robot mapping with frontier exploration, and swarm mapping.

5   Environment Mapping

As an initial application case study, we present experimental results from an environment mapping task to demonstrate the applicability of the Pi-puck ROS platform for swarm experimentation. The focus here is on the integration of the Pi-puck drivers with existing third-party ROS packages, to highlight the compatibility and ease of use of the platform. The currently available experimental environments are quite small, limiting the maximum swarm size due to practical working limitations, but experimentation with larger environments and more robots is planned once possible.

Single-Robot Mapping. Single-robot SLAM has previously been implemented by GCtronic, both using a real e-puck and an e-puck model simulated in Webots [11]. This was achieved via the e-puck's Bluetooth ROS driver, the

Webots ROS interface, and the OpenSLAM GMapping package. However, this approach used the e-puck's analogue IR sensors, so the mapping range was limited to around 40 mm. Longer-range mapping has since been achieved through the use of ToF sensor modules via an Arduino interface and custom hardware [16].

To test the compatibility of the Pi-puck ROS drivers with other ROS packages, we integrated them with the GMapping package [23] (running on a separate computer) and the default ROS navigation stack. We used a single Pi-puck to map the two arenas shown in Fig. 5, controlled both via an obstacle avoidance controller, and with the explore_lite frontier exploration package [6]. Both of these environments were successfully mapped, as shown in Fig. 5.

Multi-robot Mapping. Kegeleirs et al. [8] recently investigated the effect of different random walk behaviours on an e-puck swarm's ability to map simple environments. Their work was implemented using ROS Indigo support for the e-puck's Gumstix Overo COM turret and the GMapping package, along with the multirobot_map_merge package [6] for combining the maps produced by each robot. Results from their initial experiments in the ARGoS robot simulator [20] were quite promising, but unfortunately the maps produced by the real e-puck robots were far less faithful to the true environment. This can be attributed to the limited range and high noise of the base e-puck's IR sensors, as well as compound odometry errors.

To test the ability of the Pi-puck ROS drivers to work in an environment where multiple robots are operating together on the same ROS network, a small swarm of four robots was used to perform mapping (without localisation) using GMapping while avoiding obstacles. The four individual maps were then combined in real-time using the multirobot_map_merge package. The Pi-puck swarm was able to successfully map the environments (as shown in Fig. 5), thanks to the improved IMU on the expansion board and the longer-range, higher-accuracy ToF sensor modules. Initial robot positions were randomised, not known to the swarm, and were not coordinated in any way.

6   Conclusion

The Pi-puck is an open-source extension for the e-puck and e-puck2 robot platforms that expands their capabilities by interfacing with the Raspberry Pi – a popular single-board computer. This paper has detailed the latest hardware revision of the Pi-puck, as well as the software infrastructure developed to support it, including Raspbian and ROS integration. This affords access to the Debian and ROS ecosystems, allowing for easy use of standard algorithms for tasks such as navigation and SLAM.

We hope that the hardware presented in this paper will facilitate experimentation with swarm algorithms that were previously either not possible or inconvenient to implement, and that the evolving software infrastructure continues to support the efforts of other researchers. Full documentation and source code for the Pi-puck platform and associated extensions is available online[10], in addition to the resources on the GCtronic Wiki [4].

References

1. Brutschy, A., et al.: The TAM: abstracting complex tasks in swarm robotics research. Swarm Intell. **9**(1), 1–22 (2015)
2. Garattoni, L., Francesca, G., Brutschy, A., Pinciroli, C., Birattari, M.: Software infrastructure for e-puck (and TAM). Technical report, TR/IRIDIA/2015-004, Université Libre de Bruxelles (2015)
3. Garrido-Jurado, S., Muñoz Salinas, R., Madrid-Cuevas, F., Medina-Carnicer, R.: Generation of fiducial marker dictionaries using mixed integer linear programming. Pattern Recogn. **51** (2015)
4. GCtronic: GCtronic Wiki. https://www.gctronic.com/doc/index.php?title=GCtronic_Wiki
5. Gutiérrez, Á., Campo, A., Dorigo, M., Donate, J., Monasterio-Huelin, F., Magdalena, L.: Open e-puck range & bearing miniaturized board for local communication in swarm robotics. In: International Conference on Robotics and Automation, pp. 3111–3116. IEEE (2009)
6. Hörner, J.: Map-merging for multi-robot system. Bachelor's thesis, Charles University in Prague, Faculty of Mathematics and Physics, Prague (2016)
7. Jones, S., Studley, M., Hauert, S., Winfield, A.F.T.: A two teraflop swarm. Front. Robot. AI Multi-Robot Syst. **5**(11), 1–19 (2018)
8. Kegeleirs, M., Garzón Ramos, D., Birattari, M.: Random walk exploration for swarm mapping. In: Althoefer, K., Konstantinova, J., Zhang, K. (eds.) TAROS 2019. LNCS (LNAI), vol. 11650, pp. 211–222. Springer, Cham (2019). https://doi.org/10.1007/978-3-030-25332-5_19
9. Liu, W., Winfield, A.F.T.: Open-hardware e-puck Linux extension board for experimental swarm robotics research. Microprocess. Microsyst. **35**(1), 60–67 (2011)
10. Madgwick, S.O.H., Harrison, A.J.L., Vaidyanathan, R.: Estimation of IMU and MARG orientation using a gradient descent algorithm. In: IEEE International Conference on Rehabilitation Robotics (2011)
11. Michel, O.: Cyberbotics Ltd., Webots: professional mobile robot simulation. Int. J. Adv. Rob. Syst. **1**(1), 5 (2004)
12. Millard, A.G., et al.: The Pi-puck extension board: a Raspberry Pi interface for the e-puck robot platform. In: International Conference on Intelligent Robots and Systems (IROS), pp. 741–748. IEEE (2017)
13. Millard, A.G., et al.: ARDebug: an augmented reality tool for analysing and debugging swarm robotic systems. Front. Robot. AI Multi-Robot Syst. **5**(87), 1–6 (2018)
14. Millard, A.G., Joyce, R., Gray, I.: Human-swarm interaction via e-ink displays. In: ICRA Human-Swarm Interaction Workshop (2020)
15. Mondada, F., et al.: The e-puck, a robot designed for education in engineering. In: Conference on Autonomous Robot Systems and Competitions, vol. 1, pp. 59–65 (2009)
16. Moriarty, D.: Swarm Robotics - Mapping Using E-Pucks: Part II. https://medium.com/@DanielMoriarty/swarm-robotics-mapping-using-e-pucks-part-ii-ac15c5d62e3

---

[10] https://pi-puck.readthedocs.io.

17. Nedjah, N., Junior, L.S.: Review of methodologies and tasks in swarm robotics towards standardization. Swarm Evol. Comput. **50**, 100565 (2019)
18. Pimoroni: Inky pHAT EPD Display for Raspberry Pi. https://shop.pimoroni.com/products/inky-phat
19. Pinciroli, C., Beltrame, G.: Buzz: an extensible programming language for heterogeneous swarm robotics. In: International Conference on Intelligent Robots and Systems (IROS), pp. 3794–3800. IEEE (2016)
20. Pinciroli, C., et al.: ARGoS: a modular, parallel, multi-engine simulator for multi-robot systems. Swarm Intell. **6**(4), 271–295 (2012)
21. Quigley, M., et al.: ROS: an open-source Robot Operating System. In: ICRA Workshop on Open Source Software (2009)
22. Rezeck, P.A., Azpurua, H., Chaimowicz, L.: HeRo: an open platform for robotics research and education. In: Latin American Robotics Symposium (LARS) and Brazilian Symposium on Robotics (SBR), pp. 1–6. IEEE (2017)
23. ROS Contributors: gmapping - ROS Wiki. http://wiki.ros.org/gmapping
24. Trenkwalder, S.M., Lopes, Y.K., Kolling, A., Christensen, A.L., Prodan, R., Groß, R.: OpenSwarm: an event-driven embedded operating system for miniature robots. In: 2016 IEEE/RSJ International Conference on Intelligent Robots and Systems (IROS), pp. 4483–4490. IEEE (2016)
25. West, A., Arvin, F., Martin, H., Watson, S., Lennox, B.: ROS integration for miniature mobile robots. In: Giuliani, M., Assaf, T., Giannaccini, M.E. (eds.) TAROS 2018. LNCS (LNAI), vol. 10965, pp. 345–356. Springer, Cham (2018). https://doi.org/10.1007/978-3-319-96728-8_29
26. Zhang, Y., Zhang, L., Wang, H., Bustamante, F.E., Rubenstein, M.: SwarmTalk - towards benchmark software suites for swarm robotics platforms. In: Proceedings of the 19th International Conference on Autonomous Agents and MultiAgent Systems, pp. 1638–1646 (2020)

Zealots Attack and the Revenge of the Commons: Quality vs Quantity in the Best-of-n

Giulia De Masi[1,2]([⊠])(iD), Judhi Prasetyo[3,4](iD), Elio Tuci[4](iD),
and Eliseo Ferrante[1,5](iD)

[1] Technology Innovation Institute, Abu Dhabi, UAE
[2] CNHS, Zayed University, Dubai, UAE
giuliademasi@gmail.com
[3] Middlesex University Dubai, Dubai, UAE
[4] Université de Namur, Namur, Belgium
[5] Vrije Universiteit Amsterdam, Amsterdam, Netherlands

Abstract. In this paper we study the effect of inflexible individuals with fixed opinions, or zealots, on the dynamics of the best-of-n collective decision making problem, using both the voter model and the majority rule decision mechanisms. We consider two options with different qualities, where the lower quality option is associated to a higher number of zealots. The aim is to study the trade-off between option quality and zealot quantity for two different scenarios: one in which all agents can modulate dissemination of their current opinion proportionally to the option quality, and one in which this capability is only possessed by the zealots. In both scenarios, our goal is to determine in which conditions consensus is more biased towards the high or low quality option, and to determine the indifference curve separating these two regimes. Using both numerical simulations and ordinary differential equation models, we find that: i) if all agents can modulate the dissemination time based on the option quality, then consensus can be driven to the high quality option when the number of zealots for the other option is not too high; ii) if only zealots can modulate the dissemination time based on the option quality, whil e all normal agents cannot distinguish the two options and cannot differentially disseminate, then consensus no longer depends on the quality and is driven to the low quality option by the zealots.

1 Introduction

Collective decision making is a process whereby a population of agents makes a collective decision based only on local perception and communication. Originally inspired by the behavior of social insects [2,4], collective decision making is considered an important problem connected to more elaborated collective behaviors in swarms robotics [28], such as site selection or collective motion [3].

The best-of-n problem [28] is a special case where agents have to chose the best option among n possible alternatives with potentially different qualities.

© Springer Nature Switzerland AG 2020
M. Dorigo et al. (Eds.): ANTS 2020, LNCS 12421, pp. 256–268, 2020.
https://doi.org/10.1007/978-3-030-60376-2_20

The option quality may be known to swarm members [29], or may need to be discovered [23–25]. An option can be considered *best* because it minimizes the cost required to be evaluated or because its intrinsic quality is the highest [28]. In the latter case, a method to achieve the optimal collective decision is to let each agent advertise an option for a duration that is proportional to its quality, a mechanism called "modulation of positive feedback" [8,29,30].

In this paper we focus on the best-of-n problem with $n = 2$ options in presence of stubborn individuals, henceforth called *zealots*. Zealots are individuals that have a fixed opinion that never changes. We introduce differential option quality and differential zealot quantity in an antagonistic setting: the two options are associated to different values of quality and zealot quantities; and the number of zealots is higher for the option that has a lower quantity, hence it is not obvious which option will prevail. Two specific cases are compared: i) all agents are able to measure the quality of the two options and disseminate for a time proportional to the quality; ii) only zealots and not the normal agents are able to measure the quality of the different options, and disseminate for a time proportional to the quality of their opinion, while normal agents disseminate for a time that is independent from the quality. This last scenario is referring for example to the case where, in a swarm of robots used for monitoring task, only some of them (zealots) have additional sensors and they can perceive the quality of the two options. In this case, the number of zealots can be a design parameter or a constraint depending on the problem: fully equipped robots with many sensors are more expensive due to a larger payload.

Using computer simulations and ordinary differential equations (ODEs) models, we ask the following question: Is the swarm consensus state more biased towards the option represented by more zealots or the one represented by the highest quality? We are particularly interested in identifying the "indifference curve" separating the two regions identified by the consensus state being more biased towards one or the other option. We investigate whether the indifference curve behaves differently across two scenarios. Finally, we determine whether these results are affected by two decision mechanisms: the voter models, whereby agents change their opinion copying the opinion of a random neighbor, and the majority model where instead agents adopt the opinion of the local majority.

The remaining of the paper is organized as follows. In Sect. 2, we discuss the state of art. In Sect. 3, we describe the collective decision-making model utilized in this study. In Sect. 4 we discuss the results obtained. In Sect. 5, we conclude the paper and discuss future developments.

2   State of the Art

The best-of-n problem is inspired by biological studies of swarms of ants and bees [9,15,26]. As extensively discussed in [28], the quality and cost of the options can further characterize the nature of the best-of-n decision-making problem. In the current paper, quality and not cost is the main factor driving consensus.

Another important element that bears upon the decision-making dynamics is the presence of zealots within the swarm. The influence of zealots has been abundantly studied in physics, but introduced within swarms only recently. In the following, we will first review the few contributions focusing on zealots within swarms, and then review some of the work done within physics.

In the context of swarms, a recent study [21] illustrated the impact of zealots in the context of dynamic environments, where the option qualities can drastically change over time. Here, the presence of a small number of zealots enables the swarm to always select the option with the best quality even after abrupt changes, while without zealots, the swarm is not able to adapt and the consensus remains frozen. The authors in [5] introduced three types of malicious agents that can affect resilience of a swarm: contrarians, wishy-washy, and zealots. They performed a preliminary study on their effect on the best-of-n with four mechanisms: voter, majority, cross inhibition, and k-unanimity (q-voter). In [22], the authors also looked at the effects of malicious adversarial zealots in a data communication manipulation scenario, proposing a probabilistic decision-making rule to increase resilience. A very recent extension has been applied and evaluated the same scheme to a simulated swarm robotics scenario [14].

In the context of physics, the author in [6] introduced zealots in a model of pairwise social influence for opinion dynamics, and showed a rich phase diagram of the possible dynamics when only a small percentage of zealots is present. In the context of Internet social networks, the best placement of zealots that maximizes the impact on the consensus dynamics of the population is studied in [13]. The study shows that a small number of zealots can significantly influence the overall opinion dynamics and induce the entire population to reach a large consensus over disputed issues, such as Brexit. In [17], the authors studied the role of zealots in a social system using the naming game as decision mechanism. They show that even a very small minority can drive the opinion of a large population, if committed agents are more active than the others. However, this effect can be hindered if nodes with the same opinion are more connected with each other than with nodes with different opinion, producing a polarization inside the network.

The authors of [11,18] studied the impact of zealots in a social network, considering different degrees of zealotry. The focus of [11] is studying the effect of zealotry on the convergence time of the system. In [18], despite having used the majority rule instead than the voter, the authors were able to find similar results as in [7,21], in which introducing equal number of zealots on both option sides prevents the network from reaching a consensus state. Similarly, in [32], the presence of zealots is proven to prevent the formation of consensus, introducing instabilities and fluctuations in a binary voter model of a small-world network. A recent study illustrated in [1] aimed at studying the influence of zealots on "politically polarized" state vs consensus state and found that higher "influence of zealots" produces more polarization, shorter time to polarization, and conversely less consensus and longer to impossible time to consensus.

In [31], the authors showed the presence of a tipping point at which a minority of zealots is able to swing the initial majority opinion in a network. The study described in [16] focused on zealots with the voter model to perform peer-to-peer opinion influence, however differently from our work zealots were nodes of a complex network. In [10], a scenario with zealots with the majority rule was studied. The outcome of the system was spontaneous symmetry breaking when zealot numbers were symmetrical for the two options, while consensus towards one option emerged even with minimal unbalance in the number of zealots. In these studies options did not have an intrinsic quality.

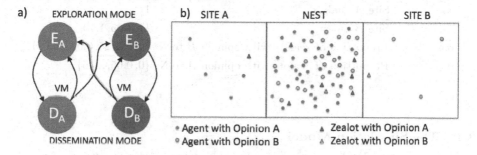

Fig. 1. (a) Probabilistic finite state machine. States represent dissemination and exploration states. Solid lines denote deterministic transitions, while dotted lines stochastic transitions. The symbol VM indicates the model (Voter/Majority) used at the end of the dissemination state. (b) The simulation arena.

To summarize, zealotry has been abundantly studied in physics, typically in fixed interaction topologies, and only recently introduced in the context of swarms, in dynamic local interaction topologies. Compared to the latest work in swarms [5,14,21,22], to the best of our knowledge, in this paper we study for the first time the interplay between different option quality and different zealot quantity, by extending the preliminary study in [19], in which the voter model only was considered and all the agents were able to disseminate differentially with quality.

3   The Model

In the best-of-n problem, a swarm of agents has to reach a collective decision among n possible alternatives. In this paper, the $n = 2$ opinions considered are labelled A and B and have intrinsic quality values ρ_A and ρ_B. The best collective decision is made if consensus is for the option with highest quality: formally, a large majority $M \leq N(1 - \delta)$ of agents agrees on the same option, where δ is a small number chosen by the experimenter. $\delta = 0$ corresponds to *perfect consensus*. Variants of the best-of-n are: the two options may have differential access times or costs [28], option quality may change over time [19,21], or the swarm may have a heterogeneous nature [21]. In the latter, a special case consists in the swarm composed by two different types of agents: zealots, agents with a fixed unchangeable opinion A or B; and normal agents, initialized with opinion

A or B, but able to change their opinion by applying a decision mechanism that relies on the observation of other agents in local proximity.

Table 1. Model parameters used in simulations

Parameter	Description	Values
N	Swarm size	{100, 1000}
ρ_A	Site A quality	1
ρ_B	Site B quality	$\{1, 1.05, 1.10, .., 2\}$
σ_B	Proportion of zealots with opinion B to N	$\{0, 0.0125, 0.025, 0.05\}$
σ_A	Proportion of zealots with opinion A to N	$\{0, 0.05, ..0.5\}$

3.1 The Simulation Model

Similarly to [20], the behaviour of the agents is controlled by the probabilistic finite state machine (FSM) shown in Fig. 1a. The FSM has four possible states: dissemination state of opinion A (D_A), dissemination state of opinion B (D_B), exploration state of opinion A (E_A), and exploration state of opinion B (E_B). Agents are located in a rectangular arena divided in a central part called the *nest* and lateral (left and right) parts called the *sites*, each associated to A or B, respectively (see Fig. 1b). All agents are initialized inside the nest, and move toward the site associated with their opinion to explore that option, for an exponentially distributed amount of time (sampled independently per agent) with mean time q, independent of the current opinion. After exploration, agents have measured the site quality and travel back to the nest after having switched to the dissemination state associated with their current opinion (D_A if they were in E_A, D_B if they were in E_B).

In the dissemination state at the nest, to meet the well-mixed criterion as much as possible [12], agents perform a correlated random walk. Each agent locally broadcasts his opinion continuously, and this message is sensed by other agents in local proximity that are in the process of applying the decision mechanism (before transitioning back to the exploration state). The time spent by the agent disseminating its opinion is exponentially distributed with mean proportional to the site quality they have last visited $g \cdot \rho_i, i \in \{a, b\}$. We considered two different cases in this paper. In the first, both normal agents and zealots with opinion A disseminate proportional to ρ_A, and both normal agents and zealots with opinion B disseminate proportional to ρ_B. In the second case, only zealots disseminate proportional to quality (ρ_A or ρ_B), while normal agents disseminate independently from the quality proportionally to $\rho = 1$. This second case is novel in this paper and was introduced to determine whether modulation of positive feedback is effective through zealots only.

At the end of dissemination, normal agents and zealots behave in two different ways. Normal agents can change their opinion based on the opinions of other agents within a specified spatial radius (in our simulations set to 10 units). The voter model or the majority rule is applied: In the case of voter model, the agent switches its opinion to the one of a random neighbors within the interaction radius [30]; while in majority rule, the agent switches its opinion to the one of the majority of its neighbors ($G = 2$ neighbors [29]).

3.2   ODEs Model

We adapted the model proposed in [21] which extends the ones in [29,30]. The variables e_A, e_B, d_A, d_B model the sub-population of agents exploring site A, exploring site B, disseminating in the nest opinion A and disseminating in the nest opinion B, respectively. The variables modeling sub-populations of zealots are constant. They are denoted with $e_{AS}, e_{BS}, d_{AS}, d_{BS}$. The total proportion of agents with opinion A and B are respectively $x_A = e_A + d_A + e_{AS} + d_{AS}$ and $x_B = e_B + d_B + e_{BS} + d_{BS}$. The total number of agents is conserved $x_A + x_B = 1$.

The system of 8 ODEs with 8 state variables is given by:

$$\dot{d}_A = -\frac{1}{\rho_{AN}g}d_A + \frac{1}{q}e_A \qquad\qquad \dot{d}_{AS} = -\frac{1}{\rho_A g}d_{AS} + \frac{1}{q}e_{AS} \quad (1)$$

$$\dot{d}_B = -\frac{1}{\rho_{BN}g}d_B + \frac{1}{q}e_B \qquad\qquad \dot{d}_{BS} = -\frac{1}{\rho_B g}d_{BS} + \frac{1}{q}e_{BS} \quad (2)$$

$$\dot{e}_A = -\frac{1}{q}e_A + \frac{p_{AA}}{\rho_{AN}g}d_A + \frac{p_{BA}}{\rho_{BN}g}d_B \qquad \dot{e}_{AS} = -\frac{1}{q}e_{AS} + \frac{1}{\rho_A g}d_{AS} \quad (3)$$

$$\dot{e}_B = -\frac{1}{q}e_B + \frac{1}{\rho_{AN}g}p_{AA}d_A + \frac{1}{\rho_{BN}g}p_{BA}d_B \qquad \dot{e}_{BS} = -\frac{1}{q}e_{BS} + \frac{1}{\rho_B g}d_{BS} \quad (4)$$

Equations on the left column describe the dynamics of normal agents, while equations on the right column describe the dynamics of zealots. In Eq. 1-left, the proportion of agents disseminating opinion A increases because of agents returning from the exploration of A at rate $\frac{1}{q}$, and decreases because of agents terminating dissemination at rate $\frac{1}{\rho_{AN}g}$. Similarly, Eq. 2-left describe the rate of increase of the number of agents disseminating opinion B. In Eq. 3-left the number of agents exploring site A decreases because of agents finishing exploration at rate $\frac{1}{q}$, and increases because of two contributions: i) agents that had previously opinion A and kept the same opinion after the application of the voter/majority model and ii) agents that had previously opinion B but switch to A as a result of the voter/majority model. Similarly, Eq. 4-left describes how agents exploring site B vary. The rates p_{AA}, p_{AB}, p_{BA}, and p_{BB} describe the probabilistic outcome of the two decision mechanisms and are described next. Note that qualities in the left column equations are indicated with ρ_{AN} and ρ_{BN} as placeholders. These correspond to the site qualities $\rho_{AN} = \rho_A$, $\rho_{BN} = \rho_B$ when all agents disseminate differentially, while $\rho_{AN} = \rho_{BN} = \rho = 1$ when only zealots disseminate differentially. The dynamic of zealots is described in a very similar way by the equations on the right column. The only difference consists in the impossibility that a zealot to change its opinion after any interaction, thus the terms that depend on the decision mechanisms are omitted. For the zealot case, the dissemination always takes place proportional to ρ_A and ρ_B.

Regarding the decision mechanism, for the voter model the probability that the outcome of the decision is A (resp. B) is the probability that, when observing a random agent disseminating, that random agent is disseminating A (resp. B). This is given by the ratio of agents disseminating A with respect to the total number of agents disseminating: $p_{AA} = p_{BA} = \frac{d_A + d_{AS}}{d_A + d_{AS} + d_B + d_{BS}}$ (resp. $p_{BB} = p_{AB} = \frac{d_B + d_{BS}}{d_A + d_{AS} + d_B + d_{BS}}$.).

For the majority model, where each agent switches its opinion to the one hold by the majority of its G neighbors, the two probabilities are simply given by the cumulative sum of probabilities distributed according to a hypergeometric distribution modeling how many neighbors have each of the two opinions [29]. As in [29], we used: $p_{AA} = \sum_{\frac{G}{2}}^{G+1} \frac{G!}{r!(G-r)!} p^r (1-p)^{G-r}$ and $p_{BA} = \sum_0^{\frac{G}{2}} \frac{G!}{r!(G-r)!} p^{G-r} (1-p)^r$.

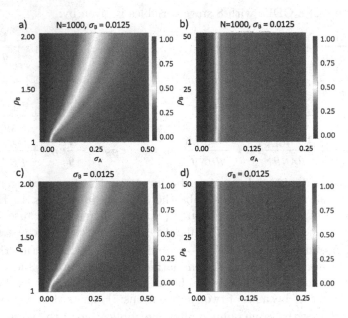

Fig. 2. Consensus heatmaps for the voter model in simulations (first row) and with ODEs (second row), for all agents performing differential dissemination (a and c) and for only performing differential dissemination (b and d). In all cases $\sigma_B = 0.0125$, and $N = 1000$ in the simulations. The colour scale represents the consensus for A. Dark blue colors indicate perfect consensus to the best opinion B, dark red colors indicate perfect consensus to the worst opinion, A, while the white color shows the indifference curve (consensus state around 0.5). (Color figure online)

4   Experimental Evaluation

The experiments were conducted using a simulation tool originally developed by [30]. The simulated arena is a rectangular, two-dimensional space. The collision of the agents is not modeled, however, previous results show that real robot experiments could be accurately reproduced [27].

In each experiment, σ_A (resp. σ_B) is the proportion of zealots committed to A (resp. B). In every run, we first initialize the zealots according to σ_A and σ_B. Afterwards, we set 50% of the remaining (normal) agents to opinion A and the remaining (normal) agents to opinion B. We fix $N = 1000$ agents and $\sigma_B = 0.0125$, as preliminary [19] as well as current study shows that these parameters do not affect the results. The nest size to 316×316 and two sites have the same size of the nest. As zealots need to be more numerous for the option with the lower quality, we set $\sigma_A \geq \sigma_B$ and $\rho_A \leq \rho_B$. Table 1 reports all parameter values.

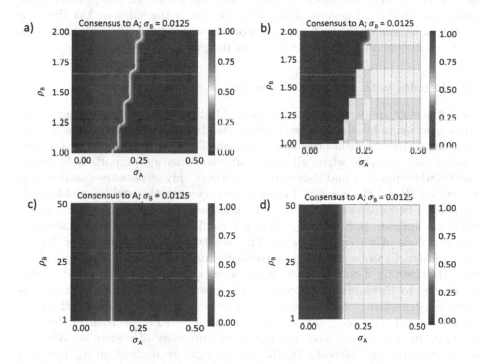

Fig. 3. Consensus heatmap obtained from ODE solution of majority model ($\sigma_B = 0.0125$). Two cases are considered: all agents disseminate for a time proportional to the quality of the option (Panel a and b) or only zealots disseminate for a time proportional to the quality of the option (Panel c and d). The colour scale represents N_A/N. Blue cells indicate perfect consensus (agreement to the best opinion, B). Red cells mean consensus to the worst opinion, A. Tiled cells in (b) and (d) indicate the lack of a second stable equilibrium. (Color figure online)

4.1   Results with the Voter Model

In Fig. 2, we report the heatmaps obtained from simulations and ODE corresponding to the two cases where all the agents disseminate proportionally to the quality (panels a and c) and where only zealots disseminate proportionally to the quality (panels b and d). The simulations results (panels a and b) are reproduced very well by the ODE predictions (c and d, respectively). When all the agents are aware or can measure the qualities of the two options, the consensus to the best option B, represented in blue color, can still be reached despite the increasing number of zealots of the opposite opinion. Only for very high number of zealots (larger than 30% of the total agents), consensus is driven to the worst option A. The indifference curve here is diagonal and depends on both parameters ρ_B and σ_A. The quality of the best option B has a predominant effect with respect to the quantity of zealots for the worst option A.

On the contrary, if only zealots can measure the quality and disseminate differentially, consensus is never driven to the best option B, except for the case where the number of zealots of the worst option is the same or less than the number of zealots of the best option. In other words, the indifference curve is in this case vertical and only depends on the parameter σ_A.

4.2   Results with the Majority Rule

Given the very good results obtained from ODE that accurately reproduce the multi-agents simulations, we used ODEs to study how the system behaves when using the majority rule as decision mechanism. These are shown in Fig. 3. Panels a and b show the case where all agents are disseminating proportionally to the quality, while panels c and d show the case where only zealots are disseminating proportionally to the quality. Figure 3 reports only the stable equilibria. In both cases, two different regimes can be observed. For every value σ_A, a stable equilibria appears (left panels), while a second stable equilibrium exists only for low values of σ_A (right panels). This additional stable equilibrium for the worst option A can be explained by the faster and less accurate dynamics of the majority rule [29]. Looking to the first stable equilibirium (left panels), results are similar to those of voter model: If all agents disseminate differentially, we observe a dependency on ρ_B, while if only zealots disseminate differentially the results depend only on σ_A. However, compared to voter decision mechanism, the majority rule seems more resilient to the quantity of zealots A: When all agents disseminate differentially, the system is more resilient to σ_A for lower values of ρ_B, while for higher values of ρ_B the voter and the majority behave in a similar way; additionally, also when only zealots disseminate differentially the system can converge to the best option for higher values of σ_A using the majority compared to the voter.

We also visualize the bifurcation diagram (Fig. 4) for the case where all agents disseminate proportionally to the quality (left column) and the case where zealots only disseminate proportionally to the quality (right column). Every row represents a different value of $\rho_B = 1, 1.5, 2$ respectively. The consensus

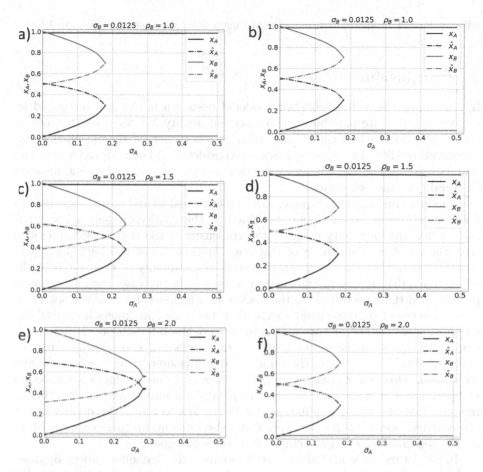

Fig. 4. Bifurcation diagram for majority model with all agents disseminating differentially (left column) and with only zealots disseminating differentially (right column) for different values of ρ_B: $\rho_B = 1$ (first row), $\rho_B = 1.5$ (second row), $\rho_B = 2$ (third row). $\sigma_B = 0.0125$ in all plots. Stable equilibria are represented by a continuous line, while unstable equilibria are represented by a dashed line and indicated with an ˆ.

state for A, denoted by x_A, is studied for increasing σ_A. Here, we confirm the presence of two stable equilibria for low values of σ_A. At some point, a saddle-node bifurcation occurs, and only one stable equilibrium survives. However, we observe that the position of the saddle-node bifurcation moves to the right with respect to σ_A only for the case where all agents disseminate differentially, and not for the case where only zealots do so. We believe these dynamics are interesting because this potentially means that the system is irreversible: if initially the number of zealots A is low, consensus will very likely be for B. However, if σ_A increases, consensus will abruptly change to A after the bifurcation. From that point onwards, reducing again σ_A will not recover consensus to B but the system will be locked in the A consensus state even for progressively lower and lower values of σ_A.

5 Conclusions

In this paper the well established model of best-of-n model is investigated by focusing on the interplay between zealots and quality. We focus on the antagonistic scenario in which the number of zealots is higher for the option that has the lowest quality. Two specific cases are considered: i) both normal agents and zealots can measure the quality of the two options; ii) only zealots can measure the quality of the two options.

The main findings of this paper are: i) if both zealots and normal agents have a different dissemination time determined by the quality of their opinion, the quality has the capability to drive the consensus to the best option, provided that the number of stubborn of the worst opinion is not too high; ii) if only zealots disseminate for a time proportional to the quality of their opinion, the consensus is driven only by the number of zealots. In this case, the quality never prevails and the consensus is to the option with higher number of zealots. From a social perspective, these results show that if only an élite knows how good different alternatives are, or have means to measure this information, the consensus cannot be driven to the better quality if the number of zealots supporting the worse quality is higher than the number of zealots supporting the best quality. This means that zealots can be explicitly designed to manipulate the opinion of the population. On the contrary, it is of paramount importance, at least in our models, that the whole population has the means to assess the quality of the alternatives, because this is the only way to be resilient, up to a given extent, to zealot manipulations and to achieve the best social good.

In the future, we would like to further analyze the dynamics, especially those with the majority model that manifested interesting irreversible dynamics. We would like to relate this model with others such as those based on cross inhibition [25]. Additionally, from the engineering perspective, we would like to understand whether it is possible to design a resilient mechanism for the normal individuals to be resilient to zealots even when they cannot measure the quality, in order to revert the results obtained with zealot-only differential dissemination. These can be useful in swarm robotics applications whereby sensors necessary to estimate quality are expensive and can only be produced for a minority of the individuals.

Acknowledgments. We would like to thank Andreagiovanni Reina and Gabriele Valentini for the useful discussions on the theoretical models and the latter for the original multi-agent simulator code.

References

1. Bhat, D., Redner, S.: Nonuniversal opinion dynamics driven by opposing external influences. Phys. Rev. E **100**, 050301 (2019)

2. Bonabeau, E., Dorigo, M., Theraulaz, G.: Swarm Intelligence: From Natural to Artificial Systems. Oxford University Press, New York (1999)
3. Brambilla, M., Ferrante, E., Birattari, M., Dorigo, M.: Swarm robotics: a review from the swarm engineering perspective. Swarm Intell. **7**(1), 1–41 (2013)
4. Camazine, S., Deneubourg, J.L., Franks, N.R., Sneyd, J., Theraulaz, G., Bonabeau, E.: Self-Organization in Biological Systems. Princeton University Press, Princeton (2001)
5. Canciani, F., Talamali, M.S., Marshall, J.A.R., Bose, T., Reina, A.: Keep calm and vote on: swarm resiliency in collective decision making. In: Proceedings of Workshop Resilient Robot Teams of the 2019 IEEE International Conference on Robotics and Automation (ICRA 2019), p. 4, IEEE Press, Piscataway (2019)
6. Colaiori, F., Castellano, C.: Consensus versus persistence of disagreement in opinion formation: the role of zealots. J. Stat. Mech. Theory Exp. **2016**(3), 1–8 (2016)
7. De Masi, G., Ferrante, E.: Quality-dependent adaptation in a swarm of drones for environmental monitoring. In: 2020 Advances in Science and Engineering Technology International Conferences (ASET). IEEE Press, Piscataway (2020, to appear)
8. Font Llenas, A., Talamali, M.S., Xu, X., Marshall, J.A.R., Reina, A.: Quality-sensitive foraging by a robot swarm through virtual pheromone trails. In: Dorigo, M., Birattari, M., Blum, C., Christensen, A.L., Reina, A., Trianni, V. (eds.) ANTS 2018. LNCS, vol. 11172, pp. 135–149. Springer, Cham (2018). https://doi.org/10.1007/978-3-030-00533-7_11
9. Franks, N.R., Pratt, S.C., Mallon, E.B., Britton, N.F., Sumpter, D.J.T.: Information flow, opinion polling and collective intelligence in house-hunting social insects. Philos. Trans. R. Soc. B Biol. Sci. **357**(1427), 1567–1583 (2002)
10. Galam, S., Jacobs, F.: The role of inflexible minorities in the breaking of democratic opinion dynamics. Physica A **381**(1–2), 366–376 (2007)
11. Ghaderi, J., Srikant, R.: Opinion dynamics in social networks with stubborn agents: equilibrium and convergence rate. Automatica **50**(12), 3209–3215 (2014)
12. Hamann, H.: Opinion dynamics with mobile agents: contrarian effects by spatial correlations. Front. Robot. AI **5**, 63 (2018)
13. Hunter, D.S., Zaman, T.: Optimizing opinions with stubborn agents under time-varying dynamics (2018)
14. Maitre, G., Tuci, E., Ferrante, E.: Opinion dissemination in a swarm of simulated robots with stubborn agents: a comparative study. In: A. Hussain, et al. (ed.) IEEE Congress on Evolutionary Computation, CEC 2020 (within IEEE World Congress on Computational Intelligence (WCCI) 2020). IEEE Press, Piscataway (2020, to appear)
15. Marshall, J.A.R., Bogacz, R., Dornhaus, A., Planqué, R., Kovacs, T.,Franks, N.R.: On optimal decision-making in brains and social insect colonies. J. R. Soc. Interface **6**(40), 1065–1074(2009)
16. Masuda, N.: Opinion control in complex networks. New J. Phys. **17**, 1–11 (2015)
17. Mistry, D., Zhang, Q., Perra, N., Baronchelli, A.: Committed activists and the reshaping of status-quo social consensus. and Related Interdisciplinary TopicsPhys. Rev. E Stat. Nonlin. Soft Matter Phys. **92**(4), 1–9 (2015)
18. Mukhopadhyay, A., Mazumdar, R.: Binary opinion dynamics with biased agents and agents with different degrees of stubbornness. In: 28th International Teletraffic Congress (ITC28), vol. 01, pp. 261–269. IEEE, Piscataway (2016)
19. Prasetyo, J., De Masi, G., Tuci, E., Ferrante, E.: The effect of differential quality and differential zealotry in the best-of-n problem. In: Coello, C.A.C., et al. (ed.) Proceedings of the Twenty-second International Conference on Genetic and Evolutionary Computation (GECCO 2020). ACM, New York, NY (2020, to appear)

20. Prasetyo, J., De Masi, G., Ferrante, E.: Collective decision making in dynamic environments. Swarm Intell. **13**(3), 217–243 (2019). https://doi.org/10.1007/s11721-019-00169-8

21. Prasetyo, J., De Masi, G., Ranjan, P., Ferrante, E.: The best-of-n problem with dynamic site qualities: achieving adaptability with stubborn individuals. In: Dorigo, M., Birattari, M., Blum, C., Christensen, A.L., Reina, A., Trianni, V. (eds.) ANTS 2018. LNCS, vol. 11172, pp. 239–251. Springer, Cham (2018). https://doi.org/10.1007/978-3-030-00533-7_19

22. Primiero, G., Tuci, E., Tagliabue, J., Ferrante, E.: Swarm attack: a self-organized model to recover from malicious communication manipulation in a swarm of simple simulated agents. In: Dorigo, M., Birattari, M., Blum, C., Christensen, A.L., Reina, A., Trianni, V. (eds.) ANTS 2018. LNCS, vol. 11172, pp. 213–224. Springer, Cham (2018). https://doi.org/10.1007/978-3-030-00533-7_17

23. Reina, A., Dorigo, M., Trianni, V.: Towards a cognitive design pattern for collective decision-making. In: Dorigo, M., et al. (eds.) ANTS 2014. LNCS, vol. 8667, pp. 194–205. Springer, Cham (2014). https://doi.org/10.1007/978-3-319-09952-1_17

24. Reina, A., Miletitch, R., Dorigo, M., Trianni, V.: A quantitative micro-macro link for collective decisions: the shortest path discovery/selection example. Swarm Intell. **9**(2–3), 75–102 (2015)

25. Reina, A., Valentini, G., Fernández-Oto, C., Dorigo, M., Trianni, V.: A design pattern for decentralised decision making. PLoS ONE **10**(10), e0140950 (2015)

26. Seeley, T.D.: Honeybee Democracy. Princeton University Press, Princeton (2010)

27. Valentini, G., Brambilla, D., Hamann, H., Dorigo, M.: Collective perception of environmental features in a robot swarm. In: Dorigo, M., et al. (eds.) ANTS 2016. LNCS, vol. 9882, pp. 65–76. Springer, Cham (2016). https://doi.org/10.1007/978-3-319-44427-7_6

28. Valentini, G., Ferrante, E., Dorigo, M.: The best-of-n problem in robot swarms: formalization, state of the art, and novel perspectives. Front. Robot. AI **4**, 9 (2017)

29. Valentini, G., Ferrante, E., Hamann, H., Dorigo, M.: Collective decision with 100 Kilobots: Speed versus accuracy in binary discrimination problems. Auton. Agents Multi-Agent Syst. **30**(3), 553–580 (2016)

30. Valentini, G., Hamann, H., Dorigo, M.: Self-organized collective decision making: The weighted voter model. In: Lomuscio, A., Scerri, P., Bazzan, A., Huhns, M. (eds.) Proceedings of the 13th International Conference on Autonomous Agents and Multiagent Systems, AAMAS 2014, IFAAMAS, pp. 45–52 (2014)

31. Xie, J., Sreenivasan, S., Korniss, G., Zhang, W., Lim, C., Szymanski, B.K.: Social consensus through the influence of committed minorities. and Related Interdisciplinary TopicsPhys. Rev. E Stat. Nonlin. Soft Matter Phys. **84**(1), 1–9 (2011)

32. Yildiz, E., Ozdaglar, A., Acemoglu, D., Saberi, A., Scaglione, A.: Binary opinion dynamics with stubborn agents. ACM Trans. Econ. Comput. **1**(4), 19:1–19:30 (2013)

Short Papers

Short Papers

AutoMoDe-Arlequin: Neural Networks as Behavioral Modules for the Automatic Design of Probabilistic Finite-State Machines

Antoine Ligot, Ken Hasselmann, and Mauro Birattari[✉]

IRIDIA, Université libre de Bruxelles, Brussels, Belgium
mbiro@ulb.ac.be

Abstract. We present `Arlequin`, an off-line automatic design method that produces control software for robot swarms by combining behavioral neural-network modules generated via neuro-evolution. The neural-network modules are automatically generated once, in a mission-agnostic way, and are then automatically assembled into probabilistic finite-state machines to perform various missions. With `Arlequin`, our goal is to reduce the amount of human intervention that is required for the implementation or the operation of previously published modular design methods. Simultaneously, we assess whether neuro-evolution can be used in a modular design method to produce control software that crosses the reality gap satisfactorily. We present robot experiments in which we compare `Arlequin` with `Chocolate`, a state of the art modular design method, and `EvoStick`, a traditional neuro-evolutionary swarm robotics method. The preliminary results suggest that automatically combining neural-network modules into probabilistic finite-state machines is a promising approach to the automatic conception of control software for robot swarms.

1 Introduction

Swarm robotics is an approach to controlling groups of autonomous robots [13]. A robot swarm is a decentralized system in which individual robots do not have predefined roles and act solely based on the local information collected through their sensors or shared by nearby peers. A collective behavior in a swarm emerges from the interactions between the robots, and between the robots and the environment. These interactions depend on how the system evolves and are therefore unknown at design time. Designing the individual behavior of the robots to obtain the desired collective behavior is a challenging task as there is no general methodology to do so [8].

For specific missions in specific cases, experts can use principled manual design methods to obtain the desired collective behavior [1,2,7,25,26,31,37,42,44]. In the general case, however, experts usually proceed by trial and error. An alternative to manual design exists: optimization-based design, which consists in searching among a set of possible individual behaviors the one that maximizes a

© Springer Nature Switzerland AG 2020
M. Dorigo et al. (Eds.): ANTS 2020, LNCS 12421, pp. 271–281, 2020.
https://doi.org/10.1007/978-3-030-60376-2_21

mission-depend objective function that measures the performance of the swarm. These methods can be classified as online or offline [9,18]: in the first case, the optimization is performed while the robots operate in the environment; in the second one, it is performed before deployment, typically using computer simulations. The work presented in this paper belongs in offline automatic design.

A popular approach to the offline automatic design of robot swarms is neuro-evolutionary swarm robotics [47], in which individual behaviors are artificial neural networks whose weights, and possibly their topologies, are fine-tuned by an evolutionary algorithm [5,12,18,46,48]. Unfortunately, neuro-evolutionary swarm robotics suffers from a major drawback: it typically does not cope well with the so-called *reality gap*, that is, the intrinsic difference between simulation and reality [10,30,46]. As a result, the performance of the generated control software is likely deceiving in reality and drops significantly with respect to the one observed in simulation [16,41]. Despite the effort made to handle the reality gap [6,17,28–30,32,39], none of the ideas explored so far appears to be the ultimate solution [18,35,46]. Other approaches to the offline automatic design of robot swarms, based on modularity, have been proposed: they generate control software by assembling low-level behavioral modules [14,15,20]. In this paper, we present a novel automatic modular design method: Arlequin. This method belongs to the AutoMoDe family [19,20,27,33,45]. The novelty of Arlequin is that, contrarily to the previous instances of AutoMoDe that automatically combine behavioral modules conceived by hand, it automatically combines behavioral modules that were themselves automatically generated *a priori* via neuro-evolutionary swarm robotics. With Arlequin, our goal is two-fold: (i) to conceive a method that requires less human expertise during its implementation than the current instances of AutoMoDe, and (ii) further corroborate the conjecture of Francesca et al. [20] that lead to the creation of AutoMoDe.

Francesca et al. [20] conjectured that the reality gap problem faced in evolutionary swarm robotics bears a resemblance to the generalization problem of machine learning, and that the performance drop observed when porting control software to physical robots is due to a sort of *overfitting* of the conditions experienced during the design. According to the *bias/variance tradeoff* [22,49], the expected generalization error of a learning algorithm can be decomposed into a bias and a variance factor. High-complexity learning algorithms have high variance and low bias, whereas low-complexity ones have low variance and high bias. For an increasing level of complexity, the generalization error typically first decreases then increases again. To minimize the generalization error, one must find the optimal level of complexity of the learning algorithm. Based on this reasoning, Francesca et al. conjectured that the difficulty of evolutionary swarm robotics to cross the reality gap is due to an excessively high representational power that entails a sort of overfitting of the idiosyncrasies of simulation [16,41]. The authors therefore created AutoMoDe to have a higher bias than the neuro-evolutionary approaches in order to decrease the representational capability of the control architecture, and to hopefully reduce the performance drop experienced. In AutoMoDe, the bias is injected by restricting to the control software to be a combination of pre-existing modules. So far, the empirical evidence indicates that manually conceiving modules in simulation and validating them on physical robots can effectively limit the overall performance drop caused by the module level. With Arlequin, we investigate whether the principles of modularity also hold true when the behavioral modules are automatically generated by a neuro-evolutionary method. That is, we investigate whether the bias injected by restricting the control software produced to be combination of neural-network modules is enough to cross the reality gap satisfactorily.

We created Arlequin to be similar in many aspects to Chocolate, a previously presented instance of AutoMoDe [19]. Indeed, the two methods only differ in the behavioral modules used. We did so to single out the aspect we wish to investigate: the relative advantages and disadvantages of generating behavioral modules automatically. Chocolate has at his disposal six hand-coded behavioral modules, which are replaced by six neural-network modules in Arlequin. To generate these neural-network modules, we inferred an objective function describing each of the six hand-coded behavioral modules of Chocolate, and fed these objective functions to a neuro-evolutionary design method called Evo-Stick [20]. Similarly to the modules of Chocolate, the neural-network modules are generated once, independently of the specific missions Arlequin will then solve. We evaluate the performance of Arlequin on two missions involving 20 e-puck robots. To assess whether the conjecture of Francesca et al. on the bias/variance tradeoff also holds true when the predefined behavioral modules are generated automatically via neuro evolution, we compare the performance of Arlequin with the ones of EvoStick and Chocolate.

2   AutoMoDe-Arlequin

Arlequin generates control software for a version of the e-puck [40]—a small, circular, two-wheeled robot—equipped with a range-and-bearing board [24], a ground sensor module, and an Overo Gumstix board [21]. We considered a subset of the capabilities of the robot. In particular, the control software that can be generated has access to the ground sensor module to detect the color of the ground situated below the robot (i.e., black, gray, or white); the infrared sensor module to detect the presence of nearby obstacles and of a light source; the range-and-bearing module to detect the presence of peers within a range of approximately 0.7 m and to infer a vector V_d indicating their direction of attraction; and the wheels actuators to move the robot.

Arlequin generates control software by automatically combining predefined modules into probabilistic finite-state machines. The modules comprise six low-level behaviors (i.e., simple actions performed by the robot) and six conditions (i.e., situations experienced by the robot). The low-level behaviors are associated to states of the probabilistic finite-state machine, whereas the conditions are associated to transitions. The low-level behavior associated with the active state is executed as long as the conditions associated with all its outgoing transitions are evaluated as false. Once a condition associated with an outgoing transition is evaluated as true, the active state is updated and the corresponding low-level behavior is executed. Arlequin has many commonalities with AutoMoDe-Chocolate [19]. The two methods adopt irace [4,38] as optimization algorithm to select and combine the different modules into a probabilistic finite-state machine. The two methods also impose the same constraints on the probabilistic finite-state machines produced: they can comport up to four states with up to four outgoing transitions per states. Finally, Arlequin and Chocolate have at their disposal the same hand-coded condition modules. We refer the reader to the original description of these conditions [20].

Arlequin and Chocolate differ in the predefined behavioral modules adopted: Chocolate combines hand-coded parametric modules, whereas Arlequin combines neural-network modules generated by EvoStick [20]. EvoStick is a relatively simple implementation of the classical neuro-evolutionary robotics approach: it generates control software in the form of neural networks whose synaptic weights are obtained via an evolutionary process. In EvoStick, the produced neural networks are fully connected, do not contain hidden layers, and have 25 input and 2 output nodes. The neural networks are therefore characterized by a total of 50 parameters, each being a real value in $[-5, 5]$. The 25 input nodes are organized as follows: 3 are dedicated to the readings of the ground sensors, 8 to the readings of the proximity sensors, 8 to the readings of the light sensors, 5 to the readings of the range-and-bearing sensors (4 for the scalar projections of the vector V_d pointing to the neighboring peers on four unit vectors, and 1 for the number of detected robots), and one serves as bias. The 2 output nodes control the speed of the left and right wheels of the robot. EvoStick uses populations of 100 individuals and evaluates each individual 10 times per generation.

To obtain behaviors that are similar to the six hand-coded low-level behaviors of Chocolate via neuro-evolution, we inferred an objective function for each of them. We fed these objective functions to EvoStick to generate control software for a swarm of 20 simulated e-puck, and considered simulation runs of 120 s. For each of the low-level behaviors, EvoStick generated 10 instances of control software. We then evaluated each instance of control software 20 times in simulation using different initial conditions, and selected the ones with the highest average performance to be used as low-level behaviors for Arlequin. The design budget allocated to EvoStick is 20 000 execution runs, which corresponds to 20 generations. The six hand-coded low-level behaviors of Chocolate and the corresponding objective functions we devised to obtain the automatically generated modules of Arlequin are described in Sect. 2.1.

2.1   Low-Level Behaviors

Exploration: In Chocolate, the robot moves straight until an obstacle is perceived by its front proximity sensors, then turns on the spot for a random number of steps drawn in $\{0, 1, ..., \pi\}$. The parameter $\pi \in \{0, 1, ..., 100\}$ is meant to be afterwards tuned by the optimization algorithm on a per-mission basis. In Arlequin, the environment is discretized into a two-dimensional grid G, and the objective function considered rewards the number of cells visited individually. The objective function, to be maximized, is $\sum_{r=1}^{N} \sum_{i=1}^{X} \sum_{j=1}^{Y} G_r[i][j]$, where $G_r[i][j] = 1$ if robot r visited cell $G_r(i,j)$ at least once, 0 otherwise; N is the number of robots in the swarm; and $X = 20$ and $Y = 20$ are the numbers of rows and columns in grid G, respectively. *Stop:* In Chocolate, the robot does not move. In Arlequin, the objective function penalizes the displacement of the individual robots. The objective function, to be minimized, is $\sum_{t=1}^{T} \sum_{r}^{N} ||P_r(t) - P_r(t-1)||$, where $P_r(t)$ is the position of robot r at time t, and T is the duration of the experimental run. *Phototaxis:* In Chocolate, the robot moves towards the light, if perceived. Otherwise, the robot moves straight. In Arlequin, the objective function penalizes the distance between the individual robots and the light. The objective function, to be minimized, is $\sum_{t=1}^{T} \sum_{r=1}^{N} ||P_r(t) - P_{light}||$, where $P_r(t)$ and P_{light} are the positions of robot r at time t and of the light, respectively. *Anti-phototaxis:* In Chocolate, the robot moves away from the light, if perceived. Otherwise, the robot moves straight. In Arlequin, the objective function rewards the distance between the individual robots and the light. The objective function, to be maximized, is $\sum_{t=1}^{T} \sum_{r=1}^{N} ||P_r(t) - P_{light}||$, where $P_r(t)$ and P_{light} are the positions of robot r at time t and of the light, respectively. *Attraction:* In Chocolate, the robot moves towards the neighboring peers (V_d), if perceived. Otherwise, the robot moves straight. A parameter $\alpha \in [1,5]$ controls the speed of convergence towards the detected peers and is meant to be afterwards tuned by the optimization algorithm on a per-mission basis. In Arlequin, the objective function penalizes the distance between each pair of robots within the swarm. The objective function, to be minimized, is $\sum_{t=1}^{T} \sum_{i=1}^{N-1} \sum_{j=i+1}^{N} ||P_i(t) - P_j(t)||$, where $P_i(t)$ and $P_j(t)$ are the positions of robot i and j, respectively. *Repulsion:* In Chocolate, the robot moves away from the neighboring peers, if perceived. Otherwise, it moves straight. A parameter $\alpha \in [1,5]$ controls the speed of divergence and is meant to be afterwards tuned by the optimization algorithm on a per-mission basis. In Arlequin, the objective function rewards, for each individual robot, the distance from its closest peer. The objective function, to be minimized, is $\sum_{t=1}^{T} \sum_{r=1}^{N} ||P_r(t) - P_{r_{min}}(t)||$, where $P_r(t)$ is the position of robot r and $P_{r_{min}}(t)$ is the one of the robot closest to robot r at time t.

3   Experiments

We generated control software with Arlequin, Chocolate, and EvoStick for two missions: FORAGING and AGGREGATION-XOR [20]. We considered a swarm of 20 e-puck robots that operate in a dodecagonal arena of 4.91 m^2 delimited by walls. For each mission, the design budget allowed to each method is 200 000 simulation runs. For each mission, we executed each design method 10 times and collected the best instance of control software produced by each execution. We assessed the performance of each instance of control software twice: once in simulation and once on physical robots [3]. We present the results in the form of notched boxplots: the notches represent the 95% confidence interval on the position of the median. If the notches of two boxes do not overlap, the difference between the respective medians is significant [11]. All simulation runs were performed with ARGoS [43], which allowed us to directly port the control software generated to the physical robots without any modifications. All the control software generated, the raw data collected, and the experimental runs recorded are available online as supplementary material [36] (Fig. 1).

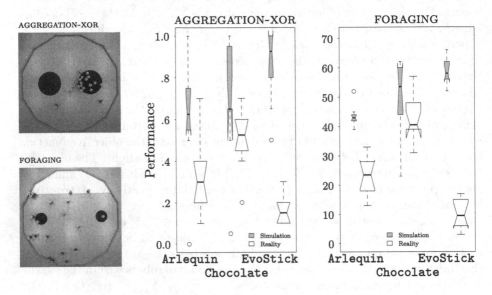

Fig. 1. The arenas and the results of the experiments.

AGGREGATION-XOR. The robots must aggregate on one of the two black areas. After 180 s, the performance measured by the function $F_A = \max(N_l, N_r)/N$, where N_l and N_r are the number of robots located on each of the two black area; and N is the total number of robots. In simulation, Arlequin and Chocolate show similar performance, but Arlequin is outperformed by EvoStick. In reality, Arlequin and EvoStick suffer from a significant performance drop, with EvoStick suffering from the reality gap the most. Indeed, the drop experienced by Arlequin is at most 0.48, whereas the one experienced by EvoStick is at least 0.55, which makes the performance drop experienced by Arlequin significantly lower than the one experienced by EvoStick (95% confidence computed with a paired Wilcoxon test). Chocolate shows similar performance in simulation and in reality. The performance drop experienced by the three methods when crossing the reality gap is such that, in reality, Arlequin outperforms EvoStick, but is outperformed by Chocolate.

FORAGING. The robots must retrieve objects from two source areas (black circles) and deposit them in a nest (white area). The objects are virtual: a robot is deemed to carry an object after it enters one of the source areas and to retrieve the object when it then enters the nest. A light source is placed behind the nest.

The performance measured by the function $F_F = N_o$, where N_o is the total number of objects retrieved after 180 s. In simulation, Arlequin is outperformed by EvoStick and Chocolate. In reality, the three methods suffer from a significant performance drop, with EvoStick suffering the most, followed by Arlequin, then Chocolate. The drop of Arlequin is at most 26, whereas the one of EvoStick is at least 42, which makes the drop of Arlequin significantly lower than the one of EvoStick (95% confidence computed with a paired Wilcoxon test). As a result, Arlequin outperforms EvoStick, but is outperformed by Chocolate.

4 Conclusions

We presented Arlequin, a novel instance of AutoMoDe that differs from the previously presented ones by the nature of the predefined behavioral modules to be combined: Arlequin uses neural network modules generated via neuro-evolution, whereas the others use hand-coded ones. The behavioral modules of Arlequin were generated via EvoStick, a neuro-evolutionary method. We compared the performance of the control software generated by Arlequin with the one of Evo-Stick and Chocolate on two missions. In both missions, the control software produced by Arlequin suffered from a significant performance drop. However, the control software generated by EvoStick suffered from a significantly larger drop than the one produced by Arlequin, and as a result, Arlequin outperformed EvoStick in reality. This corroborates the conjecture of Francesca et al. [20]: restricting the control software to be a combination of low-level, simple behaviors yields better results in reality than the traditional neuro-evolutionary approach, despite being the other way around in simulation. Our results show that this holds true also when the low-level behaviors are neural networks.

Future work will explore different ways of generating and selecting the pool of modules to be combined into probabilistic finite-state machines (i.e., select the modules on the basis of their performance assessed in *pseudo-reality* [34, 35] or on physical robots, generate them with the *transferability approach* [32]). Future work will also be dedicated to further reducing the human expertise required during the implementation of Arlequin. Recently, Gomes and Christensen [23] proposed an approach to conceive low-level behaviors in a completely automated fashion. Their approach is based on *repertoires* of behaviors obtained in a task-agnostic fashion with a diversity algorithm. We wish to investigate how one could automatically produce control software for swarm robotics by combining behavioral modules selected from these repertoires.

Acknowledgements. The experiments were conceived by the three authors and performed by AL and KH. The article was drafted by AL and revised by the three authors. The research was directed by MB.

The project has received funding from the European Research Council (ERC) under the European Union's Horizon 2020 research and innovation programme (grant agreement No 681872). MB acknowledges support from the Belgian *Fonds de la Recherche Scientifique* – FNRS.

References

1. Beal, J., Dulman, S., Usbeck, K., Viroli, M., Correll, N.: Organizing the aggregate: languages for spatial computing. In: Marjan, M. (ed.) Formal and Practical Aspects of Domain-Specific Languages: Recent Developments, pp. 436–501. IGI Global, Hershey (2012). https://doi.org/10.4018/978-1-4666-2092-6.ch016
2. Berman, S., Kumar, V., Nagpal, R.: Design of control policies for spatially inhomogeneous robot swarms with application to commercial pollination. In: IEEE International Conference on Robotics and Automation, ICRA, Piscataway, NJ, USA, pp. 378–385. IEEE (2011). https://doi.org/10.1109/ICRA.2011.5980440
3. Birattari, M.: On the estimation of the expected performance of a metaheuristic on a class of instances. How many instances, how many runs? Technical report TR/IRIDIA/2004-01, IRIDIA, Université libre de Bruxelles, Belgium (2004)
4. Birattari, M., Yuan, Z., Balaprakash, P., Stützle, T.: F-race and iterated F-race: an overview. In: Bartz-Beielstein, T., Chiarandini, M., Paquete, L., Preuss, M. (eds.) Experimental Methods for the Analysis of Optimization Algorithms, pp. 311–336. Springer, Heidelberg (2010). https://doi.org/10.1007/978-3-642-02538-9_13
5. Bongard, J.C.: Evolutionary robotics. Commun. ACM **56**(8), 74–83 (2013)
6. Bongard, J.C., Lipson, H.: Once more unto the breach: co-evolving a robot and its simulator. In: Pollack, J.B., Bedau, M.A., Husbands, P., Watson, R.A., Ikegami, T. (eds.) Artificial Life IX: Proceedings of the Conference on the Simulation and Synthesis of Living Systems, pp. 57–62. MIT Press, Cambridge (2004)
7. Brambilla, M., Brutschy, A., Dorigo, M., Birattari, M.: Property-driven design for swarm robotics: a design method based on prescriptive modeling and model checking. ACM Trans. Auton. Adapt. Syst. **9**(4), 17:1–17:28 (2014). https://doi.org/10.1145/2700318
8. Brambilla, M., Ferrante, E., Birattari, M., Dorigo, M.: Swarm robotics: a review from the swarm engineering perspective. Swarm Intell. **7**(1), 1–41 (2013). https://doi.org/10.1007/s11721-012-0075-2
9. Bredeche, N., Haasdijk, E., Prieto, A.: Embodied evolution in collective robotics: a review. Front. Robot. AI **5**, 12 (2018). https://doi.org/10.3389/frobt.2018.00012
10. Brooks, R.A.: Artificial life and real robots. In: Varela, F.J., Bourgine, P. (eds.) Towards a Practice of Autonomous Systems. Proceedings of the First European Conference on Artificial Life, pp. 3–10. MIT Press, Cambridge (1992)
11. Chambers, J.M., Cleveland, W.S., Kleiner, B., Tukey, P.A.: Graphical Methods For Data Analysis. CRC Press, Belmont (1983)
12. Doncieux, S., Mouret, J.-B.: Beyond black-box optimization: a review of selective pressures for evolutionary robotics. Evol. Intell. **7**(2), 71–93 (2014). https://doi.org/10.1007/s12065-014-0110-x
13. Dorigo, M., Birattari, M., Brambilla, M.: Swarm robotics. Scholarpedia **9**(1), 1463 (2014). https://doi.org/10.4249/scholarpedia.1463
14. Duarte, M., et al.: Evolution of collective behaviors for a real swarm of aquatic surface robots. Plos One **11**(3), e0151834 (2016). https://doi.org/10.1371/journal.pone.0151834
15. Duarte, M., Oliveira, S.M., Christensen, A.L.: Evolution of hierarchical controllers for multirobot systems. In: Sayama, H., Rieffel, J., Risi, S., Doursat, R., Lipson, H. (eds.) Artificial Life 14. Proceedings of the Fourteenth International Conference on the Synthesis and Simulation of Living Systems, pp. 657–664. MIT Press, Cambridge (2014). https://doi.org/10.7551/978-0-262-32621-6-ch105

16. Floreano, D., Husbands, P., Nolfi, S.: Evolutionary robotics. In: Siciliano, B., Khatib, O. (eds.) Springer Handbook of Robotics, pp. 1423–1451. Springer, Heidelberg (2008). https://doi.org/10.1007/978-3-540-30301-5_62

17. Floreano, D., Mondada, F.: Evolution of plastic neurocontrollers for situated agents. In: Maes, P., Matarić, M.J., Meyer, J.A., Pollack, J.B., Wilson, S.W. (eds.) From Animals to Animats 4: Proceedings of the Fourth International Conference on Simulation of Adaptive Behavior (SAB), pp. 402–410. MIT Press, Cambridge (1996)

18. Francesca, G., Birattari, M.: Automatic design of robot swarms: achievements and challenges. Front. Robot. AI **3**(29), 1–9 (2016). https://doi.org/10.3389/frobt.2016.00029

19. Francesca, G., et al.: AutoMoDe-chocolate: automatic design of control software for robot swarms. Swarm Intell. **9**(2–3), 125–152 (2015). https://doi.org/10.1007/s11721-015-0107-9

20. Francesca, G., Brambilla, M., Brutschy, A., Trianni, V., Birattari, M.: AutoMoDe: a novel approach to the automatic design of control software for robot swarms. Swarm Intell. **8**(2), 89–112 (2014). https://doi.org/10.1007/s11721-014-0092-4

21. Garattoni, L., Francesca, G., Brutschy, A., Pinciroli, C., Birattari, M.: Software infrastructure for e-puck (and TAM). Technical report TR/IRIDIA/2015-004, IRIDIA, Université libre de Bruxelles, Belgium (2015)

22. Geman, S., Bienenstock, E., Doursat, R.: Neural networks and the bias/variance dilemma. Neural Comput. **4**(1), 1–58 (1992). https://doi.org/10.1162/neco.1992.4.1.1

23. Gomes, J., Christensen, A.L.: Task-agnostic evolution of diverse repertoires of swarm behaviours. In: Dorigo, M., Birattari, M., Blum, C., Christensen, A.L., Reina, A., Trianni, V. (eds.) ANTS 2018. LNCS, vol. 11172, pp. 225–238. Springer, Cham (2018). https://doi.org/10.1007/978-3-030-00533-7_18

24. Gutiérrez, A., Campo, A., Dorigo, M., Donate, J., Monasterio-Huelin, F., Magdalena, L.: Open e-puck range & bearing miniaturized board for local communication in swarm robotics. In: Kosuge, K. (ed.) IEEE International Conference on Robotics and Automation, ICRA, Piscataway, NJ, USA, pp. 3111–3116. IEEE (2009). https://doi.org/10.1109/ROBOT.2009.5152456

25. Hamann, H.: Swarm Robotics: A Formal Approach. Springer, Cham (2018). https://doi.org/10.1007/978-3-319-74528-2

26. Hamann, H., Wörn, H.: A framework of space-time continuous models for algorithm design in swarm robotics. Swarm Intell. **2**(2–4), 209–239 (2008). https://doi.org/10.1007/s11721-008-0015-3

27. Hasselmann, K., Robert, F., Birattari, M.: Automatic design of communication-based behaviors for robot swarms. In: Dorigo, M., Birattari, M., Blum, C., Christensen, A.L., Reina, A., Trianni, V. (eds.) ANTS 2018. LNCS, vol. 11172, pp. 16–29. Springer, Cham (2018). https://doi.org/10.1007/978-3-030-00533-7_2

28. Jakobi, N.: Evolutionary robotics and the radical envelope-of-noise hypothesis. Adapt. Behav. **6**(2), 325–368 (1997). https://doi.org/10.1177/105971239700600205

29. Jakobi, N.: Minimal simulations for evolutionary robotics. Ph.D. thesis, University of Sussex, Falmer, UK (1998)

30. Jakobi, N., Husbands, P., Harvey, I.: Noise and the reality gap: the use of simulation in evolutionary robotics. In: Morán, F., Moreno, A., Merelo, J.J., Chacón, P. (eds.) ECAL 1995. LNCS, vol. 929, pp. 704–720. Springer, Heidelberg (1995). https://doi.org/10.1007/3-540-59496-5_337

31. Kazadi, S.: Model independence in swarm robotics. Int. J. Intell. Comput. Cybern. **2**(4), 672–694 (2009). https://doi.org/10.1108/17563780911005836

32. Koos, S., Mouret, J.B., Doncieux, S.: The transferability approach: crossing the reality gap in evolutionary robotics. IEEE Trans. Evol. Comput **17**(1), 122–145 (2013). https://doi.org/10.1109/TEVC.2012.2185849
33. Kuckling, J., Ligot, A., Bozhinoski, D., Birattari, M.: Behavior trees as a control architecture in the automatic modular design of robot swarms. In: Dorigo, M., Birattari, M., Blum, C., Christensen, A.L., Reina, A., Trianni, V. (eds.) ANTS 2018. LNCS, vol. 11172, pp. 30–43. Springer, Cham (2018). https://doi.org/10.1007/978-3-030-00533-7_3
34. Ligot, A., Birattari, M.: On mimicking the effects of the reality gap with simulation-only experiments. In: Dorigo, M., Birattari, M., Blum, C., Christensen, A.L., Reina, A., Trianni, V. (eds.) ANTS 2018. LNCS, vol. 11172, pp. 109–122. Springer, Cham (2018). https://doi.org/10.1007/978-3-030-00533-7_9
35. Ligot, A., Birattari, M.: Simulation-only experiments to mimic the effects of the reality gap in the automatic design of robot swarms. Swarm Intell. **14**(1), 1–24 (2019). https://doi.org/10.1007/s11721-019-00175-w
36. Ligot, A., Hasselmann, K., Birattari, M.: AutoMoDe-Arlequin: neural networks as behavioral modules for the automatic design of probabilistic finite state machines: supplementary material (2020). http://iridia.ulb.ac.be/supp/IridiaSupp2020-005/index.html
37. Lopes, Y.K., Trenkwalder, S.M., Leal, A.B., Dodd, T.J., Groß, R.: Supervisory control theory applied to swarm robotics. Swarm Intell. **10**(1), 65–97 (2016). https://doi.org/10.1007/s11721-016-0119-0
38. López-Ibáñez, M., Dubois-Lacoste, J., Pérez Cáceres, L., Birattari, M., Stützle, T.: The irace package: iterated racing for automatic algorithm configuration. Oper. Res. Perspect. **3**, 43–58 (2016). https://doi.org/10.1016/j.orp.2016.09.002
39. Miglino, O., Lund, H.H., Nolfi, S.: Evolving mobile robots in simulated and real environments. Artif. Life **2**(4), 417–434 (1995). https://doi.org/10.1162/artl.1995.2.4.417
40. Mondada, F., et al.: The e-puck, a robot designed for education in engineering. In: Gonçalves, P., Torres, P., Alves, C. (eds.) Proceedings of the 9th Conference on Autonomous Robot Systems and Competitions, pp. 59–65. Instituto Politécnico de Castelo Branco, Castelo Branco (2009)
41. Nolfi, S., Floreano, D.: Evolutionary Robotics: The Biology, Intelligence, and Technology of Self-Organizing Machines. MIT Press, Cambridge (2000)
42. Pinciroli, C., Beltrame, G.: Buzz: a programming language for robot swarms. IEEE Softw. **33**(4), 97–100 (2016). https://doi.org/10.1109/MS.2016.95
43. Pinciroli, C., et al.: ARGoS: a modular, parallel, multi-engine simulator for multi-robot systems. Swarm Intell. **6**(4), 271–295 (2012). https://doi.org/10.1007/s11721-012-0072-5
44. Reina, A., Valentini, G., Fernández-Oto, C., Dorigo, M., Trianni, V.: A design pattern for decentralised decision making. PLOS ONE **10**(10), e0140950 (2015). https://doi.org/10.1371/journal.pone.0140950
45. Salman, M., Ligot, A., Birattari, M.: Concurrent design of control software and configuration of hardware for robot swarms under economic constraints. PeerJ Comput. Sci. **5**, e221 (2019). https://doi.org/10.7717/peerj-cs.221
46. Silva, F., Duarte, M., Correia, L., Oliveira, S.M., Christensen, A.L.: Open issues in evolutionary robotics. Evol. Comput. **24**(2), 205–236 (2016). https://doi.org/10.1162/EVCO_a_00172
47. Trianni, V.: Evolutionary Swarm Robotics. Springer, Berlin (2008). https://doi.org/10.1007/978-3-540-77612-3

48. Trianni, V.: Evolutionary robotics: model or design? Front. Robot. AI **1**, 13 (2014). https://doi.org/10.3389/frobt.2014.00013
49. Wolpert, D.: On bias plus variance. Neural Comput. **9**, 1211–1243 (1997). https://doi.org/10.1162/neco.1997.9.6.1211

Coalition Formation Problem: A Group Dynamics Inspired Swarming Method

Mickaël Bettinelli[✉], Michel Occello, and Damien Genthial

Univ. Grenoble Alpes, Grenoble INP, LCIS, 26000 Valence, France
mickael.bettinelli@lcis.grenoble-inp.fr

Abstract. The coalition formation problem arises when heterogeneous agents need to be gathered in groups in order to combine their capacities and solve an overall goal. But very often agents are different and can be distinguished by several characteristics like desires, beliefs or capacities. Our aim is to make groups of agents according to several characteristics. We argue that a swarming method inspired by group dynamics allows groups to be formed on the basis of several characteristics and makes it very robust in an open system context. We evaluate this approach by making groups of heterogeneous cognitive agents and show that our method is adapted to solve this problem.

1 Introduction

Agents in a multi-agent system (MAS) face complex problems and do not always have all the capabilities to solve them alone. Thus, agents need to share theirs capacities in cooperative groups in order to reach the overall goal of their system. They need to find the best suited agents to compose their group in order to maximize the overall performance for the task resolution. This problem is called the coalition formation problem and has been addressed in many forms. To illustrate with a realistic example, in case of a large scale natural disaster [10] robots with different capabilities may rescue victims and they need to make coalitions owning the complete capabilities to rescue people. In this work we present an approach inspired by the group dynamics field in the Humanities and Social Sciences (HSS) using a swarming model with heterogeneous agents. Related work is quoted in Sect. 2 and 3 introduces our approach and defines the problem. A swarming model inspired from the literature is presented in Sect. 4. The integration of group dynamics features into a swarming context is described in Sect. 5 and experimental results shown in Sect. 6.

2 Related Work

The coalition formation problem is a broad problem addressed from different point of views as multi-agent system [16], robotic [6] or swarming [9]. It can be

Grenoble INP—Institute of Engineering Univ. Grenoble Alpes.

© Springer Nature Switzerland AG 2020
M. Dorigo et al. (Eds.): ANTS 2020, LNCS 12421, pp. 282–289, 2020.
https://doi.org/10.1007/978-3-030-60376-2_22

derived into very similar problems called Task allocation problem [12] or Knapsack problem [2]. Swarming approaches [1] are one method used to address this problem. They often use an optimization method technique called Ant Colony Optimization (ACO) [19] in order to find optimal coalitions but in this work we focus on swarming based on social potential field. Social potential field [13] is a distributed method used to make swarm from individuals by using attraction and repulsion forces between each individual. Social potential field is used to bring out a global behavior to individuals. [13] uses it for autonomous multi-robot control and [15] uses it into the multi-agent system field to make agents patrol on a terrain with obstacles. Among the swarming literature, some works are using heterogeneous agents or robots [6,7,17] that have different characteristics from others: they may have type, abilities or dynamics. [7] uses swarming with heterogeneous agents in order to achieve a self-organization of a MAS by modifying force fields depending on the type of agents. The swarming method used in this paper is inspired from this work. Yet, this work makes possible the self-organisation of agents into a system based on one characteristic (the type of agents) in order to make groups of homogeneous agents. The reality is often more complex and we would like to reuse this method to tackle the coalition formation problem where agents come together to try to achieve an overall goal based on more factors than a type or a weight.

3   The Coalition Formation Problem

In a large scale natural disaster scenario, robots can be stuck or destroyed because of debris. Thus the process of coalition formation has to be dynamic and robust. It has to be able to take into account the variability of agents and the openness of the system. This is why we choose a swarming method with social potential fields to make coalitions: their formation is processed distributively by each agent allowing to add or remove agents from the system without stopping it. In addition, agents characteristics can be modified at any time making this swarming method very robust.

As seen in the Sect. 2 some works focus on segregation of agents into heterogeneous swarming models. To the best of our knowledge, work that focuses on sorting heterogeneous swarms uses heterogeneous agents that differ only in one characteristic. But in a complex scenario we need agents to own a lot of characteristics as physical characteristics (e.g. battery state, sensors, etc.), capacities (moving, taking objects, etc.), mental characteristics (e.g. personal goal, desires, learning skills, etc.), etc. Thus, we propose a social approach of heterogeneous swarming able to make coalitions from several factors. Because humans are well suited to form groups and work efficiently in them, we draw inspiration from the group dynamics field [3] in HSS and use it with the swarming method as a new approach for the coalition formation problem. The group dynamics is a field that describes what small groups are, how they are formed and how they are maintained. This approach makes the group formation possible by taking into account several factors making heterogeneous swarming more relevant.

In the decentralized task allocation problem, we consider a set of N individuals $A = \{a_0, a_1, ..., a_n\}$ in a 1-dimension Euclidean space. Agents have heterogeneous characteristics $C = \{c_0, c_1, ..., c_n\}$ and desires $D = \{d_0, d_1, ..., d_n\}$. A characteristic c_i is meeting a desire d_i. Each individual a_n has its own set of characteristics and desires such that $C_{a_n} = \{c_{a_n}^i, c_{a_n}^j, ..., c_{a_n}^m\}$ and $D_{a_n} = \{d_{a_n}^k, d_{a_n}^l, ..., d_{a_n}^o\}$. Each individual is a point, unaware of its dimension, that knows the characteristics of all the other agents. The objective of the agents is to form groups with other agents who best meet their desires such that $C_{a_j} \subset D_{a_i}$. The distance between two agents in the space is an attraction metric representing the attraction value that two individuals have for each other. It depends on the extent to which the agent's desires are satisfied. The shorter the distance, the stronger the attraction between the agents. A *group* is a set of agents g_i for which all the agents have an attraction value for each other below a given threshold such that $g_i = \{a_i, a_j, ..., a_m\}$ where $\forall a_i, a_j \in g_i$, $attraction(a_i \rightarrow a_j) < threshold$ and $attraction(a_j \rightarrow a_i) < threshold$. Processing rules of the attraction are described in the next section through a swarm intelligence method. We assume here that agents can be part of only one group. In other word, $\forall g_i, g_j, g_i \cap g_j = \emptyset$.

4   Control Law Definition

4.1   Attraction/Repulsion Function

Agents attraction update is processed following this equation:

$$\dot{x}^i = \sum_{j=1, j \neq i}^{N} f\left(x^i - x^j\right) \qquad f(y) = -y\left(\frac{a}{\|y\|} - \frac{b}{\|y\|^4}\right) \tag{1}$$

where x^i and x^j are the position of individuals i and j into the Euclidean space. $f(y)$ being the attraction/repulsion function inspired from [14] where a and b are two constants and $\|y\|$ the Euclidean norm given by $\|y\| = \sqrt{y^T y}$ which is the distance between two agents in the Euclidean space. We also made two parameter functions to choose the right a and b constants

$$f_a(z) = \frac{0.05^z}{2} \qquad\qquad f_b(z) = 20^{(3+z)} \cdot z^3 \tag{2}$$

z being a bias representing social factors described in Sect. 4.2. f_a and f_b were both made after empirical tests. They allow to keep a good ratio between a and b and are well managing the attraction/repulsion function whatever the z value is. However, these functions were not designed to be optimal.

Note that because the attraction/repulsion function is very similar to the one from [14], we do not give the theoretical proof of stabilization of this function in this document. However, you can find it in an extended version on ArXiv and HAL.

4.2   Social Factors Integration

In group dynamics, the Group Formation subfield focuses on the processes that generates bonds of attraction between members of groups. The group formation process is a complex phenomenon implying numerous dimensions. Among these dimensions, the attraction principles takes a large part. There are two types of attraction, the social attraction and the personal attraction. Social attraction is an attraction for a group whereas the personal attraction is "based on idiosyncratic preferences grounded in personal relationships" [5]. Because in swarming individuals are not aware of groups they are making, we focus on the personal attraction allowing to predict whether an individual is attracted to another one or not. The following principles are based on personal attraction [3,11]:

- proximity principle (p): proximity allows individuals to increase the number of their interactions. We see here the proximity principle as a distance between individuals.
- similarity principle (s): individuals like people who are similar to them [4,18]. In our system, the similarity is a distance between mind states of agents.
- complementarity principle (c): individuals like other whose qualities complement their own. We represent it by the complementarity of the agents' capacities.
- reciprocity principle (r): liking tends to be mutual.
- physical attractiveness principle (a): individuals are more attracted to people who have a great physical attractiveness [4]. In our system the physical attractiveness is seen as the adequacy between the characteristics of an individual and desires of others.
- minimax principle (m): individuals are attracted to people that offer them maximum reward and minimal cost [4].

We want agents to be able to assess the attraction they have for other agents in order to allow to form groups into the system. To do so, we build the Eq. 3 that integrates these principles. However, group dynamics is only an inspiration source to build our equation. Even if we try to be consistent with the literature, we are not claiming that this equation can be used to predict the attraction between two real people.

$$z = (0.5 + p) \cdot average(a, s, c, m) \tag{3}$$

5   Experimentation

In order to evaluate our mechanisms, we integrated them in an agent model. As our approach is socially inspired we have chosen the cognitive agent architecture Soar [8] to which we add specific features involving control laws. We characterize an agent by four modules that can be seen as sets of information.

- personal characteristics (P): are physical or mental characteristics agents have (e.g. battery state, weight, shape),

- capacities (C): are skills of the agent, actions it can execute on its environment, virtual processing it can do, or perceptions it can get from the environment (*e.g.* moving, taking objects, *etc.*),
- beliefs (B): are the facts agents have about their environment (*e.g.* acquaintances characteristic, attraction for acquaintances, its group quality),
- desires (D): are objectives or situations that the agent would like to accomplish or bring about (*e.g.* making a high quality group, resolving its personal goal, help its group mates);

In this experiment, we are using a 1-dimensions matrix of three floats between 0 and 1 to represent information of each module of our agents. The size of matrices does not have any importance for the proper functioning of the system. In order to use these information, we need to integrate the attraction principles from the Sect. 4.2 to our experimentation. Each principle undergoes a filter function used to scale each result between 0 and 1.

Proximity principle is the distance between two agents, represented as:

$$||p|| = x^i - x^j \qquad\qquad p = filter(\sqrt{u^T u}, d) \qquad (4)$$

where x^i and x^j are the position of agents i and j into the Euclidean space and d a parameter to be adjusted.

Similarity, Physical attractiveness and Minimax principle are processed in the same way as the Proximity principle. The similarity principle is a distance between the beliefs of two agents. The physical attractiveness principle is a distance between physical characteristics of one agent and the personal desires of another agent. The minimax principle is a distance between capacities of one agents and the desires of another one. Let $\lceil M \rceil$ be the number of elements of a N dimensions matrix M.

$$||u|| = x_B^i - x_B^j \qquad\qquad s = filter(\sqrt{u^T u}, \lceil B \rceil) \qquad (5)$$

where x_b^i and x_b^j are the Belief matrix of agents i and j. The Physical attractiveness principle and the minimax principle are processed exactly the same equation replacing the matrix used depending on the above description.

Complementarity principle is a distance between skills of two agents:

$$||c|| = x_C^i - x_C^j \qquad\qquad c = 1 - filter(\sqrt{u^T u}, \lceil C \rceil) \qquad (6)$$

where x_s^i and x_s^j are the skills matrix of agents i and j and $\lceil C \rceil$ the number of elements in the Capacities matrix.

Reciprocity principle is the mean of z between two individuals.

$$r = mean(z^{ij}, z^{ji}) \qquad (7)$$

where z^{ij} and z^{ji} are a real number representing the attraction of i to j and j to i. Previous equations undergoing the filter function defined as

$$filter(x, x_{max}) = min(\frac{x}{x_{max}}, 1) \qquad (8)$$

Finally, each agent process the control law from the Sect. 4 in 1-dimension for each other agent of the system.

6   Evaluation

In order to make each agent characteristics attractive to others, we pseudo-randomly generate populations of individuals. If the generation was fully random the characteristics of individuals could not be appropriate to be attractive to other agents and the number of groups would be unpredictable. Though, agents characteristics are not known in advance and it allows us to better illustrate the process of group formation.

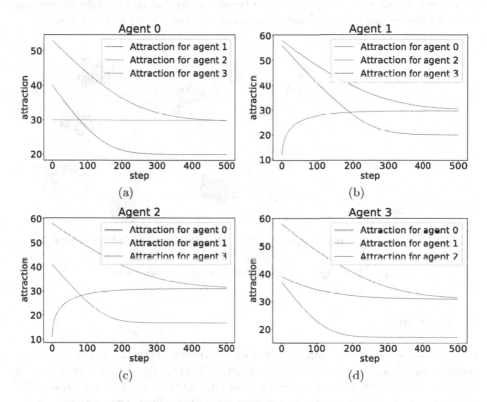

Fig. 1. Attraction an agent i has for other agents of the system.

Figures 1 illustrate the attraction each agent have for each of their acquaintances depending on the number of steps of the simulation. Firstly this simulation confirms experimentally the proof of stabilization of the attraction. Secondly, we can see that agents converge towards different attraction values. For example, agent 0 has a strong attraction for agents 1 and lower ones for agents 2 and 3. These differences are explained by a difference of characteristics that made agents 2 and 3 unattractive to agent 0. Finally, as illustrated, each agent has an attraction value over all the others allowing an outside viewer to visualize these links on a graph. To do so, we process a Gaussian mixture model clustering on attraction values allowing us to make clusters and to find agents for which they have the strongest attraction. Taking the cluster where attraction has the lowest average and linking agent to each agent of this cluster makes possible to build a graph representing groups as illustrated on the Fig. 2(a). Repeating the same process on a larger agent population (fifty agents, six groups), we obtain the Fig. 2(b). Moreover, each individual is capable to assess dynamically the attraction for its acquaintances. It means that even if groups are stabilized for a specific state of the system they can changed if an agent characteristic is modified. Note that this visualization is only a representation of the attraction agent have for each other. Links and groups made on these graphs can be modified depending on how clusters are built or what you consider being a strong attraction.

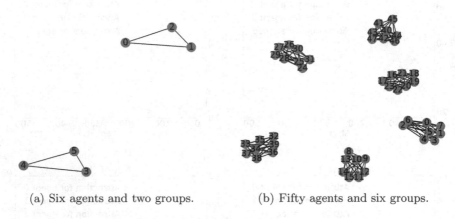

(a) Six agents and two groups.   (b) Fifty agents and six groups.

Fig. 2. Graph showing formed groups on two cases.

7   Conclusion

This work tackles the coalition formation problem using a group dynamics inspired swarming approach. Its dynamicity allows to form groups of agents in an open system with a decentralized manner making this method very robust. In addition, the group dynamics inspiration from HSS allows the system to make swarms with heterogeneous agents made by a high number of characteristics.

We showed that our approach is robust and adapted to the coalition formation problem. However, this problem can be enhanced by some realistic constraints as overlapping groups or by making agents unaware of all the other agents. This work will be used into a decision support system for the remanufacturing and the repurposing of post-used products for the Circular project (ANR-15-IDEX-02). This project focuses on developing the necessary technologies and conditions to make new circular industrial systems able to transform post-used products into new products. In this context, formed groups will represent the new products proposed by the system.

References

1. Dos Santos, D.S., Bazzan, A.L.: Distributed clustering for group formation and task allocation in multiagent systems: a swarm intelligence approach. Appl. Soft Comput. **12**(8), 2123–2131 (2012)
2. Feng, Y., Wang, G.-G., Deb, S., Lu, M., Zhao, X.-J.: Solving 0–1 knapsack problem by a novel binary monarch butterfly optimization. Neural Comput. Appl. **28**(7), 1619–1634 (2015). https://doi.org/10.1007/s00521-015-2135-1
3. Forsyth, D.R.: Group dynamics. Cengage Learning (2010)
4. Henningsen, D.D., Henningsen, M.L.M., Booth, P.: Predicting social and personal attraction in task groups. Groupwork **23**(1), 73–93 (2013)
5. Hogg, M.A., Hains, S.C.: Friendship and group identification: a new look at the role of cohesiveness in groupthink. Eur. J. Soc. Psychol. **28**(3), 323–341 (1998)
6. Irfan, M., Farooq, A.: Auction-based task allocation scheme for dynamic coalition formations in limited robotic swarms with heterogeneous capabilities. In: 2016 International Conference on Intelligent Systems Engineering (ICISE), pp. 210–215. IEEE (2016)
7. Kumar, M., Garg, D.P., Kumar, V.: Segregation of heterogeneous units in a swarm of robotic agents. IEEE Trans. Autom. Control **55**(3), 743–748 (2010)
8. Laird, J.E., Congdon, C.B.: The soar user's manual version 9.5. 0. Technical report, Computer Science and Engineering Department, University of Michigan (2015)
9. Liu, S.h., Zhang, Y., Wu, H.y., Liu, J.: Multi-robot task allocation based on swarm intelligence. J. Jilin Univ. (Eng. Technol. Ed.) **1**, 123–129 (2010)
10. Mouradian, C., Sahoo, J., Glitho, R.H., Morrow, M.J., Polakos, P.A.: A coalition formation algorithm for multi robot task allocation in large-scale natural disasters. In: 2017 13th International Wireless Communications and Mobile Computing Conference (IWCMC), pp. 1909–1914. IEEE (2017)
11. Newcomb, T.M.: Some varieties of interpersonal attraction (1960)
12. Rauniyar, A., Muhuri, P.K.: Multi-robot coalition formation problem: task allocation with adaptive immigrants based genetic algorithms. In: 2016 IEEE International Conference on Systems, Man, and Cybernetics (SMC), pp. 000137–000142. IEEE (2016)
13. Reif, J.H., Wang, H.: Social potential fields: a distributed behavioral control for autonomous robots. Robot. Auton. Syst. **27**(3), 171–194 (1999)
14. Shi, H., Xie, G.: Collective dynamics of swarms with a new attraction/repulsion function. Math. Probl. Eng. **2011** (2011)
15. Shvets, E.: Stochastic multi-agent patrolling using social potential fields. In: ECMS, pp. 42–49 (2015)
16. Souidi, M.E.H., Piao, S.: A new decentralized approach of multiagent cooperative pursuit based on the iterated elimination of dominated strategies model. Math. Probl. Eng. **2016** (2016)
17. Szwaykowska, K., Romero, L.M.y.T., Schwartz, I.B.: Collective motions of heterogeneous swarms. IEEE Trans. Autom. Sci. Eng. **12**(3), 810–818 (2015)
18. Walster, E., Aronson, V., Abrahams, D., Rottman, L.: Importance of physical attractiveness in dating behavior. J. Pers. Soc. Psychol. **4**(5), 508 (1966)
19. Wu, H., Li, H., Xiao, R., Liu, J.: Modeling and simulation of dynamic ant colony's labor division for task allocation of UAV swarm. Phys. A Stat. Mech. Appl. **491**, 127–141 (2018)

Collective Gradient Perception in a Flocking Robot Swarm

Tugay Alperen Karagüzel[1]($\boxtimes$), Ali Emre Turgut[1], and Eliseo Ferrante[2]

[1] Department of Mechanical Engineering, Middle East Technical University,
Ankara, Turkey
tugay.karaguzel@gmail.com, aturgut@metu.edu.tr
[2] Vrije Universiteit Amsterdam, Amsterdam, The Netherlands
e.ferrante@vu.nl

Abstract. Animals can carry their environmental sensing abilities beyond their own limits by using the advantage of being in a group. Some animal groups use this collective ability to migrate or to react to an environmental cue. The environmental cue sometimes consists of a gradient in space, for example represented by food concentration or predators' odors. In this study, we propose a method for collective gradient perception in a swarm of flocking agents where single individuals are not capable of perceiving the gradient but only sample information locally. The proposed method is tested with multi-agent simulations and compared to standard collective motion methods. It is also evaluated using realistic dynamical models of autonomous aerial robots within the Gazebo simulator. The results suggest that the swarm can move collectively towards specific regions of the environment by following a gradient while solitary agents are incapable of doing it.

1 Introduction

Many different species move in a collective fashion in order to survive [15]. Species provide various critical advantages from collective motion. Sharing the location of food [3], avoiding predators [10] and locating the source of nutrient [8] are some examples of these advantages. In these examples, the direction of motion of a living creature is often influenced by environmental cues. For some animals, this directed motion is a group behavior centered around their emergent sensing [6]. By definition [2], emergent sensing is the social interactions that facilitate comparisons across scalar measurements made by individuals. These comparisons lead to a collective computation of the environmental gradient. An example of emergent sensing can be seen in golden shiners [1]: While a single golden shiner fish is not capable of sensing the light gradient in order to follow it, a school of golden shiners is able to gather in the darkest regions of the environment. Golden shiners achieve this by taking social cues into account in addition to environmental cues. This combination leads to a collective computation of the light gradient [11]. The fascinating collective abilities of animal groups have been inspiring the robotics field for several decades. Swarm robotics has emerged as

© Springer Nature Switzerland AG 2020
M. Dorigo et al. (Eds.): ANTS 2020, LNCS 12421, pp. 290–297, 2020.
https://doi.org/10.1007/978-3-030-60376-2_23

a scientific field through inspiration from these studies. The collective motion of robot swarms is one of the major research areas [4,13]. In [4], a swarm of mobile ground robots flocks in a pre-defined goal direction by only relying on local sensing and on the on-board computation capabilities of robots. The authors in [16] present a swarm of autonomous aerial robots wandering in the air as a flock. Flying robots construct a wireless communication network to share and obtain information about global positions and velocities. The algorithms used in [5] and [16] show how the results of nature-inspired approaches can be effective on artificial systems. Emergent sensing in artificial systems has also been investigated in different studies. In [12], autonomous robots that are incapable of sensing and locating an underwater RF source accomplish this task by forming a group. The work in [14] presents a collective decision making method applied to a group of simple mobile ground robots that makes them able to perceive the environment and to decide what is the environmental feature with the highest relative frequency.

In this paper we propose a novel method for performing collective gradient sensing, which extends an existing collective motion method [4,13], in a robot swarm. Every agent in that swarm can sense other agents, the environment boundaries and a local and scalar environmental cue within a limited range. The perception and following of the gradient by the swarm is achieved by agents modifying their desired distances to other agents correlated with their local environmental perceptions.

2   Methodology

2.1   Standard Collective Motion (SCM)

The swarm consists of N agents which can freely move in a bounded 2-dimensional environment. At each time instant, the focal agent has a direction of motion, which is called heading, and it can only move along this heading direction according to its linear speed. The heading direction changes smoothly by the angular velocity of the agent. The focal agent i calculates the linear speed and the angular speed as a function of a virtual force vector that combines all the local information perceived from its neighbors as:

$$f_i = \alpha p_i + \beta h_i + \gamma r_i \tag{1}$$

p_i is proximal control vector, h_i is alignment control vector and r_i is avoidance vector. α, β and γ are corresponding weight coefficients for p_i, h_i and r_i. Proximal control and avoidance are always enabled [13], while we test collective motion both with and without alignment control, as in [13].

At each time instant, p_i is calculated for m perceived neighbors within the proximal control perception range D_p of the focal agent i.

$$p_i = \sum_{m \in N} p_i^m(d_i^m, \sigma_i) \angle e^{j\phi_i^m} \tag{2}$$

$p_i^m(d_i^m, \sigma_i)$ is the magnitude of proximal control vector and $\angle e^{j\phi_i^m}$ is the angle of perceived agent m. The magnitude of proximal control vector is calculated as follows:

$$p_i^m(d_i^m, \sigma_i) = -\epsilon \left[2 \frac{\sigma_i^4}{(d_i^m)^5} - \frac{\sigma_i^2}{(d_i^m)^3} \right] \tag{3}$$

d_i^m is the relative distance of agent m perceived by i, ϵ is the strength coefficient of proximal control vector and σ_i is the coefficient for desired distance of i to other agents. The relation between desired distance of i and σ_i is: $d_{des}^i = 2^{1/2}\sigma_i$. We indicate both d_i^m and σ_i as variables because the former changes as a function of the neighbors distance and the second can change according to the desired distance modulation method explained in Sect. 2.2.

The alignment control vector h_i is found by summing the headings of neighbors within the alignment control communication range D_a with heading of the focal agent i, and normalizing the result.

$$h_i = \frac{\angle e^{j\theta_0} + \sum_{m \in N} \angle e^{j\theta_m}}{||\angle e^{j\theta_0} + \sum_{m \in N} \angle e^{j\theta_m}||} \tag{4}$$

The heading of agent m located in the communication range D_a of i is $\angle e^{j\theta_i^m}$. The heading of the focal agent i is $\angle e^{j\theta_0}$. Headings are calculated with respect to a common frame of reference for all agents and shared in that way. In a real application, this frame of reference can be implemented either by a digital compass ("common north") [13] or with a shared directional signal (such as a light source) [5].

The avoidance vector r_i is calculated for every perceived environment boundary within D_r. The magnitude of avoidance vector [9] is calculated for the all perceived boundary b among all B environment boundaries as follows and all boundary avoidance vectors are summed for resultant effect:

$$r_i^b = k_{rep} \left(\frac{1}{L_b} - \frac{1}{L_0} \right) \left(\frac{p_i^b}{L_b^3} \right) \tag{5}$$

k_{rep} is the strength coefficient of avoidance vector, L_0 is the relaxation threshold of the function and L_b is the perceived distance to boundary b. The unit vector p_i^b designates the direction of boundary b in agent i's frame of reference.

As in [4], the linear and angular velocities of the focal agent is calculated by first projecting the resultant virtual force vector f_i on the two orthogonal axes of agent i's local frame of reference. In local frame of reference, the x-axis is parallel to the focal agent heading, and the usual right-hand rule is used. The linear and angular speeds are calculated as: $U_i = K_1 f_x$ and $\omega_i = K_2 f_y$, respectively. The linear speed U_i is determined by multiplying the local x component of f_i (f_x) by the linear speed gain K_1. The angular speed is determined by multiplying y component of f_i (f_y) by angular gain K_2. The linear speed is bounded between 0 and U_{max} and angular speed of the agent i is bounded between $-\omega_{max}$ and ω_{max}.

2.2   Desired Distance Modulation (DM)

The 2-dimensional environment that agents are located in is assumed to contain a scalar value at every point of it. Agents can only perceive the value at their instantaneous positions. The proposed method, which is the contribution of this paper, suggests that if every agent changes the innate coefficient for desired distance σ_i according to the local perceived value G°, the swarm can collectively exhibit a taxis behavior towards decreasing gradient values. The function linking the local perceived value by agent i with the coefficient σ_i, which is an input for proximal control vector, for desired distance is:

$$\sigma_i = \sigma_{max} - \left(\frac{G^\circ}{G_{max}}\right)^{0.1}(\sigma_{max} - \sigma_{min}) \tag{6}$$

σ_{max} and σ_{min} are the maximum and minimum values of σ_i, respectively. G_{max} is the maximum local value contained in the environment.

3   Multi-agent Simulations

In multi-agent simulations, we consider a swarm of N agents with no inertial properties. Their positions and headings (both continuous variables) are updated by discrete integration. We consider a squared arena with 20 *units* sides. To implement the gradient, the arena is divided into a square grid and every piece of the grid, with edge size of 0.025 *units*, contains a value between 0 and 255. For any environment, the values of the grids are designed such that the variation is smooth in all directions. The two environments used in multi-agent simulations are reported in Fig. 1. An agent can only perceive the value corresponding to its current grid cell. The perception of the grid value is instantaneous and the agent does not keep the value of previously seen grids.

We study the effect of two components of the proposed method on the ability of the swarm to achieve gradient perception and compare them with standard collective motion (SCM) method. Two components are desired distance modulation (DM) and heading alignment (HA). DM is considered with and without HA to reveal individual effect of DM in addition to combined effect of DM and HA on the collective behavior. In SCM, HA is included and desired distance is constant and same for every agent. We consider two different gradient models. For a chosen gradient model, the experiment is repeated 512 times for DM, $HA + DM$ and SCM. In each repetition, agents are initialized around a randomly chosen center of mass with random positions and orientations. The control step dt is set to 0.1 s. A single experiment consists of 80000 time steps. The instantaneous values from the environment are assumed to be perceived and collected only once every 100 time steps. In addition to local values, average of all agent positions is found with same rate in order to visualize the group trajectories. We define and use the *average local value* to measure the success of the swarm in perceiving and following the gradient towards *decreasing values*. The average local value is calculated as: first, the instantaneously perceived local

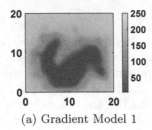
(a) Gradient Model 1

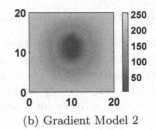
(b) Gradient Model 2

Fig. 1. Gradient models used in multi-agent simulations. Color bars show grid values.

values of all the agents in the swarm are averaged; second, the group average is averaged again over the experiment duration, and a single value is obtained for each run.

3.1 Results

In Fig. 2, we report results for two different gradient models for all the methods (SCM, DM and $HA+DM$), by using colored regions to represent the distribution of data with box plots at the center of each region. The SCM method fails to minimize the average local value in all gradient models having a median value between approximately 75 and 150. However, DM and $HA+DM$ methods perform well in all cases having median values approximately below 10.

When we consider the distributions, SCM method shows a high spread whereas both DM and $HA+DM$ methods show a very low spread. There is not a noticeable performance difference of DM and $HA+DM$ in these experiments. In Fig. 3 the average trajectories of the three methods in Gradient Model 1 are plotted. The SCM method shows an homogeneous distribution (Fig. 3a). DM and $HA+DM$ methods show a clear concentration on the region with low local values (Fig. 3b and Fig. 3c), DM method being more concentrated on the central regions.

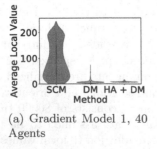
(a) Gradient Model 1, 40 Agents

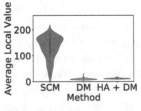
(b) Gradient Model 2, 40 Agents

Fig. 2. Distribution of average local values with multi-agent simulations with following parameter values: $N = 40$, $D_p = 2.0\ units$, $\alpha = 2.0$, $\sigma_{max} = 0.6$, $\sigma_{min} = 0.2$, $\epsilon = 12.0$, $D_a = 2.0\ units$, $\beta = 2.0$, $D_r = 0.5\ units$, $\gamma = 1.0$, $k_{rep} = 80.0$, $L_0 = 0.5$, $K_1 = 0.5$, $K_2 = 0.06$, $U_{max} = 0.05\ units/t$, $\omega_{max} = \pi/4\ rad/t$, $G_{max} = 255$, $dt = 0.1$

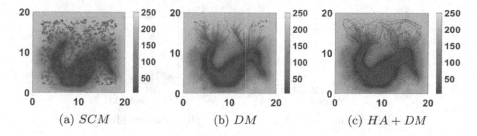

Fig. 3. Average trajectories of center of mass of 40 agents on Gradient Model 1

4   Physics-Based Simulations

Physics-based simulations are conducted with dynamical models[1] of Crazyflie[2] nano quadcopters. The Gazebo simulator is employed for dynamical calculations and a modified version of the real Crazyflie firmware (see Footnote 1) is used as SITL (Software in the Loop). The software framework is constructed on ROS (Robot Operating System). The velocity commands are produced and published by corresponding scripts running on ROS [7]. The SITL firmware receives velocity commands and applies them via low-level controller software. Velocity commands are not perfectly tracked, as in the real platform. The same gradient models studied in multi-agent simulations are used in physics-based simulations. Given the simplicity of the Crazyflie platform and the lack of proximal and environmental sensors, we emulated the required sensors by feeding the corresponding information to robots in a way that meets the constraints of the proposed method. The emulation also includes communication for alignment control. The calculations for linear and angular speeds are repeated 100 times in a simulated second ($dt = 0.01$). Calculated linear speed is transformed to a velocity command fed to SITL firmware. The robot swarm in Gazebo consists of $N = 6$ quadcopters. Boundary of the flight arena is a square with an edge size of 10 *units*. DM and $HA + DM$ methods are implemented and experiments are repeated 20 times.

4.1   Results

Figure 4 presents the results of the average local values using DM and $HA+DM$ methods for two gradient models. Colored regions represent the distribution of data and box plots are depicted at the center of each region. The results are qualitatively in accordance with the multi-agent simulations. However, in physics-based simulations, there is a slight increase in median values and the distribution is more spread. There could be several reasons for this discrepancy. In physics-based simulations, robots are modeled realistically considering all the dynamics, hence velocity tracking of robots is imperfect. In addition to that

---

[1] https://github.com/wuwushrek/sim_cf.

[2] https://www.bitcraze.io/products/old-products/crazyflie-2-0/.

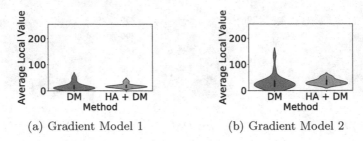

(a) Gradient Model 1   (b) Gradient Model 2

Fig. 4. Distribution of average local values of physics-based simulations with following parameter values: $N = 6$, $D_p = 2.0\ units$, $\alpha = 3.0$, $\sigma_{max} = 0.8$, $\sigma_{min} = 0.4$, $\epsilon = 12.0$, $D_a = 2.0\ units$, $\beta = 1.0$, $D_r = 0.5\ units$, $\gamma = 1.0$, $k_{rep} = 50.0$, $L_0 = 0.5$, $K_1 = 0.5$, $K_2 = 0.03$, $U_{max} = 0.25\ units/t$, $\omega_{max} = 4\pi\ rad/t$, $G_{max} = 255$, $dt = 0.01$

position information of the robots is also noisy. Therefore, robots oscillate more to keep their distances with the neighbors at the desired level increasing the spread of local value distribution.

5   Conclusion

In this study, we extended the collective motion method introduced in [4,13] by adding a local perception capability to each robot. Assuming that an environment has a gradient of some sort such as food concentration, through agent-based and physics-based simulations, we showed that robots locally sensing this gradient while moving collectively as a group were able to move towards specific regions of the environment. Robots achieve this by simply changing the desired distance parameter defined in the original model [4,13] based on the local information about the gradient. The effect of heading alignment control (HA) together with desired distance modulation (DM) on success of gradient perception and following is evaluated with systematic experiments on multi-agent simulations and physics based simulations. Last but not least, the improving effect of HA on the gradient perception can be observed on simulations with dynamical robot models as well. Although the mean values are comparably close to each other for $HA + DM$ and DM, distribution of $HA + DM$ demonstrates a more consistent and successful collective gradient perception.

References

1. Berdahl, A., Torney, C.J., Ioannou, C.C., Faria, J.J., Couzin, I.D.: Emergent sensing of complex environments by mobile animal groups. Science **339**(6119), 574–576 (2013)
2. Berdahl, A.M., et al.: Collective animal navigation and migratory culture: from theoretical models to empirical evidence. Philos. Trans. R. Soc. B Biol. Sci. **373**(1746), 20170009 (2018)

3. Couzin, I.D., Krause, J., Franks, N.R., Levin, S.A.: Effective leadership and decision-making in animal groups on the move. Nature **433**(7025), 513–516 (2005)
4. Ferrante, E., Turgut, A.E., Huepe, C., Stranieri, A., Pinciroli, C., Dorigo, M.: Self-organized flocking with a mobile robot swarm: a novel motion control method. Adapt. Behav. **20**(6), 460–477 (2012)
5. Ferrante, E., Turgut, A.E., Stranieri, A., Pinciroli, C., Birattari, M., Dorigo, M.: A self-adaptive communication strategy for flocking in stationary and non-stationary environments. Nat. Comput. **13**(2), 225–245 (2013). https://doi.org/10.1007/s11047-013-9390-9
6. Grünbaum, D.: Schooling as a strategy for taxis in a noisy environment. Evol. Ecol. **12**(5), 503–522 (1998)
7. Hönig, W., Ayanian, N.: Flying Multiple UAVs Using ROS. In: Koubaa, A. (ed.) Robot Operating System (ROS). SCI, vol. 707, pp. 83–118. Springer, Cham (2017). https://doi.org/10.1007/978-3-319-54927-9_3
8. Kearns, D.B.: A field guide to bacterial swarming motility. Nat. Rev. Microbiol. **8**(9), 634–644 (2010). https://doi.org/10.1038/nrmicro2405
9. Khaldi, B., Cherif, F.: A virtual viscoelastic based aggregation model for self-organization of swarm robots system. In: Alboul, L., Damian, D., Aitken, J. (eds.) TAROS 2016. LNCS, vol. 9716, pp. 202–213. Springer, Cham (2016). https://doi.org/10.1007/978-3-319-40379-3_21
10. Olson, R.S., Hintze, A., Dyer, F.C., Knoester, D.B., Adami, C.: Predator confusion is sufficient to evolve swarming behaviour. J. R. Soc. Interface **10**(85), 20130305 (2013)
11. Puckett, J.G., Pokhrel, A.R., Giannini, J.A.: Collective gradient sensing in fish schools. Sci. Rep. **8**(1), 7587 (2018)
12. Shaukat, M., Chitre, M.: Adaptive behaviors in multi-agent source localization using passive sensing. Adapt. Behav. **24**(6), 446–463 (2016)
13. Turgut, A.E., Çelikkanat, H., Gökçe, F., Şahin, E.: Self-organized flocking in mobile robot swarms. Swarm Intell. **2**(2–4), 97–120 (2008)
14. Valentini, G., Brambilla, D., Hamann, H., Dorigo, M.: Collective perception of environmental features in a robot swarm. In: Dorigo, M., Birattari, M., Li, X., López-Ibáñez, M., Ohkura, K., Pinciroli, C., Stützle, T. (eds.) ANTS 2016. LNCS, vol. 9882, pp. 65–76. Springer, Cham (2016). https://doi.org/10.1007/978-3-319-44427-7_6
15. Vicsek, T., Zafeiris, A.: Collective motion. Phys. Rep. **517**(3–4), 71–140 (2012)
16. Vásárhelyi, G., Virágh, C., Somorjai, G., Nepusz, T., Eiben, A.E., Vicsek, T.: Optimized flocking of autonomous drones in confined environments. Sci. Robot. **3**(20) (2018)

Fitting Gaussian Mixture Models Using Cooperative Particle Swarm Optimization

Heinrich Cilliers[1] and Andries P. Engelbrecht[1,2]($\boxtimes$) (iD)

[1] Computer Science Division, Stellenbosch University, Stellenbosch, South Africa
[2] Department of Industrial Engineering and Computer Science Division, Stellenbosch University, Stellenbosch, South Africa
{19035837,engel}@sun.ac.za

Abstract. Recently, a particle swarm optimization (PSO) algorithm was used to fit a Gaussian mixture model (GMM). However, this algorithm incorporates an additional step in the optimization process which increases the algorithm complexity and scales badly to a large number of components and large datasets. This study proposes a cooperative approach to improve the scalability and complexity of the PSO approach and illustrates its effectiveness compared to the expectation-maximization (EM) algorithm and the existing PSO approach when applied to a number of clustering problems.

1 Introduction

A GMM is a model consisting of a number of Gaussians[1], where each Gaussian is characterised by three different parameters. Fitting a GMM is done by finding suitable estimates for each Gaussian's parameters. These estimates are usually found by using the EM algorithm [9]. However, EM is sensitive to its initialization and is prone to yielding sub-optimal solutions. Ari and Aksoy [1] designed an algorithm to fit GMMs using PSO, referred to as AA in this paper. However, AA has some inefficiencies. Firstly, AA incorporates a correspondence identification step in the optimization process, which increases computational complexity. Secondly, the variant of PSO scales badly to larger datasets and a large number of components. Lastly, the mixture coefficients are not optimized directly by the PSO which could lead to sub-optimal results.

This study proposes a cooperative PSO approach to GMM, where (1) all parameter estimates are directly estimated using PSO, (2) the correspondence identification step is omitted, (3) and multiple swarms are used to estimate parameters cooperatively. This cooperative PSO approach is empirically compared to the EM and the AA algorithm on a number of data clustering problems. The results show that the cooperative approach outperforms both EM and AA algorithms w.r.t. overall clustering quality.

The remainder of this paper is as follows: Sect. 2 discusses GMMs briefly, followed by Sects. 3 and 4 which contain details of the implementations of the

---

[1] Also referred to as components.

© Springer Nature Switzerland AG 2020
M. Dorigo et al. (Eds.): ANTS 2020, LNCS 12421, pp. 298–305, 2020.
https://doi.org/10.1007/978-3-030-60376-2_24

proposed approaches. Section 5 contains a discussion of the empirical procedure used to evaluate and compare the algorithms. Section 6 contains the results of the empirical analysis. Finally, Sect. 7 contains the conclusion of this study.

2   Gaussian Mixture Model

Mixture models [6] are probabilistic models representing the distribution over a general population as a mixture of distributions over a number of sub-populations. A GMM is a mixture model whereby the distribution over the entire population, assuming K sub-populations, is modeled by K Gaussians. A sample from this distribution, denoted by **x**, has the following probability density function (p.d.f.):

$$f(\boldsymbol{x}) = \sum_{k=1}^{K} \pi_k \cdot f_N \left(\boldsymbol{x} \mid \boldsymbol{\mu}_k, \Sigma_k \right) \tag{1}$$

$$f_N(\boldsymbol{x} \mid \boldsymbol{\mu}_k, \Sigma_k) = (2\pi)^{-\frac{D}{2}} |\Sigma_k|^{-\frac{1}{2}} exp \left[-\frac{1}{2} (\boldsymbol{x} - \boldsymbol{\mu}_k)^T \Sigma_k^{-1} (\boldsymbol{x} - \boldsymbol{\mu}_k) \right] \tag{2}$$

where the k-th component's mixture coefficient, mean vector and covariance matrix is given by π_k, $\boldsymbol{\mu}_k$ and Σ_k respectively. In equation (2) f_N and D denote the multivariate normal p.d.f. and the dataset dimensionality, respectively. All covariance matrices must be positive definite and all mixture coefficients must sum to one and be a value in $(0, 1)$.

3   Gaussian Mixture Modeling Particle Swarm Optimization

Assume a D-dimensional dataset and K components. The first proposed algorithm, referred to as the Gaussian mixture modeling PSO (GMMPSO), is an adapted version of AA [1], based on the inertia weight PSO [10]. Similar to AA, GMMPSO uses the log-likelihood of the data as the objective function, i.e.

$$L = \sum_{i=1}^{N} \log \left\{ \sum_{k=1}^{K} \hat{\pi}_k \cdot f_N \left(\mathbf{x}_i | \hat{\boldsymbol{\mu}}_k, \hat{\Sigma}_k \right) \right\} \tag{3}$$

where the estimates for the mixture coefficient, mean vector and covariance matrix of the k-th component is given by $\hat{\pi}_k$, $\hat{\boldsymbol{\mu}}_k$ and $\hat{\Sigma}_k$, respectively. Similar to AA, GMMPSO parameterizes each D-dimensional covariance matrix by D positive real values and $\tau = \frac{D(D-1)}{2}$ real values, representing Eigenvalues and Givens rotation angles[2,3], respectively. Mean vectors are simply parameterized by D real values representing the vector elements. To parameterize mixture coefficients, consider the concept of normalization. Normalizing a set of K positive

---

[2] The angle of a rotation in a plane described by two axis.
[3] For D dimensions there are τ planes wherein a rotation can be applied.

values $\{\rho_k\}_{k=1}^K$ by their sum yields a set of K values which satisfy the mixture coefficient constraints. Thus, the K mixture coefficients are parameterized by K positive real numbers. Now, almost similar to a particle from AA, a particle from GMMPSO will have the form

$$\boldsymbol{p}_i = \begin{bmatrix} \boldsymbol{C}_1^T \ldots \boldsymbol{C}_K^T \end{bmatrix}, \text{ where } \boldsymbol{C}_k^T = \begin{bmatrix} \rho_k \ \mu_{k1} \ldots \mu_{kD} \ \lambda_{k1} \ldots \lambda_{kD} \ \phi_{k1} \ldots \phi_{k\tau} \end{bmatrix} \quad (4)$$

where ρ_k, μ_{kd}, λ_{kd} and ϕ_{ke} denote the unnormalized mixture coefficient, d-th dimension of the mean vector, d-th Eigenvalue and e-th Givens rotation angle of the k-th component. Evaluation of a particle is done after a reconstruction step, where all parameter estimates are calculated from their respective parameterizations. Firstly, mixture coefficient estimates are calculated by using the aforementioned normalization method. Secondly, the k-th mean vector estimate is constructed from the μ_{kd} values. Lastly, the k-th covariance matrix is constructed using Eigen decomposition and rotation matrices. First, let Λ_k be a D-dimensional diagonal matrix with $\{\lambda_{kd}\}_{d=1}^D$ on its diagonal. Then let V_k be the product of the basic rotation matrices[4] calculated from $\{\phi_{ke}\}_{e=1}^\tau$ by applying the method described by Duffin and Barret [4] to each ϕ_{ke}. Finally, calculate the estimate as $\hat{\Sigma}_k = V_k \Lambda V_k^T$. The log-likelihood can now be calculated.

Particles are initialized as follows: Let $x_{min,d}$ and $x_{max,d}$ represent the minimum and maximum value of the d-th dimension from the given dataset, respectively. Also, let η be the largest Eigenvalue of the sample covariance matrix calculated from the dataset. Then ρ_k, μ_{kd}, λ_{kd} and ϕ_{ke} are sampled from a uniform distribution over the bounds $[0, 1]$, $[x_{min,d}, x_{max,d}]$, $[0, \eta]$ and $[-\pi, \pi]$, respectively for all k and d.

4   Cooperative GMMPSO

The PSO variant used by AA and GMMPSO does not scale well to larger datasets and a large number of components. Van den Bergh and Engelbrecht [5] designed a cooperative PSO variant which improved scalability significantly; thus, a cooperative approach is considered here. Inspired by the CPSO-S_K [5] algorithm, two cooperative GMMPSO (CGMMPSO) variants are proposed.

The first algorithm, referred to as CGMMPSO-K, optimizes the parameters of a K-GMM[5] by assigning a component's parameters to a single sub-swarm. Thus, K swarms are utilized during the optimization procedure. The i-th particle from the k-th subswarm, denoted by $\boldsymbol{p}_{ik}$, takes on the form $\boldsymbol{p}_{ik} = \boldsymbol{C}_k^T$.

The second cooperative variant, referred to as CGMMPSO-K^{++}, uses an additional swarm dedicated to the mixture coefficients of the K-GMM; thus, $K+1$ swarms are used. An arbitrary particle from a component swarm looks similar to that of CGMMPSO-K, excluding ρ_k. The position of the i-th particle of the mixture coefficient swarm, denoted by $\boldsymbol{p}_i^*$, takes on the form $\boldsymbol{p}_i^* = \begin{bmatrix} \rho_1 \ldots \rho_K \end{bmatrix}$.

---

[4] A matrix which applies a Givens rotation.
[5] A GMM assuming K components.

5   Experimental Setup

This section explains how the experiments were performed, and summarizes the real-world and artificial datasets used. A synopsis of the performance measures is provided and the statistical procedure is described.

5.1   General Experiment Information

Performance was measured on cluster separation, cluster compactness and overall quality. Cluster separation is evaluated by the average inter-cluster distance (J_{inter})[8]. Cluster compactness is evaluated by the average weighted intra-cluster distances, calculated as

$$J_{intra} = \frac{1}{K} \sum_{k=1}^{K} \frac{\sum_{i=1}^{N} \gamma_k(\boldsymbol{x}_i) \cdot d(\boldsymbol{x}_i, \boldsymbol{\mu}_k)}{\sum_{i=1}^{N} \gamma_k(\boldsymbol{x}_i)}, \text{ where } \gamma_k(\mathbf{x}_i) = \frac{\pi_k f_N(\boldsymbol{x}_i | \boldsymbol{\mu}_k, \Sigma_k)}{\sum_{l=1}^{K} \pi_l f_N(\boldsymbol{x}_i | \boldsymbol{\mu}_l, \Sigma_l)} \tag{5}$$

where d($\mathbf{p}$, $\mathbf{q}$) denotes the Euclidean distance between $\boldsymbol{p}$ and $\boldsymbol{q}$. Lastly, overall quality is evaluated using the Xie and Beni (XB) index [12] where the distance from $\mathbf{x}_i$ to $\boldsymbol{\mu}_k$ is weighted with $\gamma_k(\mathbf{x}_i)$. Solutions with small intra-cluster distances, small XB index and high inter-cluster distances are preferred.

The performances of the EM [9], AA [1] and new algorithms were evaluated on a set of artificial and three real-world datasets. The EM algorithm was initialized by the adapted version of the Gonzales algorithm developed by Blömer and Bujna [2]. Thirty independent runs were performed for each experiment and algorithm. All PSO algorithms were executed with the same control parameter values, i.e. an inertia weight of 0.729 and social and cognitive coefficients of 1.494 each [11]. To ensure a fair comparison between all algorithms, a swarm size of 30 was used for all swarms and sub-swarms. At first this might seem unfair to the single swarm algorithms since the cooperative swarms will have multiples of 30 particles in total. However, if the sub-swarm size of all cooperative variants are equal to the swarm size of the single swarm variants, then all dimensions in the solution vector of each algorithm will receive the same number of decision variable updates. Lastly, to ensure that the means estimated by the PSO-based algorithms stay within acceptable bounds, all means must satisfy a boundary constraint: $x_{min,d} \leq \mu_{k,d} \leq x_{max,d} \ \forall \ k$ and d. Ari and Aksoy did not state how mixture coefficients were exactly calculated; thus, for all experiments on AA the mixture coefficients were calculated as

$$\pi_k = \frac{1}{N} \sum_{i=1}^{N} \frac{f(\mathbf{x}_i | \boldsymbol{\mu}_k, \Sigma_k)}{\sum_{l=1}^{K} f(\mathbf{x}_i | \boldsymbol{\mu}_l, \Sigma_l)} \tag{6}$$

A wins and losses approach as used by Georgieva [7] was used to statistically analyze the overall performance of the algorithms over all datasets. A Mann-Whitney U test was performed on the results from the 30 independent runs for each dataset and all combinations of the algorithms. If a statistically significant difference exists, the resulting U statistic was used to determine the winning and losing algorithm. There is no winning or losing algorithm if no statistically significant difference was found. An algorithm's score was calculated as $\#wins - \#losses$. The algorithm with the highest score is considered the better algorithm overall. All statistical tests were performed with a significance level of 5%.

5.2   Datasets

The real-world datasets, acquired from the UCI Machine learning repository[6], include the Iris (4 features, 150 samples), Seeds (7 features, 210 samples) and Travel Reviews (TR) (11 features, 980 samples) datasets. Since the true number of clusters are unknown, all algorithms were applied four times to each dataset with varying numbers of clusters and iterations. Table 2 lists the configuration for each experiment performed on each dataset. Various artificial datasets were generated to evaluate the overall effectiveness of all algorithms. These datasets varied in the number of features (D), true number of clusters (K), number of samples (N), cluster shape (spherical or elliptical), degree of separation between clusters (C), and presence of outliers. Table 1 lists the four characteristics of the nine main configurations of the artificial data and the number of iterations performed by all algorithms on each configuration. Thirty-six artificial data experiments were generated from the different combinations of the main configurations, cluster shape and outlier presence. All algorithms were applied to the artificial datasets with the true number of clusters known a priori.

Artificial datasets were generated as follows: Mixture coefficients, denoted by $\{\pi_k\}_{k=1}^K$, were generated by normalising a set of K integer values sampled from $U(1, 10)$[7]. The covariance matrices, denoted by $\{\Sigma_k\}_{k=1}^K$, were generated in one of two ways depending on the shape. For spherical clusters Σ was calculated as $s \cdot I_D$, where $s \sim U(0.5, 20.0)$ and I_D denotes the D-dimensional identity matrix. For elliptical clusters the covariance matrices were calculated using $\Sigma_k = V_k \Lambda_k V_k^T$. V_k is the orthogonal matrix yielded from the QR decomposition of a random matrix, i.e. a matrix with all its entries sampled from $U(0, 1)$. Λ_k is a diagonal matrix with each diagonal entry sampled from $U(0.5, 20.0)$. Mean vectors, denoted by $\{\mu_k\}_{k=1}^K$, were generated by creating a mean at the origin, then sequentially generating $K - 1$ means by using $\mu_k = \mu_l + \delta_k \cdot \theta_k$, where $\delta_k \sim U(0, 1)$, l is a random integer in the bound $(1, k-1)$ and θ is a random unit vector, i.e. $\theta_j \sim U(-1, 1)$; the vector was then normalised. Afterwards, the means were moved away from each other until a pairwise separation [3] constraint was fulfilled for all pairs. Now, for all k, $N \cdot \pi_k$ samples were generated according to a multivariate Gaussian distribution with mean vector μ_k and covariance matrix Σ_k. Lastly, outliers were generated by mutating 5% of the total number of samples into outliers. The mutation was performed by moving a sample on the

---

[6] Dua, D. and Graff, C. (2019). UCI Machine Learning Repository [http://archive.ics. uci.edu/ml]. Irvine, CA: University of California, School of Information and Computer Science.

[7] $U(a, b)$ denotes a uniform distribution over the interval (a, b).

Table 1. Artificial dataset configurations

Configuration	D	K	N	C	# Iterations
1	2	2	200	3	500
2	3	3	300	3	1500
3	4	4	400	3	3000
4	5	5	500	3	5000
5	3	2	200	2	1000
6	5	3	300	2	3000
7	7	5	500	2	5000
8	5	2	200	1	2500
9	10	3	400	1	5000

Table 2. Real-world dataset configurations

Dataset	# Clusters	# Iterations
Iris	2	2500
Iris	3	3000
Iris	4	3500
Iris	5	4000
Seeds	2	3500
Seeds	3	4250
Seeds	4	5000
Seeds	5	6000
TR	2	5000
TR	3	6000
TR	4	7000
TR	5	8000

Table 3. Cluster compactness scores per algorithm (sorted in descending order w.r.t. score)

Algorithm	# Wins	# Losses	Score
EM	106	28	78
CGMMPSO-K	90	23	67
CGMMPSO-K^{++}	81	27	54
GMMPSO	32	97	−65
AA	13	147	−134

Table 4. Cluster separation scores per algorithm (sorted in descending order w.r.t. score)

Algorithm	# Wins	# Losses	Score
AA	107	17	90
GMMPSO	58	30	28
CGMMPSO-K	34	45	−11
CGMMPSO-K^{++}	27	49	−22
EM	9	94	−85

edge of the mixture away from its respective mean till its Mahalanobis distance is between five and nine units.

6 Results

Tables 3, 4 and 5 summarize the scores achieved by each algorithm for the J_{intra}, J_{inter} and XB measures, respectively. GMMPSO and AA showed better cluster separation compared to the other algorithms. The AA algorithm yielded clusters with the best separation, followed by the GMMPSO with the second best and CGMMPSO-K with the third best score. Despite EM having the best cluster compactness, the algorithm performed the worst w.r.t. cluster separation. Finally, regarding the overall clustering quality the CGMMPSO-K algorithm outperformed all other algorithms, followed by CGMMPSO-K^{++} in close second and EM with the third best score. GMMPSO performed the worst w.r.t. overall clustering quality.

It is evident from the results that CGMMPSO-K outperformed CGMMPSO-K^{++} on all measures, making CGMMPSO-K the best overall cooperative variant. Although GMMPSO did have better cluster separation than CGMMPSO-K, CGMMPSO-K did outperform GMMPSO w.r.t. cluster compactness and overall

Table 5. Overall clustering quality scores per algorithm (sorted in descending order w.r.t. score)

Algorithm	# Wins	# Losses	Score
CGMMPSO-K	53	24	29
CGMMPSO-K^{++}	57	31	26
EM	58	43	15
AA	48	59	−11
GMMPSO	11	70	−59

clustering quality. Thus, CGMMPSO-K is the best of the three new algorithms. Note the contrasting results of the EM and AA algorithms. EM performed the best w.r.t. cluster compactness and AA performed the best w.r.t. cluster separation. However, EM outperformed AA w.r.t. overall clustering quality.

Finally, EM performed better than CGMMPSO-K only w.r.t. cluster compactness. However, EM performs worse on both cluster separation and overall clustering quality. Thus, CGMMPSO-K is regarded as the best overall algorithm for the considered data clustering problems.

7   Conclusion

This study proposed three particle swarm optimization (PSO) algorithms to estimate parameters for a Gaussian mixture model (GMM). The performance of the proposed algorithms, the EM algorithm and AA algorithm was evaluated on a number of artificial and real-world datasets. Their respective performances were compared and the results revealed that CGMMPSO-K performed the best overall. The performance of this approach must still be evaluated on larger datasets and a larger number of components to properly determine the extent of the scalability improvement. It would be beneficial to investigate an approach to dynamically determine the optimal number of components of a GMM, using PSO.

Acknowledgements. The authors would like to thank the Centre for High Performance Computing for the use of their resources to run the simulations used in this study. As well as the UCI Machine Learning Repository for the real-world datasets used.

References

1. Ari, C., Aksoy, S.: Maximum likelihood estimation of Gaussian mixture models using particle swarm optimization. In: Proceedings of the International Conference on Pattern Recognition, pp. 746–749 (2010)

2. Blömer, J., Bujna, K.: Adaptive seeding for Gaussian mixture models. In: Bailey, J., Khan, L., Washio, T., Dobbie, G., Huang, J.Z., Wang, R. (eds.) PAKDD 2016. LNCS (LNAI), vol. 9652, pp. 296–308. Springer, Cham (2016). https://doi.org/10. 1007/978-3-319-31750-2_24

3. Dasgupta, S.: Learning mixture of Gaussians. In: Proceedings of the 40th Annual Symposium on Foundations of Computer Science, pp. 634–644 (1999)

4. Duffin, K., Barrett, W.: Spiders: a new user interface for rotation and visualization of n-dimensional point sets. In: Proceedings of the 1994 IEEE Conference on Scientific Visualization, pp. 205–211 (1994)

5. Engelbrecht, A.P., Van den Bergh, F.: A cooperative approach to particle swarm optimization. Proc. IEEE Trans. Evol. Comput. **8**, 225–239 (2004)

6. Gensler, S.: Finite mixture models. In: Homburg, C., Klarmann, M., Vomberg, A. (eds.) Handbook of Market Research. Springer, Cham (2017). https://doi.org/10. 1007/978-3-319-05542-8_12-1

7. Georgieva, K.: A Computational Intelligence Approach to Clustering of Temporal Data. Master's thesis, University of Pretoria, South Africa (2015)

8. Georgieva, K.S., Engelbrecht, A.P.: Dynamic differential evolution algorithm for clustering temporal data. In: Lirkov, I., Margenov, S., Waśniewski, J. (eds.) LSSC 2013. LNCS, vol. 8353, pp. 240–247. Springer, Heidelberg (2014). https://doi.org/ 10.1007/978-3-662-43880-0_26

9. Redner, R., Walker, H.: Mixture densities, maximum likelihood and the EM algorithm. SIAM Rev. **26**, 195–239 (1984)

10. Shi, Y., Eberhart, R.: A modified particle swarm optimizer. In: Proceedings of the 1998 IEEE International Conference on Evolutionary Computation, pp. 69–73 (1998)

11. Shi, Y., Eberhart, R.: Comparing inertia weights and constriction factors in particle swarm optimization. In: Proceedings of the 2000 Congress on Evolutionary Computation, vol. 1, pp. 84–88 (2000)

12. Xie, X., Beni, G.: A validity measure for fuzzy clustering. IEEE Trans. Pattern Anal. Mach. Intell. **13**, 841–847 (1991)

Formation Control of UAVs and Mobile Robots Using Self-organized Communication Topologies

Weixu Zhu[1](✉)(iD), Michael Allwright[1](iD), Mary Katherine Heinrich[1](iD),
Sinan Oğuz[1](iD), Anders Lyhne Christensen[2](iD),
and Marco Dorigo[1](iD)

[1] IRIDIA, Université Libre de Bruxelles, Brussels, Belgium
{weixu.zhu,michael.allwright,mary.katherine.heinrich,sinan.oguz,
mdorigo}@ulb.ac.be
[2] SDU Biorobotics, The Mærsk Mc-Kinney Møller Institute,
University of Southern Denmark, Odense, Denmark
andc@mmmi.sdu.dk

Abstract. Formation control in a robot swarm targets the overall swarm shape and relative positions of individual robots during navigation. Existing approaches often use a global reference or have limited topology flexibility. We propose a novel approach without these constraints, by extending the concept of 'mergeable nervous systems' to establish distributed asymmetric control via a self-organized wireless communication network. In simulated experiments with UAVs and mobile robots, we present a proof-of-concept for three sub-tasks of formation control: formation establishment, maintenance during motion, and deformation. We also assess the fault tolerance and scalability of our approach.

1 Introduction

We target the control of mobile multi-robot formations—in other words, the maintenance of a possibly adaptive shape during navigation, including both shape outline and relative positions of individuals. Formation control is more frequently studied in control theory than swarm robotics (cf. distinction pointed out by [20]). In swarm robotics, physical coordination with non-physical connections has been studied in flocking (e.g., [7]), where an amorphous group forms during motion, and in self-assembly without physical connections, which has been demonstrated for immobile shapes that are definite and static [18] or amorphous and adaptive [19]. In these approaches, flexibility of individual robot positions has been used as a feature, similar to formation-containment control (e.g., [6]) in control theory, which maintains an overall convex hull.

Formation control—maintaining both overall shape and individual relative positions—merits further study in swarm robotics. We propose an approach based on the existing 'mergeable nervous systems' (MNS) [12] concept.

© Springer Nature Switzerland AG 2020
M. Dorigo et al. (Eds.): ANTS 2020, LNCS 12421, pp. 306–314, 2020.
https://doi.org/10.1007/978-3-030-60376-2_25

The MNS concept combines aspects of centralized and decentralized control, via distributed asymmetric control over a communication graph formed exclusively by self-organization. Our method targets control of definite swarm shape and relative positions of individuals, in non-physically connected robots. Widely studied formation control approaches [1, 4, 11] primarily make use of formation-level central coordination, and include *leader–follower* [21] (including virtual leader [17]), *virtual structure* [9,17], and *behavior-based* [2,3]. Our proposed hybrid approach uses a virtual structure that is not only a reference coordinate frame and a target formation, but also a target topology of the communication network. Robots cede motion control to distributed leaders (i.e., parents) that are their immediate neighbors in the communication topology, rather than following a single shared leader. Similarly to *behavior-based* control, the target formation is not necessarily rigid, as the parents can adapt the motion control of their immediate followers (i.e., children) on the fly, during tasks such as obstacle avoidance.

We select the review by [11] to define the aims of our proof-of-concept experiments. Then, a comprehensive approach to formation control should include the following sub-tasks: 1) formation establishment from random positions, 2) formation maintenance during motion, and 3) formation 'deformability' [11] (i.e., updating the target formation on the fly) during obstacle avoidance. We test formation establishment (Sect. 3.1) with various shapes and sizes of target formations. For formation maintenance, we test time-and-position cooperative and reactive motion, in response to an external stimulus (Sect. 3.2). For formation deformability, we test a scenario requiring multiple updates to the target formation during obstacle-exposed navigation (Sect. 3.2). We also target beneficial features typically seen in self-organization. First, we test fault tolerance, in terms of formation recovery after robot failure (Sect. 3.3). Second, we test scalability (Sect. 3.4), in terms of convergence time during formation establishment (Sect. 3.1) and reaction time in response to an external stimulus (Sect. 3.2).

2   Methods

Our formation control approach is based on the 'mergeable nervous systems' (MNS) concept [12], previously demonstrated with physical connections among ground robots. Here, we extend the concept to non-physical connections, with self-organized wireless communication topologies in a heterogeneous swarm.

Target Topology, Target Formation, and Motion Control. In our approach, an MNS is a set of robots connected in a self-organized wireless communication network, specifically a directed rooted tree, where the root acts as the brain robot of the MNS. A self-organization process results in a network with a given target topology. This network is used to execute distributed motion control, to move robots to positions and orientations that match a given target formation. The target topology is represented in graph G, and target formation is represented by a set of attributes A associated to the links of G. For each link between a parent robot and child robot, A includes the child target position and orientation, relative to the parent, and includes the robot type of the child (either UAV or ground robot). A robot uses the full G and A as reference only if it is currently a brain. G and A are defined externally and can be updated during runtime. A non-brain robot receives a portion of the target from its parent, to use as its new reference. Specifically, robot r_n receives G'_n, the subgraph downstream from it, and the associated subset A_n.

To establish the target formation and maintain it during motion, each child cedes motion control to its parent, which directs it to the relative position and orientation indicated in A. The motion instructions communicate linear and angular velocities, with the parent as reference frame, via the following: 1) new linear velocity vector $\mathbf{v}$, magnitude in m/s; 2) new angular velocity vector $\boldsymbol{\omega}$, magnitude in rad/s; and 3) current orientation in unit quaternion $\mathbf{q}_t$, representing rotation axis and angle. To execute the instructions, the child first rotates $\mathbf{v}$ and $\boldsymbol{\omega}$ by $-\mathbf{q}_t$, resulting in new vectors $\mathbf{v}_q$ and $\boldsymbol{\omega}_q$, then begins moving in direction $\mathbf{v}_q$ at speed $\|\mathbf{v}_q\|$ m/s while rotating around $\boldsymbol{\omega}_q$ at speed $\|\boldsymbol{\omega}_q\|$ rad/s. In order to calculate instructions that will move the child towards the target, the parent senses its child's current displacement vector $\mathbf{d}_t$ and orientation $\mathbf{q}_t$, with itself as reference frame. At each step, the parent sends new motion instructions after calculating a new desired displacement $\mathbf{d}_{t+1}$ and orientation $\mathbf{q}_{t+1}$ for the child, and then calculating $\mathbf{v}$ and $\boldsymbol{\omega}$ according to Eq. 1, as follows:

$$\mathbf{v} = k_1 \left(\frac{\mathbf{d}_{t+1} - \mathbf{d}_t}{\|\mathbf{d}_{t+1} - \mathbf{d}_t\|} \right), \qquad \boldsymbol{\omega} = k_2 \cdot \|f(\mathbf{q}_{t+1}^{-1} \times \mathbf{q}_t)\|, \tag{1}$$

where k_1 and k_2 are speed constants, and where function $f(x)$ converts a quaternion to an Euler angle.

Formation Establishment and Maintenance. A target topology is established by robots forming directed communication links, becoming members of the same MNS. MNS topologies are self-organized via distributed *recruitment* operations and *handover* operations. Recruitment operations form new links. A robot tries to form new links with another robots if these two robots are not in the same network. Handover operations redistribute robots if their current topology nodes do not match the target G and A. A robot may handover its children to its parent or other children based on its G and A to change the topology of the network. Regardless of how robots are initially recruited, those at incorrect nodes will be shifted along the topology until all robots match the target G and A. In case of faulty robots, recruitment and handover operations also restore the target topology. When a robot is moving, it sends motion instruction to its children to maintain their relative positions and orientations according to A. A robot reacting to an external signal may send emergency motion instructions to its neighbors, and the instructions propagate through the MNS. The MNS is 'deformable' [11] (i.e., can switch the target formation on the fly) by updating the target G and A in the brain.

2.1   Experiment Setup

We run experiments with the multi-robot simulator ARGoS [16], using kinematic robot control. The arena is $10 \times 10 \times 2.5$ m^3, fully enclosed, and optionally includes $0.04 \times 0.04 \times 0.02$ m^3 static obstacles. The UAV model is based on the DJI F540 multi-rotor frame, which we extend with four ground-facing cameras. We limit UAV speed to 0.1 m/s, to match the 0.1 m/s maximum speed of the ground robots. UAVs maintain a 1.5 m altitude, after taking off at the start of an experiment. The ground robot model is an extended e-puck [8,13,14], with a fiducial marker (0.03×0.03 m^2 AprilTag [15]) encoding the robot ID. Obstacles also have AprilTags, encoding an obstacle identifier. At 1.5 m altitude, a UAV reliably views a ground area of 1.5×1.5 m^2, detecting positions and orientations of ground robots and obstacles. If two UAVs are connected, they detect each other via 'virtual sensing' [10]—they each infer the other's position and orientation relative to a mutually detected ground robot. Our setup assumes restriction to short-range communication. Messages can only be passed between robots if they are connected in the graph G, or if one is in the other's field of view.

We run experiments (video available[1]) for formation establishment, obstacle avoidance, fault tolerance, and scalability. We define and use nine target formations (F1–F9, see Fig. 1(a)). We conduct 100 runs per experiment and record robot positions throughout. We assess performance via 'position error' [11]—i.e., the difference between actual relative positions and those indicated in the target formation.

3   Results

3.1   Formation Establishment

We first test formation establishment from random starting positions. Establishment is considered successful if all robots merge into a single MNS and the robots achieve the given target formation. We test all nine target formations (100 runs each). Using the robots' final node allocations after experiment termination, we use Euclidean distance to calculate position error E at each timestep, as follows:

$$
E = \frac{1}{n} \sum_{i=1}^{n} E_i, \quad E_i = \mid d(\mathbf{p}_i - \mathbf{p}_1) - d(\mathbf{r}_i - \mathbf{r}_1) \mid, \tag{2}
$$

where n is the total number of robots, $\mathbf{p}_i$ is a robot's current position, $\mathbf{r}_i$ is a robot's target position, and $i = 1$ is the brain. E_i for the brain is always zero, because the brain's relative position to itself is constant. Position error E over time is given in Fig. 1(d), for all nine target formations. In all runs, the swarm successfully establishes the target formation within 400 s. The larger the swarm size, the more time it takes to converge. On average, convergence time is 12.79 s per robot (standard deviation of 5.32 s).

---

[1] http://iridia.ulb.ac.be/supp/IridiaSupp2020-006/index.html.

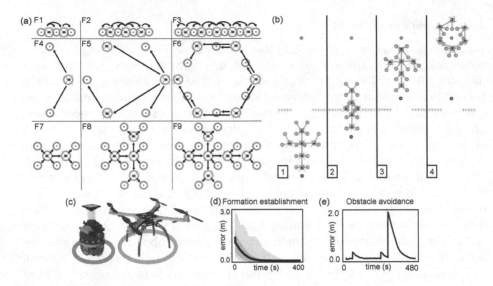

Fig. 1. (a) Target formations (F1–F9). Red circles are UAVs, blue are ground robots. (b) Formation-level obstacle avoidance scenario (screenshot from simulator); deformation. (c) Ground robot and UAV. (d) Formation establishment results. Average position error E over time for all target formations (900 runs total). Dark grey shows standard deviation; light grey shows maximum and minimum. (e) Formation-level obstacle avoidance results from example run. Position error E over time.

3.2   Formation-Level Obstacle Avoidance

We test 'deformability' [11]—i.e., whether the target formation can be updated on the fly, for instance by switching from a cross-shaped formation to a circular formation. For deformation, the brain updates the target topology and formation. Deformation is successful if the MNS establishes the new target formation after an update, such that the position error E returns to its prior level (approximately 0.1 m position error). In this experiment type, we define a wall with a narrow opening (a complex obstacle for formation control [11]) and a small box to be encircled. We use a shepherd robot as stimulus. In step 1, see Fig. 1(b), the MNS is in formation F9 and moves towards the wall because of the shepherd robot. In step 2, it switches from the cross-shaped formation F9 to a more elongated formation similar to formation F3, passing the opening. In step 3, it switches back to formation F9. In step 4, it encounters the small box, and switches to a circular formation similar to formation F6, surrounding the box. Position error E (see Eq. 2) over time is given in Fig. 1(e). Peaks occur when the target formation switches; the largest peak corresponds to the largest difference between the old and new formations. In all 100 runs, E returns to its prior stable level, after each formation switch.

3.3   Fault Tolerance

We test recovery of the topology and formation when a robot fails—i.e., its communication links break and it is arbitrarily displaced. Recovery is successful if position error returns to its prior level, from before the failure. Searching for robots is not within the scope of this paper, so the failed robot is displaced to a random position within the MNS's field of view. Our approach is tolerant even to brain failure, as any robot can be replaced by its topologically closest neighbor. With 100 runs each, we test failure of a leaf node (Fig. 2(a,d)); a non-brain inner node (Fig. 2(b,e)); and a brain (Fig. 2(c,f)). We begin the assessment of each experiment at timestep 280 s, once all robots have established formation F9. In this target formation, all leaf nodes are ground robots and all inner nodes are UAVs. From the set of robots that are candidates for failure in the respective experiment (e.g., those at leaf nodes), one robot is randomly selected as the failed robot, and is removed and displaced at timestep 300 s. Shaded plots of position error E (see Eq. 2) over time are given in Fig. 2(a–c), and scatter plots of recovery time in relation to displacement distance of the failed robot are given in Fig. 2(d–f). Results show that the closer the failed robot is to the brain topologically, the longer the time to recover (note the different scale on the y-axes of Fig. 2(a–c)), and the less direct the relationship between displacement distance and recovery time. This weaker relationship reflects the increased difficulty of recovery, when the failure is closer to the brain. The MNS succeeds in fully recovering in 99% of 300 runs. In the remaining 1%, one ground robot erroneously moves slightly out of view; as searching for robots is not part of the experiment setup, it remains out of view.

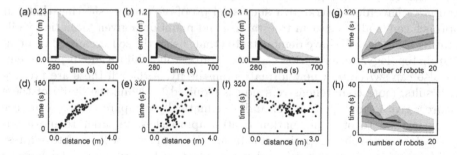

Fig. 2. (a–f) Formation recovery after three failure types: (a,d) leaf node, (b,e) non-brain inner node, and (c,f) brain. (a–c) Position error E over time, for each failure type (100 runs each). (d–f) Relationship between recovery time and displacement distance, for each failure type (100 runs each). (g–i) Scalability analysis. (g) Convergence time by number of robots. Each color line indicates average time for a shape type (shape types F1–3, F4–6, and F7–9 in Fig. 1(a)). Dark grey is standard deviation for all shape types; light grey, maximum and minimum. (h) Convergence time per robot, by number of robots. (Color indications match (g).)

3.4 Scalability

We assess scalability in terms of the initial time to converge on the target formation (in Sect. 3.1), and the reaction time during motion while the formation is being maintained (in Sect. 3.2). The total time to converge (see Fig. 2(g)) tends to increase sublinearly with increasing number of robots—in other words, the system scales slightly better than linearly. Convergence time per robot (see Fig. 2(h)) tends to decrease as the number of robots increases. These tendencies occur because merges often happen in parallel early in the establishment process. The formation shape also impacts convergence time; as this is a multidimensional variable, a comprehensive understanding would require further study.

In a physical MNS, reaction time depends on the number of robots a message passes through [12]. For our wireless MNS, we find that reaction time increases linearly according to the number of links from the stimulated robot to the furthest robot. Currently, there is no spread; one message takes one simulation step (200 ms). In real robots, message time will likely vary. In the experiments of [12], the real message rate was 100 ms (half the rate we set in simulation), using messages of comparable size. For wireless communication, a candidate for our setup would be Zigbee, with effective bit rate of 250 kbps [5].

4 Discussion and Conclusions

We have proposed a self-organized approach to formation control based on the existing concept of 'mergeable nervous systems,' which combines aspects of centralized and decentralized control. Robots in a swarm execute distributed asymmetric control via self-organized communication topologies. In simulated experiments we have demonstrated a successful proof-of-concept, showing that our approach can enable a swarm to establish and maintain a given formation while avoiding obstacles. We have demonstrated that, using the self-organized communication topology, the formation can recover after robot failure and displacement, and also can switch to a new formation on the fly. Although these are promising results, more comprehensive study is required to define the limits of these features, give formal guarantees, and systematically compare the performance of our method to other formation control approaches. In order to move our approach to real-robot experiments, future developments will need to address conditions such as sensor noise and communication latency, and add a layer of dynamic control in addition to kinematic control for the UAVs. Overall, we draw the conclusion that in the tested experimental setup our MNS-based approach is capable of fault-tolerant and scalable formation control during navigation, in a heterogeneous robot swarm comprising UAVs and ground robots.

Acknowledgements. This work is partially supported by the Program of Concerted Research Actions (ARC) of the Université libre de Bruxelles, by the Office of Naval Research Global (Award N62909-19-1-2024), by the European Union's Horizon 2020 research and innovation programme under the Marie Skłodowska-Curie grant agreement No 846009, and by the China Scholarship Council Award No 201706270186.

Marco Dorigo and Mary Katherine Heinrich acknowledge support from the Belgian F.R.S.-FNRS, of which they are a Research Director and a Postdoctoral Researcher respectively.

References

1. Anderson, B.D., Fidan, B., Yu, C., Walle, D.: UAV formation control: theory and application. In: Blondel, V.D., Boyd, S.P., Kimura, H. (eds.) Recent Advances in Learning and Control. Lecture Notes in Control and Information Sciences, vol. 371, pp. 15–33. Springer, London (2008). https://doi.org/10.1007/978-1-84800-155-8_2
2. Balch, T., Arkin, R.C.: Behavior-based formation control for multirobot teams. IEEE Trans. Robot. Autom. **14**(6), 926–939 (1998)
3. Cao, Z., Xie, L., Zhang, B., Wang, S., Tan, M.: Formation constrained multi-robot system in unknown environments. In: IEEE International Conference on Robotics and Automation (Cat. No. 03CH37422), vol. 1, pp. 735–740. IEEE (2003)
4. Chen, Y.Q., Wang, Z.: Formation control: a review and a new consideration. In: IEEE/RSJ International Conference on Intelligent Robots and Systems, pp. 3181–3186. IEEE (2005)
5. Cox, D., Jovanov, E., Milenkovic, A.: Time synchronization for ZigBee networks. In: Proceedings of the Thirty-Seventh Southeastern Symposium on System Theory, SSST 2005, pp. 135–138. IEEE (2005)
6. Dong, X., Hua, Y., Zhou, Y., Ren, Z., Zhong, Y.: Theory and experiment on formation-containment control of multiple multirotor unmanned aerial vehicle systems. IEEE Trans. Autom. Sci. Eng. **16**(1), 229–240 (2018)
7. Ferrante, E., Turgut, A.E., Huepe, C., Stranieri, A., Pinciroli, C., Dorigo, M.: Self-organized flocking with a mobile robot swarm: a novel motion control method. Adapt. Behav. **20**(6), 460–477 (2012)
8. Gutiérrez, Á., Campo, A., Dorigo, M., Donate, J., Monasterio-Huelin, F., Magdalena, L.: Open e-puck range & bearing miniaturized board for local communication in swarm robotics. In: IEEE International Conference on Robotics and Automation, pp. 3111–3116. IEEE (2009)
9. Lewis, M.A., Tan, K.H.: High precision formation control of mobile robots using virtual structures. Autonom. Robots **4**(4), 387–403 (1997). https://doi.org/10.1023/A:1008814708459
10. Liu, L., Kuo, S.M., Zhou, M.: Virtual sensing techniques and their applications. In: International Conference on Networking, Sensing and Control, pp. 31–36. IEEE (2009)
11. Liu, Y., Bucknall, R.: A survey of formation control and motion planning of multiple unmanned vehicles. Robotica **36**(7), 1019–1047 (2018)
12. Mathews, N., Christensen, A.L., O'Grady, R., Mondada, F., Dorigo, M.: Mergeable nervous systems for robots. Nature Commun. **8**, 439 (2017)
13. Millard, A.G., et al.: The Pi-puck extension board: a Raspberry Pi interface for the e-puck robot platform. In: IEEE/RSJ International Conference on Intelligent Robots and Systems (IROS), pp. 741–748. IEEE (2017)
14. Mondada, F., Bonani, et al.: The e-puck, a robot designed for education in engineering. In: Proceedings of the 9th Conference on Autonomous Robot Systems and Competitions, vol. 1, pp. 59–65. IPCB: Instituto Politécnico de Castelo Branco (2009)

15. Olson, E.: AprilTag: a robust and flexible visual fiducial system. In: Proceedings of the IEEE International Conference on Robotics and Automation (ICRA), pp. 3400–3407. IEEE, May 2011
16. Pinciroli, C., et al.: ARGoS: a modular, parallel, multi-engine simulator for multi-robot systems. Swarm Intell. 6(4), 271–295 (2012). https://doi.org/10.1007/s11721-012-0072-5
17. Ren, W., Sorensen, N.: Distributed coordination architecture for multi-robot formation control. Robot. Auton. Syst. 56(4), 324–333 (2008)
18. Rubenstein, M., Cornejo, A., Nagpal, R.: Programmable self-assembly in a thousand-robot swarm. Science 345(6198), 795–799 (2014)
19. Soorati, M.D., Heinrich, M.K., Ghofrani, J., Zahadat, P., Hamann, H.: Photomorphogenesis for robot self-assembly: adaptivity, collective decision-making, and self-repair. Bioinspir. Biomim. 14(5), 056006 (2019)
20. Valentini, G., Ferrante, E., Dorigo, M.: The best-of-n problem in robot swarms: formalization, state of the art, and novel perspectives. Front. Robot. AI 4, 9 (2017)
21. Wang, P.K.: Navigation strategies for multiple autonomous mobile robots moving in formation. J. Robot. Syst. 8(2), 177–195 (1991)

Group-Size Regulation in Self-organized Aggregation in Robot Swarms

Ziya Firat[1] , Eliseo Ferrante[2,3] , Raina Zakir[4], Judhi Prasetyo[1,4] ,
and Elio Tuci[1(✉)]

[1] Department of Computer Science, University of Namur, Namur, Belgium
{ziya.firat,judhiprasetyo}@student.unamur.be, elio.tuci@unamur.be
[2] Vrije Universiteit Amsterdam, Amsterdam, The Netherlands
e.ferrante@vu.nl
[3] Technology Innovation Institute, Masdar City, Abu Dhabi, United Arab Emirates
[4] Middlesex University Dubai, Dubai, United Arab Emirates
rainazakir@gmail.com

Abstract. In swarm robotics, self-organized aggregation refers to a collective process in which robots form a single aggregate in an arbitrarily chosen aggregation site among those available in the environment, or just in an arbitrarily chosen location. Instead of focusing exclusively on the formation of a single aggregate, in this study we discuss how to design a swarm of robots capable of generating a variety of final distributions of the robots to the available aggregation sites. We focus on an environment with two possible aggregation sites, A and B. Our study is based on the following working hypothesis: robots distribute on site A and B in quantities that reflect the relative proportion of robots in the swarm that selectively avoid A with respect to those that selectively avoid B. This is with an as minimal as possible proportion of robots in the swarm that selectively avoid one or the other site. We illustrate the individual mechanisms designed to implement the above mentioned working hypothesis, and we discuss the promising results of a set of simulations that systematically consider a variety of experimental conditions.

1 Introduction

Swarm robotics studied the design of collective behaviors in group of autonomous robots [4]. Swarm robotics takes inspiration from studies in social insect and other social animals, whereby simple individuals are able to exhibit superior collective intelligence when working in groups [2]. The overall goal is to design systems that are robust, scalable, and flexible like their natural counterparts [4]. To achieve this, swarm robotics relies on the application of the following principles: i) absence of external infrastructure and reliance on only on-board sensing and computation; ii) use of local perception and communication only; that is, each robot can sense and communicate only within a given range via on-board devices; iii) the process of self-organization, that yields from microscopic behaviors and individual interactions to macroscopic complex collective behaviors.

© Springer Nature Switzerland AG 2020
M. Dorigo et al. (Eds.): ANTS 2020, LNCS 12421, pp. 315–323, 2020.
https://doi.org/10.1007/978-3-030-60376-2_26

Collective decision-making is the ability to make a collective decisions only via local interaction and communication [26]. This is an important collective response which has been extensively studied in swarm robotics. Collective decision making can take several forms: it can either be studied explicitly [26,27] or implicitly in other collective behaviors such as collective motion (decision on a common direction of motion), and aggregation (decision on a common location for gathering in the environment). Two factors that can synergistically or antagonistically influence collective decision-making process are asymmetries in the environment, or the active modulation performed by the all or some of the swarm members [26]. In a seminal study on collective decision making [8], the authors studied collective motion models in presence of so called *implicit leaders* or *informed individuals*. These have a preferred direction of motion to guide collective motion in that direction. The rest of the swarm do not possess a preferred direction of motion, nor is able to recognize informed individuals. The main result of the paper is that even a minority of informed individuals is able to guide the swarm in the desired direction, and that larger groups require smaller proportion of informed individuals for equal levels of accuracy.

As in [8], also in swarm robotics the framework of informed individuals has been studied mainly in the context of collective motion [11,12,20]. Recently, this framework has been ported to another collective behavior, namely self-organized aggregation [13,14,19]. Self-organized aggregation [1,7,18] is inspired by the biological study of cockroaches [9,21], where probabilistic models have been proposed. The same models have been adapted and implemented on distributed robotic systems [15,16,18]. Besides being another example of collective decision-making, aggregation is a basic building block for other cooperative behaviors [10,25]. Self-organized aggregation can take place in environments that are completely homogeneous (except for boundaries and potential obstacles) where no perceivable special locations, called sites or shelters, are present, thus robots are required to aggregate anywhere in this environment [5]. Alternatively, as it is the case of the current paper, the sites where robots are required to aggregate can be specific areas in the environment that can be clearly perceived by all or some of the robots [6,17].

In this paper our objective is to go one step beyond the state of the art to further study how the framework of *informed* robots can be used as a guiding principle for self-organization. We build upon recent studies on self-organized aggregation with *informed* robots [13,14,19], where robots need to select only one site among n possible alternatives, driven by *informed* robots. Each *informed* robot has knowledge on a specific aggregation site to stop on and avoids other aggregation sites. *Non-informed* robots do not possess this information, therefore may potentially aggregate on any side. Additionally, *informed* robots are assumed to be perceivable, through sensing, by other robots (e.g., they emit a signal), while *non-informed* robots cannot be sensed at all. Differently from [13,14,19], in this paper robots are required to aggregate on both sites according to different proportions as set by the designer. To control the proportion of robots aggregating on the two sides, we design a novel aggregation method. *Informed* robots are divided in two types, each preferring one of the two sites over the other. To control the relative group size on the two sites, the proposed method only requires the presence of *informed* robots with internal sub-proportions correlating with desired global allocation for the whole swarm. We perform our study using both simulation and real-robot experiments. In simulation we considered a scenarios in a circular arena where the two aggregation sites are represented by black or white colored circles, respectively. The results of our simulations show interesting relationships between swarm size and sub-proportion of informed agents, both on quality and speed of convergence on the desired aggregation site. In the following sections, we detailed the methods of our study; we discuss the significance of our results for the swarm robotics community, and we point to interesting future directions of work.

2   The Simulation Environment

A swarm of robots is randomly placed in a circular area with the floor colored in gray except for two circular aggregation sites, one in which the floor is colored in white and one in which is black. The task of the robots is to form aggregates on both sites according to rules that prescribe which proportion of the swarm has to aggregate on the white site and which proportion on the black site. Each simulated robot is controlled by a probabilistic finite state machine (PFSM), similar to the one employed in [1,5,7,21]. The robots' controller is made of three states: Random Walk ($\mathcal{RW}$), Stay ($\mathcal{S}$), and Leave ($\mathcal{L}$). When in state $\mathcal{RW}$, the movement of the robot is characterized by an isotropic random walk, with a fixed step length (5 s, at 10 cm/s), and turning angles chosen from a wrapped Cauchy probability distribution [22]. Any robot in state $\mathcal{RW}$ is continuously performing an obstacle avoidance behavior. To perform obstacle avoidance, first the robot stops, and then it changes its headings of a randomly chosen angle uniformly drawn in $[-\pi, \pi]$ until no obstacles are perceived in the forward direction of motion. Negative angles refer to clockwise rotations, while positive to anticlockwise rotations.

In our model, we consider two type of robots: *informed* robots and *non-informed* robots. *Informed* robots systematically rest only on one site. Some of them, avoid the black site and rest only on the white site (*informed* robots for white); others avoid the white site and rest only on the black site (*informed* robots for black). *Non-informed* robots can potentially rest on both types of site. Note that, the working hypothesis of this study is that the way in which the swarm distributes among the two aggregation sites reflects the relative proportion of *informed* robots for black and for white. For example, if 50% of the *informed* robots are for black and 50% of them are for white, the swarm should generate two equal size aggregates one on the black and one on the white site. This experimental work aims at verifying this working hypothesis by systematically varying the proportion of *informed* robots within the swarm. Moreover, for each proportion of *informed* robots, we vary the relative proportion of *informed* robots for black and for white.

A *non-informed* robot systematically transits from state $\mathcal{RW}$ to state $\mathcal{S}$ anytime it reaches an aggregation site. *Informed* robots for black undergo the same state change only when they reach the black site, thus ignoring the white site. *Informed* robots for white systematically transits from state $\mathcal{RW}$ to state $\mathcal{S}$ anytime they reach a white aggregation site, thus ignoring the black site. For all types of robots, the transaction from the random walk to resting on a site happens in the following: the robots moves forward within the site for a limited number of time steps in order to avoid stopping at the border of the site thus creating barriers preventing the entrance to other robots. Then, they transitions from state $\mathcal{RW}$ to state $\mathcal{S}$.

The robots leave state $\mathcal{S}$ to join state $\mathcal{L}$ with a probability P_{leave}, which is computed in the following:

$$
P_{leave} = \begin{cases} e^{-a(k-|n-x|)}, & \text{if } n > 0; \text{ it applies to all types of robots;} \\ 1, & \text{if } n = 0; \text{ only for } \textit{non-informed} \text{ robot;} \end{cases} \tag{1}
$$

with $a = 2.0$ and $k = 18$. n is current number of *informed* robots perceived at the site, and x is the number of *informed* robots perceived at the site at the time of joining a site. Note that, for any robot n and x are local estimates based the number of *informed* robots in the perceivable neighborhood which is smaller than the entire site. P_{leave} is sampled every 20 time steps. When in state $\mathcal{L}$, a robot leaves the aggregation site by moving forward while avoiding collisions with other robots until it no longer perceives the site. At this point, the robot transitions from state $\mathcal{L}$ to state $\mathcal{RW}$. While on an aggregation site, *informed* robots count themselves in order to estimate n and x.

To model this scenario, we use ARGoS multi engine simulator [23]. The simulation environment models the circular arena as detailed above, and the kinematic and sensors readings of the Foot-bots mobile robots [3]. The robot sensory apparatus includes the proximity sensors positioned around the robot circular body, four ground sensors positioned two on the front and two on the back of the robot's underside, and the range and bearing sensor. The proximity sensors are used for sensing and avoiding the walls of the arena. The readings of each ground sensors is set to 0.5 if the sensor is on gray, to 1 if on white, and 0 if on black. A robot perceives an aggregation site when all the four ground sensors return a value different from 0.5. The range and bearing sensor is used to avoid collision with other robots and to estimate how many *informed* robots are resting on a site within sensor range (i.e., the parameters n and x in Eq. 1). With this sensor, two robots can perceive each other up to a distance of 0.8 meter.

3   Results

We run two sets of experiments (hereafter, setup 1, and setup 2), in which we varied the swarm size N, with $N = 50$ in setup 1, and $N = 100$ in setup 2. As aggregation performance are heavily influenced by swarm density [5,13,14,19]

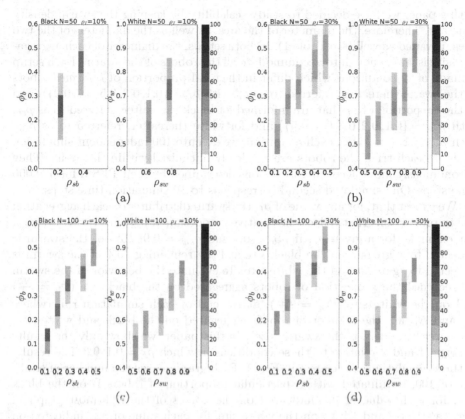

Fig. 1. Graphs in which the intensity of gray refers to the number of trials, out of 100, terminated with a particular proportion of robots on each site (i.e., Φ_b and Φ_w). The x-axes refer to the proportion of *informed* robots for black (ρ_{sb}) or for white (ρ_{sw}). The swarm size N and the total proportion of *informed* robots (ρ_I) in the swarm are: (a) $N = 50$ and $\rho_I = 0.1$; (b) $N = 50$ and $\rho_I = 0.3$; (c) $N = 100$ and $\rho_I = 0.1$; (d) $N = 100$ and $\rho_I = 0.3$. In each of these cases, the x and y-axis of the leftmost graphs refer to ρ_{sb} and Φ_b, respectively; the x and y-axis of the rightmost graphs refer to ρ_{sw} and Φ_w, respectively.

Table 1. Table showing the characteristics of each experimental condition in simulation

Setup	Swarm size	Arena diameter (cm)	Aggregation site diameter (cm)
1	50	12.9	2.8
2	100	18	4.0

in this paper we have decided to study scalability by keeping the swarm density constant. Therefore, the diameter of the area, as well as the diameters of the two sites, is varied as well (see Table 1). In both setups, the diameter of each aggregation site is large enough to accommodate all the robots of the swarm. Each setup is made of 25 conditions which differ in the total proportion of *informed* robots in the swarm (hereafter, referred to as ρ_I, with $\rho_I = \{0.1, 0.3, 0.5, 0.7, 0.9\}$), and in the proportion of ρ_I that are informed for black (hereafter, referred to as ρ_{sb}, with $\rho_{sb} = \{0.1, 0.2, 0.3, 0.4, 0.5\}$) and for white (hereafter, referred to as ρ_{sw}, with $\rho_{sw} = 1 - \rho_{sb}$). For each condition, we execute 100 independent simulation trials. In each trial, the robots are randomly initialized within the arena. They autonomously move according to actions determined by their PFSM for 300.000 time steps. One simulated second corresponds to 10 simulation time steps.

We expect that, for any value of ρ_I, the swarm distributes on each aggregation site in proportions that reflect the relative proportion of ρ_{sb} with respect to ρ_{sw}. For example, for a given ρ_I, if $\rho_{sb} = 0.1$ and $\rho_{sw} = 0.9$, 10% of the swarm is expected to aggregate on the black site and the remaining 90% of the swarm is expected to aggregate on the white site. To evaluate the behavior of the swarm we recorded the proportion of robots aggregated on the black site ($\Phi_b = \frac{N_b}{N}$) and on the white site ($\Phi_w = \frac{N_w}{N}$), at the end of each simulation run (where N_b and N_w are the number of robots aggregated on the black and white site, respectively, and N is the swarm size). In this paper, we show only the results of setup 1 and 2 relative to those conditions in which $\rho_I = 0.1, 0.3$ The results of these simulations are shown in Fig. 1. Each graph shows the number of trials, out of 100, terminated with a particular proportion of robots i) on the black site, for each value of ρ_{sb} indicated on the x-axes of the rightmost graph in Fig. 1a, 1b, 1c, and 1d; ii) on the white site, for each value of ρ_{sw} indicated on the x-axes of the leftmost graph in Fig. 1a, 1b, 1c, and 1d. The swarm size N and the total proportion of *informed* robots (ρ_I) in the swarm are: i) $N = 50$ and $\rho_I = 0.1$ in Fig. 1a; ii) $N = 50$ and $\rho_I = 0.3$ in Fig. 1b; iii) $N = 100$ and $\rho_I = 0.1$ in Fig. 1c; iv) $N = 100$ and $\rho_I = 0.3$ in Fig. 1d.

Ideally, if the swarm aggregates in way to perfectly reflect the relative proportion of *informed* robots for black and for white in the swarm, for both $N = 50$ and $N = 100$ and for all total proportion of *informed* robots in the swarm, the graphs in Fig. 1 would show only black rectangle aligned on the diagonal from bottom left to top right corners of each graph. In other words, all 100 trials in each condition of each setup would terminate with $\Phi_b = \rho_{sb}$ and $\Phi_w = \rho_{sw}$. The results of the simulations tend to slightly diverge from this ideal case. However, we can clearly see that a higher concentration of trials (the darker rectangles for each case) tend to be aligned on the above mentioned diagonal, with deviations from the ideal case that remain nevertheless close to the expected result. This is clearly observable even when the total proportion of informed robots is 0.1 of N for both setup1 with $N = 50$ and setup 2 with $N = 100$ (see Fig. 1a and 1c, respectively). Note that the results in Fig. 1 refer to the two most challenging scenarios, in which the total proportion of *informed* robots in the swarm is relatively low ($\rho_I = 0.1$ and $\rho_I = 0.3$). For higher proportion of *informed* robots in the swarm, the results of the simulations tend to get progressively closer to the best-case scenario, in which simulation trials terminates with $\Phi_b = \rho_{sb}$ and $\Phi_w = \rho_{sw}$.

In summary, the PFSM described in Sect. 2 allows to rather accurately control the way in which the robots of a swarm distribute on two different aggregation sites simply by regulating the relative proportion of *informed* robots for each site, even with a small total proportion of *informed* robots in the swarm ($\rho_I = 0.1$), and for different swarm sizes.

4   Conclusions

In this paper, we have shown that the aggregation dynamics of a swarm of robots can be controlled using the system heterogeneity. In this self organized aggregation scenario, with a swarm of robots required to operate in an arena with two aggregation sites, the system heterogeneity is represented by *informed* robots; that is, agents that selectively avoid a type of aggregation site (i.e., the black/white site) to systematically rest on the other type of site (i.e., the white/black site). The results of our simulations indicate that with a small proportion of *informed* robots a designer can effectively control the way in which an entire swarm distribute on the two aggregation sites. This is because the size of the robots' aggregates at each site tends to match the relative proportion of the two different types of *informed* robots characterizing the swarm.

We have also performed few preliminary tests with physical robots to test the effectiveness of the PFSM discussed in Sect. 2 in controlling the aggregation dynamics of a swarm of kilobot robots [24]. The results of these tests, not shown in the paper, closely match the results obtained in simulations. Nevertheless, the behavior of physical robots has been negatively affected by the frequent collisions between the robots and the arena wall which, relatively often, represented deadlock conditions with the robots unable to generate the virtuous manoeuvres necessary to recover movement. These tests have been carried out with maximum 18 kilobots. We believe that the small size of the physical robots swarm has contributed to limit the influential role of *informed* robots. However, further studies with physical robots are required to better characterize the nature of the relationship between the swarm size and the aggregation mechanisms we have discussed in this paper.

We believe that the system heterogeneity, relatively neglected in swarm robotics, can play an important role in the development of mechanisms to control the self-organized collective responses of swarms of robots. Our research agenda for the future is focused on the series of experiments based on the hypothesis that the system heterogeneity has a measurable impact on the outcomes of certain self-organized processes. We aim to identify these processes and to illustrate how they can be effectively controlled by manipulating the system heterogeneity.

References

1. Bayindir, L., Şahin, E.: Modeling self-organized aggregation in swarm robotic systems. In: IEEE Swarm Intelligence Symposium, SIS 2009, pp. 88–95. IEEE (2009)
2. Bonabeau, E., Dorigo, M., Marco, D.d.R.D.F., Theraulaz, G., et al.: Swarm Intelligence: From Natural to Artificial Systems. Oxford University Press, Oxford (1999)
3. Bonani, M., et al.: The MarXbot, a miniature mobile robot opening new perspectives for the collective-robotic research. In: IEEE/RSJ International Conference on Intelligent Robots and Systems (IROS), pp. 4187–4193 (2010)
4. Brambilla, M., Ferrante, E., Birattari, M., Dorigo, M.: Swarm robotics: a review from the swarm engineering perspective. Swarm Intell. **7**(1), 1–41 (2013)
5. Cambier, N., Frémont, V., Trianni, V., Ferrante, E.: Embodied evolution of self-organised aggregation by cultural propagation. In: Dorigo, M., Birattari, M., Blum, C., Christensen, A.L., Reina, A., Trianni, V. (eds.) ANTS 2018. LNCS, vol. 11172, pp. 351–359. Springer, Cham (2018). https://doi.org/10.1007/978-3-030-00533-7_29
6. Campo, A., Garnier, S., Dédriche, O., Zekkri, M., Dorigo, M.: Self-organized discrimination of resources. PLoS ONE **6**(5), e19888 (2010)
7. Correll, N., Martinoli, A.: Modeling and designing self-organized aggregation in a swarm of miniature robots. Int. J. Robot. Res. **30**(5), 615–626 (2011)
8. Couzin, I., Krause, J., Franks, N., Levin, S.: Effective leadership and decision making in animal groups on the move. Nature **433**, 513–516 (2005)
9. Deneubourg, J., Lioni, A., Detrain, C.: Dynamics of aggregation and emergence of cooperation. Biol. Bull. **202**(3), 262–267 (2002)
10. Dorigo, M., et al.: Evolving self-organizing behaviors for a swarm-bot. Auton. Robots **17**(2), 223–245 (2004)
11. Ferrante, E., Turgut, A.E., Huepe, C., Stranieri, A., Pinciroli, C., Dorigo, M.: Self-organized flocking with a mobile robot swarm: a novel motion control method. Adapt. Behav. **20**(6), 460–477 (2012)
12. Ferrante, E., Turgut, A.E., Stranieri, A., Pinciroli, C., Birattari, M., Dorigo, M.: A self-adaptive communication strategy for flocking in stationary and non-stationary environments. Nat. Comput. **13**(2), 225–245 (2013). https://doi.org/10.1007/s11047-013-9390-9
13. Firat, Z., Ferrante, E., Cambier, N., Tuci, E.: Self-organised aggregation in swarms of robots with informed robots. In: Fagan, D., Martín-Vide, C., O'Neill, M., Vega-Rodríguez, M.A. (eds.) TPNC 2018. LNCS, vol. 11324, pp. 49–60. Springer, Cham (2018). https://doi.org/10.1007/978-3-030-04070-3_4
14. Firat, Z., Ferrante, E., Gillet, Y., Tuci, E.: On self-organised aggregation dynamics in swarms of robots with informed robots. Neural Comput. Appl. 1–17 (2020). https://doi.org/10.1007/s00521-020-04791-0
15. Garnier, S., et al.: The embodiment of cockroach aggregation behavior in a group of micro-robots. Artif. Life **14**(4), 387–408 (2008)
16. Garnier, S., et al.: Aggregation behaviour as a source of collective decision in a group of cockroach-like-robots. In: Capcarrère, M.S., Freitas, A.A., Bentley, P.J., Johnson, C.G., Timmis, J. (eds.) ECAL 2005. LNCS (LNAI), vol. 3630, pp. 169–178. Springer, Heidelberg (2005). https://doi.org/10.1007/11553090_18
17. Garnier, S., Gautrais, J., Asadpour, M., Jost, C., Theraulaz, G.: Self-organized aggregation triggers collective decision making in a group of cockroach-like robots. Adapt. Behav. **17**(2), 109–133 (2009)

18. Gauci, M., Chen, J., Li, W., Dodd, T., Groß, R.: Self-organized aggregation without computation. Int. J. Robot. Res. **33**(8), 1145–1161 (2014)
19. Gillet, Y., Ferrante, E., Firat, Z., Tuci, E.: Guiding aggregation dynamics in a swarm of agents via informed individuals: an analytical study. In: The 2018 Conference on Artificial Life: A Hybrid of the European Conference on Artificial Life (ECAL) and the International Conference on the Synthesis and Simulation of Living Systems (ALIFE), pp. 590–597. MIT Press (2019)
20. Çelikkanat, H., Şahin, E.: Steering self-organized robot flocks through externally guided individuals. Neural Comput. Appl. **19**(6), 849–865 (2010)
21. Jeanson, R., et al.: Self-organized aggregation in cockroaches. Animal Behav. **69**(1), 169–180 (2005)
22. Kato, S., Jones, M.: An extended family of circular distributions related to wrapped Cauchy distributions via Brownian motion. Bernoulli **19**(1), 154–171 (2013)
23. Pinciroli, C., et al.: ARGoS: a modular, parallel, multi-engine simulator for multi-robot systems. Swarm Intell. **6**(4), 271–295 (2012)
24. Rubenstein, M., Ahler, C., Hoff, N., Cabrera, A., Nagpal, R.: Kilobot: a low cost robot with scalable operations designed for collective behaviors. Robot. Auton. Syst. **62**(7), 966–975 (2014). https://doi.org/10.1016/j.robot.2013.08.006. http://dx.doi.org/10.1016/j.robot.2013.08.006
25. Tuci, E., Alkilabi, M., Akanyety, O.: Cooperative object transport in multi-robot systems: a review of the state-of-the-art. Front. Robot. AI **5**, 1–15 (2018)
26. Valentini, G., Ferrante, E., Dorigo, M.: The best-of-n problem in robot swarms: Formalization, state of the art, and novel perspectives. Front. Robot. AI **4**, 9 (2017). https://doi.org/10.3389/frobt.2017.00009. https://www.frontiersin.org/article/10.3389/frobt.2017.00009
27. Valentini, G., Ferrante, E., Hamann, H., Dorigo, M.: Collective decision with 100 Kilobots: speed versus accuracy in binary discrimination problems. Auton. Agents Multi-Agent Syst. **30**(3), 553–580 (2016)

On the Effects of Minimally Invasive Collision Avoidance on an Emergent Behavior

Chris Taylor$^{(\boxtimes)}$ ⓘ, Alex Siebold ⓘ, and Cameron Nowzari ⓘ

Department of Electrical and Computer Engineering, George Mason University,
Fairfax, VA, USA
{ctaylo3,asiebold,cnowzari}@gmu.edu

Abstract. Swarms of autonomous agents are useful in many applications due to their ability to accomplish tasks in a decentralized manner, making them more robust to failures. Due to the difficulty in running experiments with large numbers of hardware agents, researchers typically resort to simulations with simplifying assumptions. While some assumptions are tolerable, we feel that two assumptions have been overlooked: one, that agents take up physical space, and two, that a collision avoidance algorithm is available to add safety to an existing algorithm. While there do exist minimally invasive collision avoidance algorithms designed to add safety while minimizing interference in the intended behavior, we show they can still cause unexpected interference. We use an illustrative example with a double-milling behavior and show, through simulations, that the collision avoidance can still cause unexpected interference and careful parameter tuning is needed.

1 Introduction

Swarms have been extensively studied and are an attractive choice for many applications due to their decentralized nature and robustness against individual failures [3,4,12]. A common goal with decentralized control in swarms is to achieve an *emergent behavior* [7,12], where the collective behavior of the swarm has properties that the behaviors of individual agents lack. This is desirable when agents lack global awareness of the higher level goal or other agents, often due to limited sensing or computation capabilities.

Since swarm experiments on real hardware are difficult, researchers generally prototype swarming algorithms in simulations with simplifying assumptions. For instance, many works assume double-integrator dynamics, noise-free sensing, no occlusions, ideal communication, or no actuation constraints [6,11,14,20]. In some cases the assumptions are acceptable and the algorithm's behavior doesn't fundamentally change when replaced with more realistic assumptions. However we feel two assumptions in particular have been overlooked. The first: that agents occupy physical space. Many works assume agents have negligible size [14,20], for instance [20] defines a collision as two agents occupying the *exact* same position,

© Springer Nature Switzerland AG 2020
M. Dorigo et al. (Eds.): ANTS 2020, LNCS 12421, pp. 324–332, 2020.
https://doi.org/10.1007/978-3-030-60376-2_27

[7,10,14] assume agents can pass right through one other, and other works make no mention of the size of agents [6,9,19].

The second overlooked assumption is that there exists a collision avoidance algorithm to add safety to an existing algorithm. In some cases this is acceptable when the behavior is already well-suited to avoid collisions [1]. However, as we show in a previous work, adding collision avoidance can disrupt certain behaviors [16]. In this work, we extend the results of [16] but use much more sophisticated minimally invasive collision avoidance techniques: control barrier certificates [5] and Optimal Reciprocal Collision Avoidance (ORCA) [17]. We show that there is still disruption in the intended behavior and it requires careful tuning to achieve good performance. We validate our results through simulations and explorations of the parameter space. See [15] for additional details, where the connection between emergence and collision avoidance is investigated further.

2   Problem Formulation

We wish to understand the effects of imposing two constraints on the swarming algorithm from Szwaykowska et al. [14]: one, that agents have non-negligible physical size, and two, that agents must avoid collisions in a realistic manner, e.g. without resorting to infinite control effort.

2.1   Individual Agent Model

To isolate just the effects of our constraints, we consider a simple agent model. Letting $r_i \in \mathbb{R}^2$ be the position of agent $i \in \{1, \ldots, N\}$ in a swarm of N agents, we consider the dynamics

$$\ddot{\mathbf{r}}_i(t) = u_i(t) \tag{1}$$

with the following two constraints at all times $t \in \mathbb{R}_{\geq 0}$:

C1. No Collisions. Letting $r > 0$ represent the outer radius of each agent, agents must obey $\|\mathbf{r}_i(t) - \mathbf{r}_j(t)\| > 2r$ for all $i, j \in \{1, \ldots, N\}, i \neq j$.

C2. Limited Acceleration. We require $\|u_i(t)\| \leq a_{\max}$.

2.2   Desired Global Behavior: Ring State

Given the model Eq. (1) and constraints, we now introduce the controller we replicate from [14]: the 'ring state'. Without collision avoidance, it is

$$\mathbf{u}_i^* = \beta(v_0^2 - \|\dot{\mathbf{r}}_i\|^2)\dot{\mathbf{r}}_i + \frac{\alpha}{N-1} \sum_{j \neq i} (\mathbf{r}_j(t - t_d) - \mathbf{r}_i(t)) . \tag{2}$$

The input $\mathbf{u}_i^*$ consists of two terms (in order): keep the agent's speed at approximately $v_0 > 0$ with gain $\beta > 0$ and attract toward the delayed position of other agents t_d seconds in the past (d for *delay*), where α adjusts the attraction strength. We also assume each agent i is also able to immediately sense the state of local agents in a set $\mathcal{N}_i$ defined by a circular sensing radius ℓ_r

$$\mathcal{N}_i = \{j \in \{1, \dots, N\}\} \backslash \{i\} \mid \|\mathbf{r}_i - \mathbf{r}_j\| \le \ell_r\}. \tag{3}$$

Using this additional information with the desired input $\mathbf{u}_i^*$, we consider different collision avoidance wrappers $\mathbf{F}_{c_r}$ parameterized by a tunable 'cautiousness' parameter $c_r > 0$. $\mathbf{F}_{c_r}$ is a function of just the desired input and the states of nearby agents in $\mathcal{N}_i$. To make our dynamics obey constraint C2, we define

$$\mathbf{clip}(\mathbf{x}, a) = \begin{cases} \mathbf{x} & \|\mathbf{x}\| < a, \\ a\frac{\mathbf{x}}{\|\mathbf{x}\|} & \text{otherwise.} \end{cases}$$

This results in the final dynamics equations of

$$\ddot{\mathbf{r}}_i = \mathbf{u}_i$$
$$\mathbf{u}_i = \mathbf{clip}[\mathbf{F}_{c_r}(\mathbf{u}_i^*, \mathbf{r}_i, \dot{\mathbf{r}}_i, \{(\mathbf{r}_j, \dot{\mathbf{r}}_j)\}_{j \in \mathcal{N}_i}), a_{\max}]. \tag{4}$$

Note that the *all-to-all* sensing in Eq. (2) is delayed by t_d seconds (e.g. as communicated over a network), whereas the *local* sensing used for collision avoidance in Eq. (4) incurs no delay (e.g. agents sense others without communication). While the work we replicate [14] explores more realistic communication constraints, we assume all-to-all, albeit delayed, communication is available since we are more interested in the effect of collisions on the intended behavior rather than limited communication.

3   Methodology

In Sect. 3.1, we describe our metrics to quantify the amount of interference caused by adding collision avoidance. In Sect. 3.2, we describe the two collision-avoidance algorithms we consider. Then, in Sect. 4, we show the results of applying collision avoidance to our algorithm of interest.

3.1   Measuring Emergent Behavior Quality

The metrics we introduce in our prior work [16] consist of the 'fatness' Φ and 'tangentness' τ. Formally, letting μ be the average position of all agents, i.e. $\mu = \frac{1}{N} \sum_{i=1}^{N} \mathbf{r}_i$, and $r_{\min}, r_{\max}$ be the minimum and maximum distance from any agent to the ring center, i.e. $r_{\min} = \min_{i \in \{1, \dots, N\}} \|\mathbf{r}_i - \mu\|$ and $r_{\max} = \max_{i \in \{1, \dots, N\}} \|\mathbf{r}_i - \mu\|$, the fatness and tangentness are defined (respectively) as

$$\Phi(t) = 1 - \frac{r_{\min}^2(t)}{r_{\max}^2(t)}, \quad \tau(t) = \frac{1}{N} \sum_{i=1}^{N} \left| \frac{\mathbf{r}_i - \mu}{\|\mathbf{r}_i - \mu\|} \cdot \frac{\dot{\mathbf{r}}_i}{\|\dot{\mathbf{r}}_i\|} \right|. \tag{5}$$

In other words, $\Phi = 0$ implies a perfectly thin ring and $\Phi = 1$ implies an entirely filled-in disc. $\tau = 0$ represents perfect alignment between agents' velocities and their tangent lines and $\tau = 1$ means all agents are misaligned. We also define $\overline{\Phi}(t), \overline{\tau}(t)$ as the average fatness, tangentness over the time interval $[t - T, t]$. We then define a single metric $\lambda \in [0, 1]$ as $\lambda = 1 - \max(\overline{\Phi}, \overline{\tau})$, where $\lambda = 1$ represents a perfect ring and $\lambda = 0$ represents maximum disorder.

3.2   Collision Avoidance

Here we summarize each collision avoidance technique that we replicate: ORCA [17] and CBC [5]. In all cases, we replicate the decentralized version of each technique where agents are only aware of the positions and velocities of other agents in their local neighbor set $\mathcal{N}_i$ but not their inputs, i.e. agent i doesn't know $\mathbf{u}_j^*$ from Eq. (2) nor $\mathbf{u}_j$ from Eq. (4), $j \neq i$.

Both strategies each have their own scalar 'cautiousness' parameter, which we have replaced with our universal tuning parameter $c_r > 0$ where increasing c_r represents an increase in how aggressively agents try to avoid each other. Intuitively, we find increasing c_r causes agents to make corrective measures earlier. Both also have a safety distance parameter D_s where each strategy guarantees $\|\mathbf{r}_i - \mathbf{r}_j\| > D_s$ to satisfy constraint C1 from Sect. 2.1. We set $D_s = 2.1r$, the agent diameter plus a 5% margin, to allow for numerical imprecision from discretization in our simulation.

Control Barrier Certificates (CBC). CBC uses 'barrier function' B_{ij} which is a function of the states $\mathbf{r}_i, \mathbf{r}_j, \dot{\mathbf{r}}_i, \dot{\mathbf{r}}_j$ of two agents and is defined such that $B_{ij} \to \infty$ as i is about to collide with j (refer to [5] for the exact definition). To synthesize a controller, [5] defines the main safety constraint as

$$\dot{B}_{ij} \leq \frac{1}{c_r B_{ij}}, \tag{6}$$

where we substitute their tunable parameter with c_r. We can express CBC in our framework as

$$\mathbf{F}_{c_r}(\mathbf{u}_i^*, \mathbf{r}_i, \dot{\mathbf{r}}_i, \{(\mathbf{r}_j, \dot{\mathbf{r}}_j)\}_{j \in \mathcal{N}_i}) = \arg\min_{\mathbf{u}_i} \|\mathbf{u}_i - \mathbf{u}_i^*\|^2$$

$$\text{subj to. } \dot{B}_{ij} \leq \frac{1}{c_r B_{ij}} \ \forall j \in \mathcal{N}_i, \tag{7}$$

$$\|\mathbf{u}_i\|_\infty \leq a_{\max}.$$

Since agents do not know each other's inputs, agent i assumes $\mathbf{u}_j^* = 0, j \neq i$, i.e. that agent j's velocity is going to be constant. Equation (7) is a quadratic program which we solve using the operator splitting quadratic programming solver (OSQP) [2,13]. If a solution does not exist, we make agent i brake, i.e. $\mathbf{u}_i = -\dot{\mathbf{r}}_i$, as [18] recommends doing.

CBC has strict requirements on the sensing radius ℓ_r in order to function, which must obey

$$\ell_r \geq D_s + \frac{1}{4a_{\max}} \left(\sqrt[3]{4c_r a_{\max}} + 2v_{\max} \right)^2,\tag{8}$$

where we choose $v_{\max} = 2v_0$ from Eq. (2) to allow flexibility in case neighboring agents violate their speed set-point v_0.

Optimal Reciprocal Collision Avoidance (ORCA). This strategy is based on the *velocity obstacle* [8,17], which is the set of all relative velocities that would result in a crash with another agent within c_r seconds in the future (we use our universal tuning parameter c_r). ORCA takes this one step further by introducing the *reciprocal* velocity obstacle, that is, a more permissive velocity obstacle that assumes the other agent is also avoiding its velocity obstacle instead of just continuing in a straight line.

Let $B(\mathbf{x}, \rho)$ represent the open ball centered at $\mathbf{x}$ of radius ρ, i.e.

$$B(\mathbf{x}, \rho) = \{\mathbf{p} \in \mathbb{R}^2 \mid \|\mathbf{p} - \mathbf{x}\| < \rho\}.$$

Leting $ORCA_{i|j}^{cr}$ denote the set of safe velocities for agent i assuming that agent j is also using the ORCA algorithm, then

$$ORCA_i^{cr} = B(\mathbf{0}, v_0) \cap \bigcap_{j \in \mathcal{N}_i} ORCA_{i|j}^{cr}\tag{9}$$

represents the set of safe velocities slower than the set-point speed v_0 for agent i when considering *all* of its local neighbors in $\mathcal{N}_i$.

In our discussion of ORCA so far, we consider safe *velocity* inputs, however our agent model in Eq. (1) assumes *acceleration* inputs. To get around this we: (1) convert each agent's desired acceleration into the new velocity that would result from it, i.e. $\dot{\mathbf{r}}_i + \Delta t \mathbf{u}_i^*$, (2) find the safe velocity $\mathbf{v}_{\text{safe}}$ using ORCA, then (3) find the input $\mathbf{u}_i$ necessary to achieve this new velocity.

In summary, the definition of ORCA expressed in our framework is

$$\mathbf{F}_{c_r}(\mathbf{u}_i^*, \mathbf{r}_i, \dot{\mathbf{r}}_i, \{(\mathbf{r}_j, \dot{\mathbf{r}}_j)\}_{j \in \mathcal{N}_i}) = \frac{\mathbf{v}_{\text{safe}} - \dot{\mathbf{r}}_i}{\Delta t}$$
$$\mathbf{v}_{\text{safe}} = \underset{\mathbf{v} \in ORCA_i^{cr}}{\arg \min} \|\mathbf{v} - (\dot{\mathbf{r}}_i + \Delta t \mathbf{u}_i^*)\|\tag{10}$$

where Δt is the timestep of the Euler integration in our simulator. Provided Δt is sufficiently small, we find varying it does not change the results.

4 Results

Here we use our behavior quality metric λ from Sect. 3.1 to assess the interference caused by ORCA and CBC. In all our experiments, we let the swarm stabilize for 2,000 simulated seconds then record metrics for an additional 2,000 s, i.e. $t = 4000, T = 2000$ from Sect. 3.1.

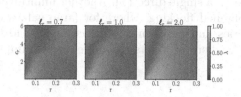

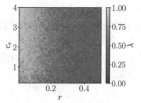

Fig. 1. Ring quality λ for ORCA as we vary agent size r, cautiousness c_r, and sensing radius ℓ_r

Fig. 2. Ring quality λ for CBC as we vary r, c_r

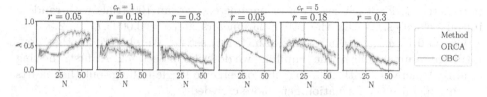

Fig. 3. A comparison of how ORCA and CBC affect the behavior quality λ as the number of agents is increased. Shaded regions show the 95% confidence intervals.

4.1 Control Barrier Certificates (CBC)

Figure 2 shows the ring quality λ as a function of agent size r and cautiousness parameter c_r, where the other parameters are fixed at $N = 20, \alpha = 0.001, t_d = 2.5, \beta = 1, v_0 = 0.16, a_{max} = 0.6$ and ℓ_r is set dynamically based on Eq. (8). For each r, c_r pair we run 10 trials with random initial velocities, then take the average λ value across the trials. We observe that CBC is fairly agnostic to the choice of c_r and only extremely large or small values seem to make much difference on the behavior quality. Predictably, the quality λ drops off as the agent size r increases, regardless of the choice of c_r. For very large values of r, CBC tends to cause agents to deadlock in a static clump since most agents cannot find a solution Eq. (7) and they switch to braking mode.

4.2 Optimal Reciprocal Collision Avoidance (ORCA)

Figure 1 shows the mean ring quality λ as a function of r, c_r. We test 60 unique values of r and c_r on each axis, 3 values of ℓ_r, and for each r, c_r, ℓ_r tuple we simulate 200 trials (we can use more than CBC since ORCA is computationally cheaper) with random initial velocities, where λ is averaged across all the trials. The other parameters are fixed at $N = 20, \alpha = 0.0008, t_d = 3.0, \beta = 2.0, v_0 = 0.12, a_{max} = 0.8$. Here we notice an unintuitive result in the darker 'bands' where λ decreases and then starts increasing again. We find this is due to agents attempting 'double-milling' behavior inside the band, where they attempt to travel in counter-rotating mills and make erratic movements to avoid each other.

Outside the band, the agents converge on a single direction. Another unintuitive result we notice with ORCA is 'unplanned flocking' where for large values of r agents appear to align their velocities and move in one direction instead of forming a ring. Contrast this to CBC, where as r gets too large agents simply deadlock.

4.3 Comparing ORCA and CBC

Figure 3 shows a comparison of the quality λ of ORCA and CBC as we vary the number of agents, for 6 combinations of the agent size r and cautiousness parameter c_r. For each selection of parameters we run 100 trials with random initial velocities. We fix the other parameters at $\alpha = 0.0005, t_d = 2.5, \beta = 1.0, v_0 = 0.12, a_{max} = 0.8$ and use Eq. (8) to set ℓ_r. We observe another non-intuitive result similar to the 'banding' we describe in Sect. 4.2. At first, the ring quality λ increases with N as agents converge to single-milling, but then starts decreasing once the formation gets more crowded.

Comparing each collision avoidance in a fair manner is difficult in general, but we notice for the parameters shown in Fig. 3 that ORCA generally performs better for larger and more numerous agents. We suspect this is due to ORCA's ability to gracefully degrade the quality of the solution and break the fewest possible constraints if the problem is infeasible. CBC instead has agents switch to a braking mode upon infeasibility.

5   Conclusion

Despite using two fairly sophisticated collision avoidance techniques as opposed to the simpler techniques in [16], we show through an illustrative example that swarming algorithms can still be disrupted by collision avoidance and we recommend that such algorithms be co-designed with collision avoidance in mind as opposed to adding it separately. We believe further research is necessary to realize existing swarming behaviors on platforms that take up non-negligible amounts of physical space and cannot collide.

Acknowledgements. This work was supported by the Department of the Navy, Office of Naval Research (ONR), under federal grants N00014-19-1-2121 and N00014-20-1-2042. The experiments were run on ARGO, a research computing cluster provided by the Office of Research Computing at George Mason University, VA. (http://orc.gmu.edu).

References

1. Arul, S.H., et al.: LSwarm: efficient collision avoidance for large swarms with coverage constraints in complex urban scenes. IEEE Robot. Autom. Lett. 4(4), 3940–3947 (2019). https://doi.org/10.1109/lra.2019.2929981
2. Banjac, G., Goulart, P., Stellato, B., Boyd, S.: Infeasibility detection in the alternating direction method of multipliers for convex optimization. J. Optim. Theory Appl. 183(2), 490–519 (2019). https://doi.org/10.1007/s10957-019-01575-y
3. Bayindir, L.: A review of swarm robotics tasks. Neurocomputing 172, 292–321 (2016). https://doi.org/10.1016/j.neucom.2015.05.116
4. Bonabeau, E., Marco, D.d.R.D.F., Dorigo, M., Theraulaz, G., et al.: Swarm Intelligence: From Natural to Artificial Systems, no. 1, Oxford University Press, Oxford (1999)
5. Borrmann, U., Wang, L., Ames, A.D., Egerstedt, M.: Control barrier certificates for safe swarm behavior. IFAC-PapersOnLine 48(27), 68–73 (2015). https://doi.org/10.1016/j.ifacol.2015.11.154
6. Cao, Y., Ren, W.: Distributed coordinated tracking with reduced interaction via a variable structure approach. IEEE Trans. Autom. Control 57(1), 33–48 (2012). https://doi.org/10.1109/TAC.2011.2146830
7. Carrillo, J., D'Orsogna, M., Panferov, V.: Double milling in self-propelled swarms from kinetic theory. Kinetic Relat. Models 2(2), 363–378 (2009). https://doi.org/10.3934/krm.2009.2.363
8. Fiorini, P., Shiller, Z.: Motion planning in dynamic environments using velocity obstacles. Int. J. Robot. Res. 17(7), 760–772 (1998). https://doi.org/10.1177/027836499801700706
9. Ghapani, S., Mei, J., Ren, W., Song, Y.: Fully distributed flocking with a moving leader for Lagrange networks with parametric uncertainties. Automatica 67, 67–76 (2016). https://doi.org/10.1016/j.automatica.2016.01.004
10. Mier-Y-Teran-Romero, L., Forgoston, E., Schwartz, I.B.: Coherent pattern prediction in swarms of delay-coupled agents. IEEE Trans. Robot. 28(5), 1034–1044 (2012). https://doi.org/10.1109/TRO.2012.2198511
11. Olfati-Saber, R.: Flocking for multi-agent dynamic systems: algorithms and theory. IEEE Trans. Autom. Control 51(3), 401–420 (2006). https://doi.org/10.1109/TAC.2005.864190
12. Reynolds, C.W.: Flocks, herds and schools: a distributed behavioral model. In: ACM SIGGRAPH Computer Graphics, vol. 21, pp. 25–34. New York (1987). https://doi.org/10.1145/37402.37406, http://portal.acm.org/citation.cfm?doid=37402.37406
13. Stellato, B., Banjac, G., Goulart, P., Bemporad, A., Boyd, S.: OSQP: an operator splitting solver for quadratic programs. ArXiv e-prints, November 2017
14. Szwaykowska, K., Schwartz, I.B., Mier-Y-Teran Romero, L., Heckman, C.R., Mox, D., Hsieh, M.A.: Collective motion patterns of swarms with delay coupling: theory and experiment. Phys. Rev. E 93(3), 032307 (2016). https://doi.org/10.1103/PhysRevE.93.032307
15. Taylor, C., Luzzi, C., Nowzari, C.: On the effects of collision avoidance on an emergent swarm behavior. arXiv preprint arXiv:1910.06412 (2019)
16. Taylor, C., Luzzi, C., Nowzari, C.: On the effects of collision avoidance on emergent swarm behavior (2020, to appear)

17. Van Den Berg, J., Guy, S.J., Lin, M., Manocha, D.: Reciprocal n-body collision avoidance. In: Pradalier C., Siegwart R., Hirzinger G. (eds.) Robotics Research. Springer Tracts in Advanced Robotics, vol. 70, pp. 3–19 (2011). Springer, Heidelberg. https://doi.org/10.1007/978-3-642-19457-3_1
18. Wang, L., Ames, A.D., Egerstedt, M.: Safety barrier certificates for collisions-free multirobot systems. IEEE Trans. Robot. **33**(3), 661–674 (2017). https://doi.org/10.1109/TRO.2017.2659727
19. Zhan, J., Li, X.: Flocking of multi-agent systems via model predictive control based on position-only measurements. IEEE Trans. Indust. Inf. **9**(1), 377–385 (2013). https://doi.org/10.1109/TII.2012.2216536
20. Zhang, H.T., Cheng, Z., Chen, G., Li, C.: Model predictive flocking control for second-order multi-agent systems with input constraints. IEEE Trans. Circ. Syst. I Regular Pap. **62**(6), 1599–1606 (2015). https://doi.org/10.1109/TCSI.2015.2418871

Set-Based Particle Swarm Optimization for Portfolio Optimization

Kyle Erwin[1] and Andries P. Engelbrecht[2(✉)]

[1] Computer Science Division, Stellenbosh University, Stellenbosch, South Africa
kyle.erwin24@gmail.com
[2] Department of Industrial Engineering and Computer Science Division,
Stellenbosh University, Stellenbosch, South Africa
engel@sun.ac.za

Abstract. Portfolio optimization is a complex real-world problem where assets are selected such that profit is maximized while risk is simultaneously minimized. In recent years, nature-inspired algorithms have become a popular choice for efficiently identifying optimal portfolios. This paper introduces such an algorithm that, unlike previous algorithms, uses a set-based approach to reduce the dimensionality of the problem and to determine the appropriate budget allocation for each asset. The results show that the proposed approach is capable of obtaining good quality solutions, while being relatively fast.

1   Introduction

In 1952, Markowitz presented the mean-variance model, a portfolio optimization model that allowed investors to optimize a portfolio based on a given level of market risk [7]. The underlying principle of the model is the diversification of investments such that the riskiness of an asset can be negated by another, unrelated safe asset. Chang et al. presented an empirical analysis on the applicability of heuristics for an adaptation of the mean-variance model that included real world constraints [1]. Much of the work done thereafter has been inspired by Chang et al. [5]. However, the performance of such approaches diminishes when large asset spaces are considered [10]. Set-based approaches have lead to improved performance in comparison to their non-set based counterparts, although the algorithms are dependent on quadratic programming. This paper proposes a set-based particle swarm optimization (SBPSO) approach, for the mean-variance portfolio optimization model, that does not require the use of quadratic programming.

The rest of this paper is organized as follows: Sect. 2 provides an overview of the concepts presented. Section 3 proposes SBPSO for portfolio optimization. Section 4 presents the empirical process used. The findings are discussed in Sect. 5. Section 6 concludes the paper.

© Springer Nature Switzerland AG 2020
M. Dorigo et al. (Eds.): ANTS 2020, LNCS 12421, pp. 333–339, 2020.
https://doi.org/10.1007/978-3-030-60376-2_28

2   Background

This section presents the necessary background for the work presented in this paper. Section 2.1 presents an overview of portfolio optimization. A genetic algorithm and it's application to portfolio optimization is discussed in Sect. 2.2. Section 2.3 describes set-based particle swarm optimization.

2.1   Portfolio Optimization

The mean–variance portfolio model is a two-objective nonlinear quadratic programming model, that minimizes risk and maximizes return given a level of risk tolerance [7]. The model is formulated as follows:

$$\text{minimize } \lambda \bar{\sigma} - (1 - \lambda)R \tag{1}$$

where $\bar{\sigma}$ is the total risk of the portfolio, R is the total return of the portfolio, and λ represents an investor's risk tolerance in the range $[0, 1]$. When λ is equal to zero, return is to be maximized regardless of the risk involved. In contrast, a λ value of one minimizes risk irrespective of profits. For the case when $0 < \lambda < 1$, an explicit trade-off between risk and return is obtained. Thus, by solving the single-objective model for varying values of λ, a range of efficient portfolios can be identified. Risk is calculated using

$$\bar{\sigma} = \sum_{i=1}^{n} \sum_{j=1}^{n} \sigma_{ij} w_i w_j \tag{2}$$

where n is the number of assets, w_i and w_j is a weighting of assets i and j, respectively, and σ_{ij} is the covariance between assets i and j. R is calculated using

$$R = \sum_{i=1}^{n} R_i w_i \tag{3}$$

where R_i is the return of asset i. Equation (1) is subject to the following constraints: The total weighting of all assets in the portfolio must be equal to one, and each asset in the portfolio must have a positive weighting or a weighting of zero.

2.2   Genetic Algorithm for Portfolio Optimization

Chang et al. found a genetic algorithm (GA) to be a more competitive approach than a tabu search and simulated annealing for the mean-variance portfolio optimization problem [1]. The GA used by Chang et al. represented solutions to the mean–variance portfolio optimization problem as individuals, where each genotype corresponds to an asset weight. The following operators were used to evolve the candidate solutions to the portfolio optimization problem: Uniform crossover between two parents, each selected by tournament selection with a tournament size of 2, and 10% mutation of one randomly chosen genotype. Equation (1) was used as the fitness function. Parameter details are given in Sect. 4.

2.3 Set-Based Particle Swarm Optimization

Particle swarm optimization (PSO) simulates the social behavior of birds within a flock, where individuals converge on a single point by exchanging locally available information [3]. Joost and Engelbrecht proposed SBPSO, a PSO algorithm based on set-theory [6]. SBPSO defines a particle's position as $\mathcal{P}(U)$, the power set of U, where U is the universe of elements defined by the problem. In general, mathematical sets do not have spatial structure, thus velocity is a set of operation pairs. An operation pair is denoted as $(\pm, e)$, with $(+, e)$ for the addition of element $e \in U$ and $(-, e)$ for the deletion of element e. Furthermore, due to the lack of spatial structure, special operators, described below, are required in order to implement the velocity and position update equations.

The addition of two velocities, $V_1 \oplus V_2$, is a mapping $\oplus : \mathcal{P}(\{+, -\} \times U)^2 \to \mathcal{P}(\{+, -\} \times U)$ implemented as the union of the two sets of operation pairs, $V_1 \cup V_2$. The difference between two positions, $X_1 \ominus X_2$, is a mapping $\ominus : \mathcal{P}(U)^2 \to \mathcal{P}(\{+, -\} \times U)$ defined as a set of operation pairs that indicate the steps required to convert $X2$ into $X1$, that is $(\{+\} \times (X_1 \setminus X_2)) \cup (\{-\} \times (X_2 \setminus X_1))$. The multiplication of a velocity by a scalar, $\eta \otimes V$, is a mapping $\eta \otimes : [0, 1] \times \mathcal{P}(\{+, -\} \times U) \to \mathcal{P}(\{+, -\} \times U)$ implemented as the random selection of $\lfloor \eta \times |V| \rfloor$ elements from V. The addition of a velocity and a position, $X \boxplus V$, is a mapping $X \boxplus V : \mathcal{P}(U) \times \mathcal{P}(\{+, -\} \times U) \to \mathcal{P}(U)$ defined as the action of applying the velocity function V to the position X, $X \boxplus V = V(X)$.

The remaining operators use the following function, $N_{\beta,X}$, to determine the number of elements to select. $N_{\beta,X}$ is implemented as $\min \{|X|, \lfloor \beta \rfloor + \mathbb{1}_\beta\}$, where β is an element of $\mathbb{R}^+$, $\mathbb{1}_\beta$ is equal to 1 if $r < \beta - \lfloor \beta \rfloor$, otherwise 0, and r is a uniformly randomly sampled value from $(0, 1)$,

The removal of elements $X(t) \cup Y(t) \cup \hat{Y}(t)$ from position $X(t)$ uses the $\odot^-$ operator, where $Y(t)$ and $\hat{Y}(t)$ are the cognitive and social guides, respectively. Denoted as $\beta \odot^- S$, where S is shorthand for $X(t) \cup Y(t) \cup \hat{Y}(t)$, is a mapping $\odot^+ : [0, |S|] \times \mathcal{P}(U) \to \mathcal{P}(\{+, -\} \times U)$. The $\beta \odot^- S$ operation is calculated using, $\beta \odot^- S = \{-\} \times (\frac{N_{\beta,S}}{|S|} \otimes S)$.

The addition of elements to position $X(t)$ outside of $X(t) \cap Y(t) \cap \hat{Y}(t)$ uses the $\odot^+$ operator. Denoted $\beta \odot^x A$, where A is shorthand for $U \setminus (X(t) \cap Y(t) \cap \hat{Y}(t))$, is a mapping $\odot^+ : [0, |A|] \times \mathcal{P}(U) \to \mathcal{P}(\{+, -\} \times U)$ implemented using, $\beta \odot^+_k A = \{+\} \times k$-Tournament Selection$(A, N_{\beta,A})$ where k is the number of individuals, and each tournament consists of $N_{\beta,A}$ elements. The element added to $X(t)$ that yields the best fitness value is selected as the winner.

Using the above equations, SBPSO defines the velocity update as

$$V_i(t+1) = c_1 r_1 \otimes (Y_i(t) \ominus X_i(t)) \oplus c_2 r_2 \otimes \hat{Y}_i(t) \ominus X_i(t))$$
$$\oplus (c_3 r_3 \odot^+_k A_i(t)) \oplus (c_4 r_4 \odot^- S_i(t)) \tag{4}$$

where r_1, r_2, r_3, and r_4 are random values, each sampled from a standard uniform distribution in the range [0,1], and c_3 and c_4 are positive acceleration constants for the addition and removal of elements, respectively. The position update, is then the application of the velocity to the position, as shown n the following equation, $X_i(t + 1) = X_i \boxplus V_i(t + 1)$.

3   Set-Based Particle Swarm Optimization for Portfolio Optimization

This paper proposes SBPSO as approach to the set-based mean-variance portfolio optimization problem. Each particle represents the assets included in a candidate portfolio. Unlike previously proposed set-based approaches, a meta-heuristic is used as the weight optimizer for the selected assets. The inertia PSO is used in this study as the weight optimizer [8]. The decision to use PSO is motivated by the fact that it is well established as a successful approach to portfolio optimization [5]. SBPSO and PSO execute in an interleaving fashion. SBPSO firstly identifies a subset of assets. The inertia PSO then optimizes the weights of each asset for a fixed number of time steps. This process repeats until the stopping condition is met. For both PSO algorithms, star network topologies are used. The $\odot^+$ operator for SBPSO is re-implemented as the random selection of $N_{\beta,A}$ elements from A. This is done to reduce the computational complexity of the proposed approach. SBPSO is expected to be a faster portfolio optimizer than the GA proposed by Chang et al., with at least similar accuracy [1].

4   Empirical Process

Five benchmarks[1] were used to assess the scalability and accuracy of the proposed SBPSO in comparison to Chang et al.'s GA. Namely, Hang Seng (Hong Kong, 31 assets), DAX 100 (Germany, 85 assets), FTSE 100 (UK, 89 assets), S&P 100 (USA, 98 assets), and Nikkei 225 (Japan, 255 assets). The the mean-variance model, equation (1), was used as the objective function. The performance of SBPSO and the GA to optimize the objective function for 50 evenly-spaced λ values in the range [0,1] was recorded over 30 independent runs. Each algorithm was allocated 5000 iterations to optimize each λ value, with SBPSO and PSO interleaving every 10 iterations. A swarm and population size of 20 was used for both algorithms. To satisfy the positive weight constraint, any negative weight is treated as zero. The weights are then normalized to ensure that the sum of all weights is equal to 1. If all weights are negative, a large objective function value is given to deter the algorithm from that area of the search space.

The parameters for each algorithm were tuned by evaluating 128 parameters sets. The parameter sets were generated using sequences of Sobol pseudo-random that spanned the parameter space for SBPSO and the GA [4]. c_1 and c_2 were generated in the range [0.00,1.00], and c_3 and c_4 in the range [0.50,5.00]. For the GA, the mutation rate was generated in the range [0.01,50], and k in the range [2]. Recommended parameters for the inertia PSO that satisfy stability conditions and guarantee that an equilibrium state will be reached were used.

---

[1] http://people.brunel.ac.uk/~mastjjb/jeb/orlib/portinfo.html.

These are $w = 0.729844$ and $c_1 = c_2 = 1.496180$. The parameter sets for each algorithm were ranked according to their average objective function value for each of the λ values. The parameter set with the lowest average ranking was selected as the optimal parameters.

The number of solutions, return, risk, generational distance (GD) [9], inverted generational distance (IGD) [2], hypervolume (HV) [11] and time were used to assess the quality, and efficiency, of the obtained solutions. The mean and standard deviation for each aforementioned performance measures was recorded. One-tailed Mann Whitney U tests with a level of significance of 95% were performed to test if an algorithm is statistically significantly better than the other.

5   Results

Table 2 presents the results obtained using the optimal parameters for each algorithm, shown in Table 1.

Table 1. Tuned parameters for each benchmark

Stock market	SPSO				GA	
	c_1	c_2	c_3	c_4	mr	k
Hang Seng	0.375000	0.375000	3.312500	4.437500	0.155469	11
DAX 100	0.117188	0.117188	3.488281	2.714844	0.052109	13
FTSE 100	0.367188	0.867188	3.363281	3.839844	0.017050	10
S & P 100	0.828125	0.484380	2.960938	2.257813	0.270313	3
Nikkei 225	0.828125	0.484380	2.960938	2.257813	0.036797	19

The GA was statistically significantly better at obtaining more optimal solutions than SBPSO for Hang Seng, Dax 100 and Nikkei 225, however, the difference in number of solutions is small. Furthermore, the profitability of solutions obtained by either algorithm are similar despite the GA, in general, having slightly higher return values. SBPSO on average was more capable of obtaining less risky portfolios than the GA, especially for the largest benchmark, i.e. Nikkei 225.

The GA was shown to be statistically significantly better than SBPSO with respect to GD for all benchmarks, except for Nikkei 225 where SBPSO is superior. Furthermore, the GA, on average, obtained better IGD results than SBPSO. Generally, there was no statistically significant difference in results obtained for HV. The standard deviations for GD, IGD and HV, in general, were smaller for the GA than for the SBPSO. However, the performance of SBPSO is in the same order of magnitude as the GA.

SBPSO was significantly faster than the GA and maintained a similar time for all benchmarks, whereas the time for GA deteriorated as the number of

Table 2. Mean and standard deviations for each performance measure

			N	R	$\bar{\sigma}$	GD	IGD	HV	Time
Hang Seng	SBPSO	$\bar{x}$	31	0.233296	0.055898	0.000980	0.000264	1.181940	12.95
		σ	1	0.010148	0.004238	0.000237	0.000014	0.002276	1.08
	GA	$\bar{x}$	31	0.237859	0.057105	0.000248	0.000229	1.182968	101.62
		σ	1	0.007075	0.002338	0.000034	0.000010	0.001691	1.50
DAX 100	SBPSO	$\bar{x}$	31	0.257062	0.028204	0.001668	0.000363	1.191110	32.84
		σ	4	0.036299	0.005491	0.000434	0.000031	0.004717	1.46
	GA	$\bar{x}$	38	0.317733	0.034459	0.000292	0.000317	1.197005	300.89
		σ	1	0.010734	0.001713	0.000156	0.000019	0.002183	2.98
FTSE 100	SBPSO	$\bar{x}$	28	0.187684	0.021696	0.000503	0.000246	1.184440	26.48
		σ	1	0.006002	0.001063	0.000216	0.000013	0.002397	3.07
	GA	$\bar{x}$	28	0.187028	0.021392	0.000421	0.000243	1.184850	376.74
		σ	1	0.007221	0.001208	0.000332	0.000026	0.002126	2.15
S& P 100	SBPSO	$\bar{x}$	41	0.318468	0.052514	0.000861	0.000273	1.198922	24.77
		σ	1	0.012195	0.003196	0.000130	0.000015	0.001561	10.64
	GA	$\bar{x}$	34	0.252994	0.038313	0.000756	0.000270	1.195580	124.68
		σ	2	0.018880	0.003900	0.000239	0.000013	0.001941	1.61
Nikkei 225	SBPSO	$\bar{x}$	39	0.129067	0.030552	0.000770	0.000247	1.195960	26.82
		σ	2	0.007218	0.002393	0.000127	0.000014	0.001923	6.02
	GA	$\bar{x}$	44	0.140672	0.036203	0.001692	0.000276	1.196826	1175.73
		σ	4	0.013242	0.003323	0.000308	0.000020	0.002132	5.25

assets increased. The average time for SBPSO for the largest benchmark was approximately four times faster than that of the GA for the smallest benchmark. The difference in computational time is most notable for Nikkei 225, the largest benchmark, where SBPSO was approximately 44 times faster than the GA. However, It should be noted that SBPSO wastes computational budget under certain circumstances. For example, if a SBPSO particle contains a single asset, the inertia PSO will waste 10 iterations to optimize the asset weight when the weight can only ever have a value of one. Secondly, if a candidate solution is repeatedly found by SBPSO, the inertia PSO does not make use of previously found best weights. Meaning that the weights have to be potentially rediscovered.

6 Conclusion

This paper proposed set-based particle swarm optimization (SBPSO) for the mean-variance portfolio optimization problem. SBPSO is used for asset selection while particle swarm optimization (PSO) is used for asset weight determination. SBPSO performed similarly to the genetic algorithm (GA) proposed by Chang et al. [1]. Furthermore, SBPSO was significantly faster than the GA – in one case, 44 times faster.

Future work will investigate why the performance of SBPSO, in some cases, is marginally worse than the GA. In order to investigated the causes, a diversity measure for the SBPSO will be developed to determine if convergence is not too fast. Also, more efficient approaches will be developed to allow the inertia PSO more time to optimize weights, instead of allowing the inertia PSO a fixed number of iterations. Also, approaches will be investigated not to loose good weightings found by the inertia PSO.

Acknowledgements. The authors acknowledge the Centre for High Performance Computing (CHPC), South Africa, for providing computational resources to this research project.

References

1. Chang, T., Meade, N., Beasley, J., Sharaiha, Y.: Heuristics for cardinality constrained portfolio optimisation. Comput. Oper. Res. **27**(13), 1271–1302 (2000). https://doi.org/10.1016/S0305-0548(99)00074-X
2. Coello Coello, C.A., Reyes Sierra, M.: A study of the parallelization of a coevolutionary multi-objective evolutionary algorithm. In: Monroy, R., Arroyo-Figueroa, G., Sucar, L.E., Sossa, H. (eds.) MICAI 2004. LNCS (LNAI), vol. 2972, pp. 688–697. Springer, Heidelberg (2004). https://doi.org/10.1007/978-3-540-24694-7_71
3. Eberhart, R., Kennedy, J.: A new optimizer using particle swarm theory. In: MHS 1995. Proceedings of the Sixth International Symposium on Micro Machine and Human Science, pp. 39–43, October 1995. https://doi.org/10.1109/MHS.1995.494215
4. Franken, N.: Visual exploration of algorithm parameter space. In: 2009 IEEE Congress on Evolutionary Computation, pp. 389–398 (2009)
5. Kalayci, C.B., Ertenlice, O., Akbay, M.A.: A comprehensive review of deterministic models and applications for mean-variance portfolio optimization. Expert Syst. Appl. **125**, 345–368 (2019). https://doi.org/10.1016/j.eswa.2019.02.011
6. Langeveld, J., Engelbrecht, A.P.: Set-based particle swarm optimization applied to the multidimensional knapsack problem. Swarm Intell. **6**(4), 297–342 (2012). https://doi.org/10.1007/s11721-012-0073-4
7. Markowitz, H.: Portfolio selection. J. Financ. **7**(1), 77–91 (1952). https://doi.org/10.1111/j.1540-6261.1952.tb01525.x
8. Shi, Y., Eberhart, R.: A modified particle swarm optimizer. In: Proceedings of the IEEE International Conference on Evolutionary Computation, pp. 69–73 (1998). https://doi.org/10.1109/ICEC.1998.699146
9. Veldhuizen, D.A.V., Lamont, G.B.: Multiobjective evolutionary algorithm research: a history and analysis. Technical reports, Department of Electrical and Computer Engineering. Graduate School of Engineering, Air Force Inst Technol, Wright Patterson, Technical Report TR-98-03 (1998)
10. Woodside-Oriakhi, M., Lucas, C., Beasley, J.: Heuristic algorithms for the cardinality constrained efficient frontier. Eur. J. Oper. Res. **213**, 538–550 (2011). https://doi.org/10.1016/j.ejor.2011.03.030
11. Zitzler, E., Thiele, L., Laumanns, M., Fonseca, C.M., da Fonseca, V.G.: Performance assessment of multiobjective optimizers: an analysis and review. IEEE Trans. Evol. Comput. **7**(2), 117–132 (2003). https://doi.org/10.1109/TEVC.2003.810758

Extended Abstracts

A Probabilistic Bipartite Graph Model for Hub Based Swarm Solution of the Best-of-N Problem

Michael A. Goodrich[✉][iD] and Puneet Jain[iD]

Department of Computer Science, Brigham Young University, Provo, UT, USA
mike@cs.byu.edu, puneetj@byu.edu

For *spatial swarms*, which are characterized by co-located agents, graph-based models complement agent-based and differential equation models, especially in providing theoretical results that apply to large-but-finite numbers of agents [4]. For *hub-based colony swarms*, where agents are often not in spatial proximity [3], except at a centralized hub, graph-based models do not appear to have received much attention. This extended abstract presents a graph-based model for a hub-based colony solving the best-of-N problem [5].

Hub-based colonies are characterized by two different kinds of entities: agents and sites (locations of interest in the world). Let $G = (V, E)$ be a bipartite graph with $V = V_{\text{agent}} \cup V_{\text{site}}$ partitioned into agent vertices and site vertices. Since G is bipartite, The edge set E has edges connecting an agent vertex only to a site vertex. A directed edge between agent a and a site s means the agent is "committed" to that site (assessing, promoting, committed to, etc.). The quality of sites is, without loss of generality, a real-valued number in $[0, 1]$.

Two probabilities determine the graph dynamics: the *attachment* probability, which determines when a new edge is formed between an agent and a site, and the *detachment* probability, which determines when an existing edge is removed. Attachment uses the preferential attachment pattern [1] and begins when an agent is randomly selected with uniform probability. If the agent is not connected, it randomly chooses a site to which it attaches with a probability proportional to the degree of the site. Detachment uses a tunable clustering pattern [2] and proceeds by selecting an edge with uniform probability from E. The probability that the edge is removed decreases linearly with site quality. Popularity-based clustering and degree-based persistence makes it likely that agents will cluster at the highest quality site, effectually solving the best-of-N problem. Figure 1 shows snapshots of an agent-based implementation of the graph dynamics.

Attachment and detachment induce graph dynamics for a discrete-time Markov process (DTMC) over a finite state space. MATLAB's `dtmc` method was used to compute a numerical solution for how the distribution evolves over time for nine agents and two sites. The quality of the best site was fixed to $\text{qual}(s_1) = 0.95$, and for the second best site was varied between $\text{qual}(s_0) \in \{0.05, 0.75\}$. Thus, the difference in qualities was $\Delta = \text{qual}(s_1) - \text{qual}(s_0) \in \{0.9, 0.2\}$.

Two initial distributions were considered: λ^{empty} placed all probability mass on the configuration with no edges, which represents a colony just beginning the best-of-N problem with no sites discovered. λ^{worst} placed all probability on the configuration with all agents connected to the second-best site, with its

© Springer Nature Switzerland AG 2020
M. Dorigo et al. (Eds.): ANTS 2020, LNCS 12421, pp. 343–344, 2020.
https://doi.org/10.1007/978-3-030-60376-2

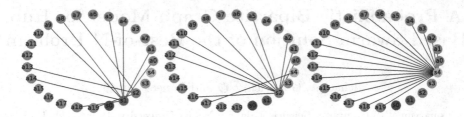

Fig. 1. Snapshots of graph configuration from agent-based simulation converging to highest quality site. From left to right $t = 50$, $t = 200$, $t = 350$.

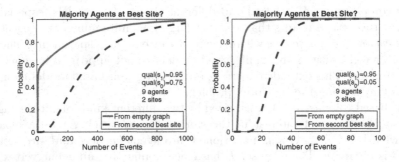

Fig. 2. Evolution of the DTMC state distribution from λ^{empty} (solid line) and from λ^{worst} (dashed line). Left: $\Delta = 0.20$, Right: $\Delta = 0.90$.

evolution representing the time taken by the colony to switch "commitment" from the inferior site to the superior site. Figure 2 shows the probability that a plurality of agents favors the superior site for a colony with 9 agents. Having similar site qualities slows convergence to the best site.

Acknowledgements. This extended abstract was partially supported by US Office of Naval Research grant N00014-18-1-2503. The results are the responsibility of the authors and not the sponsoring organization.

References

1. Barabási, A.L., Albert, R.: Emergence of scaling in random networks. Science **286**(5439), 509–512 (1999)
2. Deijfen, M., Kets, W.: Random intersection graphs with tunable degree distribution and clustering. Probab. Eng. Inf. Sci. **23**(4), 661–674 (2009)
3. Gordon, D.M.: Ant Encounters: Interaction Networks and Colony Behavior. Princeton University Press, Princeton (2010)
4. Mesbahi, M., Egerstedt, M.: Graph Theoretic Methods in Multiagent Networks. Princeton University Press, Princeton (2010)
5. Valentini, G., Ferrante, E., Dorigo, M.: The best-of-N problem in robot swarms: formalization, state of the art, and novel perspectives. Front. Robot. AI **4** (Mar 2017)

Ant Colony Optimization for K-Independent Average Traveling Salesman Problem

Yu Iwasaki and Koji Hasebe[(⊠)]

Department of Computer Science, University of Tsukuba Tennodai, Tsukuba,
Ibaraki, Japan
iwasaki@mas.cs.tsukuba.ac.jp, hasebe@cs.tsukuba.ac.jp

The traveling salesman problem (TSP) is applied to constructing a transportation route. However, it is better to have multiple routes in case of disasters or accidents. Therefore, we propose a K-Independent average traveling salesman problem (KI-Average-TSP) that minimizes the weighted sum of the average and standard deviations of the K circuits' costs in a complete graph, where the circuits are mutually independent. Compared to the study to prove propositions of mutually independent paths, this study actually constructs paths and average paths [3]. The definition of KI-TSP is presented below, where $\mathrm{cost_{avg}}$ is the average of the total cost of K circuits, $\mathrm{cost_{sd}}$ is the standard deviation of the K circuits' costs, and γ, θ are weighting parameters.

$$\min \quad \mathrm{cost_{avg}} + \gamma \cdot \mathrm{cost_{sd}}^{\theta} \tag{1}$$

$$\text{subject to} \quad \sum_{k \in K} x_{ijk} \leq 1 \quad (\forall i, j \ (i \neq j)) \tag{2}$$

$$\sum_{j \in V} \sum_{k \in K} x_{ijk} = K \quad (\forall i) \tag{3}$$

$$\sum_{j \in V} \sum_{k \in K} x_{jik} = K \quad (\forall i) \tag{4}$$

$$\sum_{i \in T} \sum_{j \in V \setminus T} x_{ijk} \geq 1 \quad (\forall k, \forall T \subset V (T \neq \phi, T \neq V)) \tag{5}$$

$$x_{ijk} \in \{0, 1\} \tag{6}$$

Figure 1 shows an example of $K = 2$ circuits in a complete graph of $N = 8$ vertices. To find approximate solutions to KI-Average-TSP, we propose K-Independent ant colony optimization (KI-ACO), an extension of the original ant colony optimization [1]. If K ants construct paths one by one as the original ACO, the paths constructed later will be longer, resulting in a larger standard deviation of K paths. Therefore, KI-ACO moves K ants along each edge at a time, so that all ants can use equally their favorable edges and reduce the standard deviation. However, as the ants move, the number of reachable vertices for the ants decreases and the circuits construction failure rate increases. As a countermeasure, we introduce two heuristics to prevent the failure of the construction.

© Springer Nature Switzerland AG 2020
M. Dorigo et al. (Eds.): ANTS 2020, LNCS 12421, pp. 345–346, 2020.
https://doi.org/10.1007/978-3-030-60376-2

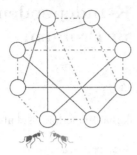

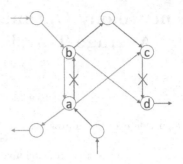

Fig. 1. An example of KI-Average-TSP where the same edge is not shared by multiple paths in a complete graph. The orange ant's path is dependent on the blue ant's one.

Fig. 2. An example of 2-best-opt which swaps $e = (a, b)$ and $e' = (c, d)$, where the edge e is used twice. New edges $e_1 = (a, c)$, $e_2 = (b, d)$ can reduce the usage count in $e = (a, b)$.

The first heuristic is the margin residual degree used in the transition probability equation, representing the number of vertices an ant can move to from its current vertex. This index is used to reduce the possibility of reusing the edge used by other ants when close to N-th movement. The second heuristic is 2-best-opt, a method based on 2-opt [2]. Figure 2 shows an example of trying to exchange an edge e used in multiple paths for another path's edge by 2-opt. We apply this method to all edges, correcting the overlapping edges greedily. After applying 2-best-opt, if K paths are independent, ants deposit pheromone calculated based on the path cost and the standard deviation. KI-ACO iterates this process until satisfying termination conditions.

To evaluate our proposed method, we conducted comparative experiments on the effects of two heuristics and pheromone updates on cost, execution time, and failure rate. From the result, we observed that KI-ACO with 2-best-opt and pheromone update improved the failure rate by about 60% and reduced the weighted sum by about 20%. Besides, margin residual degree reduced the failure rate, while increasing the execution time.

As a further study, we apply KI-ACO to a network planning problem. Also, to apply KI-TSP to more realistic problems, we are interested in a relaxed KI-TSP that allows some edges to be used in multiple paths simultaneously.

References

1. Dorigo, M., Stützle, T.: Ant colony optimization: overview and recent advances. In: Gendreau, M., Potvin, J.-Y. (eds.) Handbook of Metaheuristics. ISORMS, vol. 272, pp. 311–351. Springer, Cham (2019). https://doi.org/10.1007/978-3-319-91086-4_10
2. Johnson, D.S., McGeoch, L.A.: The traveling salesman problem: a case study in local optimization. Local Search Comb. Optim. **1**(1), 215–310 (1997)
3. Teng, Y.H., Tan, J.J., Ho, T.Y., Hsu, L.H.: On mutually independent hamiltonian paths. Appl. Math. Lett. **19**(4), 345–350 (2006)

Construction Coordinated by Stigmergic Blocks

Yating Zheng[1,2(✉)] ⓘ, Michael Allwright[2] ⓘ, Weixu Zhu[2] ⓘ, Majd Kassawat[3] ⓘ,
Zhangang Han[1] ⓘ, and Marco Dorigo[2] ⓘ

[1] School of Systems Science, Beijing Normal University, Beijing, China
zhengyating@mail.bnu.edu.cn, zhan@bnu.edu.cn
[2] IRIDIA, Université Libre de Bruxelles, Brussels, Belgium
{michael.allwright,weixu.zhu,marco.dorigo}@ulb.ac.be
[3] Universitat Jaume I, Castellon, Spain
majd@uji.es

In swarm robotics, robots coordinate their actions by communicating with their neighbors and by sensing and modifying their environment [3]. In previous work, swarms of building robots have been coordinated through stigmergy, where the observations of previous construction actions trigger further construction actions [1, 4]. In these systems, the intelligence that coordinates construction is usually embedded in the robots.

We are currently exploring how swarm construction can be realized when the intelligence that coordinates construction is distributed between the robots and the building material. In this abstract, we present a preliminary step to distributing this intelligence where the building material, in the form of building blocks, can send and receive messages from other blocks in the same structure. In our current implementation, we provide the blocks with a description of a structure. One block then takes the lead and determines where blocks are missing and should be placed. Figure 1 shows this block relaying this information by setting the colors of the light-emitting diodes on a structure during manual assembly.

In initial experiments, we have begun to explore how this setup can be used to influence the way in which construction unfolds. For example, Fig. 2 shows how the order in which robots attach blocks could be used to regulate construction so that the final structure has two blocks on the top layer on opposite sides (either front-back or left-right).

We have also started to investigate gradient following [5], where blocks use their light-emitting diodes to communicate the directions in which construction sites can be found. Figure 3 shows an example of this concept where blue blocks indicate that the nearest site is to the left and red blocks indicate that the nearest site is to the right. After both blocks have been attached, the controller sets all blocks to green to indicate that the structure is complete.

In future work, we will decentralize the approach in this abstract to realize a swarm robotics construction system where the intelligence is distributed across both the blocks and the robots. We will use the BuilderBot and Stigmergic Blocks to validate this approach in order to gain a deeper understanding as to whether these abstract concepts can be realized in a more realistic setting [2].

© Springer Nature Switzerland AG 2020
M. Dorigo et al. (Eds.): ANTS 2020, LNCS 12421, pp. 347–348, 2020.
https://doi.org/10.1007/978-3-030-60376-2

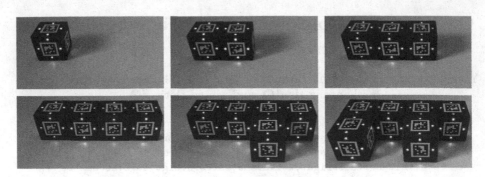

Fig. 1. Manual construction guided by the leftmost block

Fig. 2. The block in the center disables construction sites to regulate construction

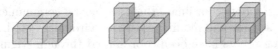

Fig. 3. The leftmost block shows how to reach a valid construction site

Acknowledgements. This work is partially supported by a Marie Skłodowska-Curie fellowship (grant agreement number 846009), by the Ministerio de Economa y Competitividad (DPI2015-69041-R), and by Universitat Jaume I (UJI-B2018-74). Yating Zheng and Weixu Zhu acknowledge support from the China Scholarship Council (grants 201806040106 and 201706270186). Marco Dorigo acknowledges support from the Belgian F.R.S.-FNRS, of which he is a research director.

References

1. Allwright, M., Bhalla, N., Dorigo, M.: Structure and markings as stimuli for autonomous construction. In: 18th International Conference on Advanced Robotics (ICAR), pp. 296–302. IEEE (2017)
2. Allwright, M., Zhu, W., Dorigo, M.: An open-source multi-robot construction system. HardwareX **5**, e00050 (2019)
3. Dorigo, M., Birattari, M., Brambilla, M.: Swarm robotics. Scholarpedia **9**(1), 1463 (2014)
4. Theraulaz, G., Bonabeau, E.: Coordination in distributed building. Science **269**(5224), 686–688 (1995)
5. Werfel, J., Nagpal, R.: Three-dimensional construction with mobile robots and modular blocks. Int. J. Robot. Res. **27**(3–4), 463–479 (2008)

Human-Swarm Teaming with Proximal Interactions

Mohammad Divband Soorati[1]([⊠]) [iD], Dimitar Georgiev[1] [iD], Javad Ghofrani[2] [iD], Danesh Tarapore[1] [iD], and Sarvapali Ramchurn[1] [iD]

[1] School of Electronics and Computer Science, University of Southampton, Southampton, UK
{m.divband-soorati,dg1g17,d.s.tarapore,sdr1}@soton.ac.uk
[2] Department of Informatics and Mathematics, Dresden University of Applied Sciences, Dresden, Germany
javad.ghofrani@gmail.com

One of the major challenges in human-swarm interaction is acquiring global information about the swarm's state and visualizing it to the human operators. The literature lacks a comprehensive study that eliminates the high communication costs of human-swarm interaction and allows the operator to continuously observe the swarm's state and control the agents [1]. This paper aims at filling this gap by proposing a minimal framework of human-swarm system with proximal interactions. We consider an example of a disaster management scenario with a swarm of simulated drones as agents, a human operator, and a disaster zone. The primary goal of the agents is to disperse and move in the area in a way that the swarm has maximum certainty about the situation in the mission zone throughout the runtime. Agents collectively explore and map the environment by disseminating their observations and incorporating their neighboring agents' maps. Instead of a global communication infrastructure between the operator and the agents, we assume that the operator can only communicate with the agents in its local neighborhood. The operator uses the incoming maps from the agents around it to estimate the swarm's state and manipulates the maps to control the swarm. Each agent updates and communicates two maps, a *belief* and a *confidence* map. A belief map is used as a map representation of the mission zone. Agents use the belief maps to store their observation of the area. A confidence map represents the distribution of the swarm as seen by individual agents. This map determines the agents' certainty level of the information in the corresponding cell of the belief map. Using the belief and the confidence maps, the agents store their representation of the environment and keep their confidence level about its correctness. As agents move and explore the mission zone, the cells in the belief maps are updated. The corresponding confidence values of the recently visited cells in the confidence map also increase. We associate time to the confidence by multiplying the confidence map with an aging factor that is smaller than one. Agents participate in a collective decision-making process by continuously sharing their maps (belief and confidence maps) and incorporating other maps in their own updates.

We create an artificial potential field where high confidence is repulsive and low confidence is attractive. The agents continuously follow the downhill gradient

© Springer Nature Switzerland AG 2020
M. Dorigo et al. (Eds.): ANTS 2020, LNCS 12421, pp. 349–350, 2020.
https://doi.org/10.1007/978-3-030-60376-2

of the potential calculated by applying Sobel filter on the confidence map. Agents move towards the least confidence area and, therefore, try to maximize their confidence about the environment. As the dynamic of the swarm is determined by the agents' confidence, the human operator can control the swarm using the values in this map to attract the agents to an area or repel them from it. In order to keep the agents inside the mission zone, the confidence of the boundaries is set to a high value. A disaster zone with an operator, 15 agents, and a moving disaster is shown in Fig. 1a. The operator continuously receives up-to-date belief maps, before (Fig. 1b) and after (Fig. 1c) the movement of the disaster area. An example of an operator controlling the swarm is also shown in Fig. 1. In this example, the operator intends to repel the agents from the center of the mission zone. Figure 1e shows an example of a manipulated confidence map that the operator disseminates. Figure 1g is the swarm distribution and shows that the swarm follows the operator's command and avoids the forbidden flight zone.

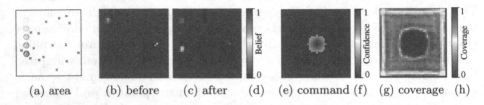

(a) area   (b) before   (c) after   (d)   (e) command (f)   (g) coverage (h)

Fig. 1. An example of a moving disaster (a) and the operator's belief maps, before (b) and after the displacement (c). An example of an operator's command (d) and its effect on the swarm's distribution (e) are shown.

Our proposed method for human-swarm teaming with proximal interactions successfully guides a swarm to explore a mission area in a dynamic environment and allows a single operator to control the swarm. There are several directions that this research can be extended and further developed. Future studies could investigate the effect of noise in sensing, positioning, communication, and actuation. We plan to implement our approach on physical unmanned aerial vehicles to evaluate the performance of the swarm in real world applications.

Acknowledgement. This project was supported by the AXA Research Fund and the Alan Turing Institute-funded project on Flexible Autonomy for Swarm Robotics.

Reference

1. Kolling, A., Walker, P., Chakraborty, N., Sycara, K., Lewis, M.: Human interaction with robot swarms: a survey. IEEE Trans. Hum. Mach. Syst. **46**(1), 9–26 (2015)

PSO Trajectory Planner for Smooth Differential Robot Velocities

Aldo Aguilar[ID], Miguel Zea[ID], and Luis A. Rivera[✉][ID]

Universidad Del Valle de Guatemala, Guatemala City, Guatemala
{agu15170,mezea,larivera}@uvg.edu.gt

A multi-agent differential robot system requires a definite algorithm to behave as a swarm with goal searching capabilities. The classic Particle Swarm Optimization (PSO) algorithm [1] is a popular tool when it comes to finding the optimal solution of a determined fitness function. The PSO is designed for swarms of particles with no mass or physical dimensions, unlike differential robots. Robots' movements are restricted, as opposed to particles, which can move in any direction at any velocity. Therefore, it cannot be directly used in goal searching applications with physical robotic swarms.

In this work, we adapt the PSO algorithm for swarms of differential robots. We propose the use of the PSO as a trajectory planner to enable the agents to collectively find the optimal path to a goal. The velocity vectors generated by the PSO can be used as pointers to new PSO positions $\mathbf{x}_{i+1}$. These multiple reference points can be easily tracked by a differential robot. Based on this idea, the PSO trajectory planner (PSO-TP) is designed as an algorithm that is tasked with simply updating the position of the reference waypoint, or PSO Marker. The PSO marker is then tracked by the kinematic controller of the robot describing a smooth and continuous path. After a certain amount of iterations of the kinematic controller tracking the current waypoint, the PSO-TP becomes active again and it updates the location of the PSO Marker. The kinematic controller starts tracking the new reference point and describes another continuous path segment. This process is repeated continuously until the PSO-TP converges to the goal. A new scaling factor η multiplying $\mathbf{v}_{i+1}$ is added to have a direct control over the length of the PSO velocity vectors, thus limiting the acceptable distances between the robot and the PSO Marker to avoid the marker surpassing the search space boundaries and causing the kinematic controllers to saturate the robots' actuators. The modified update rule for the particles position is given by:

$$\mathbf{x}_{i+1} = \mathbf{x}_i + \eta \mathbf{v}_{i+1} \qquad (1)$$

The PSO-TP needs to take into account the restrictions derived from the kinematic equations of the robotic agents and the finite dimensions of the search space. For that purpose, it is coupled with the necessary controllers to map particle velocities into smooth and continuous differential robot velocities. Since the differential robot model is non-linear and not controllable, a diffeomorphism of the original system into a simpler system is required. The implemented control transformation is based on the assumption that we can directly control the planar velocities u_1 and u_2 of a single point in a differential robot. The diffeomorphism transforms the control signals into the linear and angular velocities

© Springer Nature Switzerland AG 2020
M. Dorigo et al. (Eds.): ANTS 2020, LNCS 12421, pp. 351–352, 2020.
https://doi.org/10.1007/978-3-030-60376-2

of the differential robot, which in turn are mapped to wheel velocities using the unicycle model [4]. Four different controllers combined with the diffeomorphism were tested to determine the best in terms of convergence speed, trajectory smoothness and smooth wheel velocities. The four tested controllers were the Transformed Unicycle Controller (TUC) (Eq. 21 in [3]), Transformed Unicycle with LQR (TUC-LQR), Transformed Unicycle with LQI (TUC-LQI), and a Lyapunov-stable Pose Controller (LSPC) [2].

After selecting the proper PSO-TP parameters to achieve the desired swarm behavior, we analysed the PSO-TP implementation on differential robots using the Webots simulation environment. The simulations were performed using swarms of 10 E-Puck robots in a 2×2 m space and a sampling period of 32 ms. The Sphere and Keane benchmark functions were used as fitness functions $f(\cdot)$ to evaluate the algorithm's performance. The controllers that tended to generate straighter paths towards the goal were the TUC-LQR and the TUC-LQI. We used the minimum energy property of cubic spline interpolations to measure the smoothness of each actuator velocity signal generated by each controller [5]. The TUC-LQI controller signals presented the lowest average bending energy results indicating greater control smoothness. Furthermore, it presented a small standard deviation, which indicated little dependency on the initial positions of the E-pucks. Control saturation rate for each velocity control signal was also calculated. The TUC presented actuator saturation between 50% and 90% of the time, whereas the LSPC, TUC-LQR and TUC-LQI presented no saturation.

In conclusion, the PSO-TP was able to guide a differential robot swarm towards a goal in a finite amount of time generating smooth trajectories. The implementation of the PSO Velocity scaling factor η allowed an easier manipulation of the operating range of the PSO-TP. It effectively restricted the positions of the PSO Marker so that they were always within the search space. The TUC-LQI controller outperformed the others in terms of achieving smooth continuous differential robot velocities, while following the paths generated by the PSO trajectory planner. Videos and images demonstrating our results can be found in https://tinyurl.com/AguilarZeaRivera2020.

References

1. Kennedy, J., Eberhart, R.: Particle swarm optimization. In: ICNN'95 - International Conference on Neural Networks. vol. 4, pp. 1942–1948 (1995)
2. Malu, S.K., Majumdar, J.: Kinematics, localization and control of differential drive mobile robot. Glob. J. Res. Eng. **14**, 1–7 (2014)
3. Martins, F., Brandão, A.: Motion control and velocity-based dynamic compensation for mobile robots. Applications of Mobile Robots (Nov 2018)
4. O'Flaherty, R.: A Control Theoretic Perspective on Learning in Robotics. School of Electrical and Computer Engineering, Georgia Institute of Technology (2016)
5. Wolberg, G., Alfy, I.: An energy-minimization framework for monotonic cubic spline interpolation. J. Comput. Appl. Math. **143**(2), 145–188 (2002)

Author Index

Printed in the United States
By Bookmasters